Audel™

Electric Motors:
All New 6th Edition

Audel™

Electric Motors:
All New 6th Edition

Rex Miller
Mark Richard Miller

WILEY
Wiley Publishing, Inc.

Executive Publisher: Robert Ipsen
Publisher: Joe Wikert
Editor: Katie Feltman
Developmental Editor: Kevin Shafer
Editorial Manager: Kathryn Malm
Production Editor: Angela Smith
Text Design & Composition: Wiley Composition Services

Library of Congress Cataloging-in-Publication Data:

ISBN: 0-764-54198-6

Printed in the United States of America

10 9 8 7 6 5 4 3 2

Contents

Preface

Electric motors play an important role in furnishing power to the industrialized world. Robots make great use of small motors, as do automobiles, home appliances, and aircraft. Their versatility, dependability, and economy of operation cannot be equaled by any other form of motive power. It is estimated that electric motors are utilized in over 90 percent of industrial applications, a figure that would be higher except for the absence of power lines in some remote areas. The application of electric motors in the average home has also reached a high degree, ranging from the smallest units found in electric clocks, to the larger units in air conditioners, heating plants, etc. It is a rare individual, indeed, whose daily life is not affected in some way by an electric motor.

This book has been prepared as a practical guide to the selection, maintenance, installation, operation and repair of electric motors. Electricians, industrial maintenance personnel, and installers should find this book, with its clear descriptions and illustrations and simplified explanations, a ready source of information for the many problems they might encounter while maintaining or repairing electric motors. Both technical and non-technical persons who desire to gain knowledge of electric motors will benefit from the theoretical and practical coverage this book offers.

Acknowledgments

No book can be written without the aid of many people. It takes a great number of individuals to put together the information available about any particular technical field into a book. The field of electrical motors is no exception. Many firms have contributed to the information, illustrations, and analysis of the book.

The authors would like to thank every person involved for his or her contributions. Some of the firms that supplied technical information and illustrations are listed below:

Allis-Chalmers, Inc.

Amprobe, Inc.

Bodine Electric Co.

Brown & Sharpe Co.

Doerr Electric Corporation

Fasco Industries, Inc.

General Electric Company

Hayden Switch and Instrument, Inc.

Howard Industries, Inc.

Leeson Electric Corporation

Lennox Industries, Inc.

Norland Div. of Scott and Fetzer Company

Reliance Electric Co.

Robbins & Meyers, Inc.

Sears, Roebuck & Company

Square D Company

Stanley Tool Company

Stock Drive Products

Voorlas Manufacturing Company

Western Electric Company

Westinghouse Electric Corporation

About the Authors

Rex Miller was a Professor of Industrial Technology at The State University of New York, College at Buffalo for over 35 years. He has taught at the technical school, high school, and college level for well over 40 years. He is the author or co-author of over 100 textbooks ranging from electronics to carpentry and sheetmetal work. He has contributed more than 50 magazine articles over the years to technical publications. He is also the author of seven Civil War regimental histories.

Mark Richard Miller finished his BS degree in New York and moved on to Ball State University where he obtained an MS degree and went to work in San Antonio. He taught in high school and went to graduate school in College Station, Texas, where he completed his doctorate. He took a position at Texas A&M University in Kingsville, Texas where he now teaches in the Industrial Technology Department as a Professor and Department Chairman. He has co-authored seven books and contributed many articles to technical magazines. His hobbies include refinishing a 1970 Plymouth Super Bird and a 1971 Road-Runner. He is also interested in playing guitar, a hobby he pursued while in college as the lead guitarist for a band called The Rude Boys.

Part I

Motor Basics

Chapter 1

Motor Principles

In order to obtain a clear concept of the principles on which the electric motor operates, it is necessary first to understand the fundamental laws of magnetism and magnetic induction. It is not necessary to have a great number of expensive laboratory instruments to obtain this knowledge. Instead, children's toy magnets, automobile accessories, and so on, will suffice. It is from these principles that the necessary knowledge about the behavior of permanent magnets and the magnetic needle can be obtained.

In our early schooldays, we learned that the earth is a huge permanent magnet with its north magnetic pole somewhere in the Hudson Bay region, and that the compass needle points toward the magnetic pole. The compass is thus an instrument that can give an indication of magnetism.

The two spots on the magnet that point one to the north and the other to the south are called the poles: one is called the north-seeking pole (N) and the other the south-seeking pole (S).

Magnetic Attraction and Repulsion

If the south-seeking, or S, pole of a magnet is brought near the S pole of a suspended magnet, as in Figure 1-1, the poles repel each other. If the two N poles are brought together, they also repel each other. But if an N pole is brought near the S pole of the moving

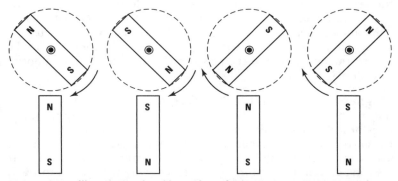

Figure 1-1 Illustrating that like poles of permanent magnets repel each other and unlike poles attract each other.

magnet, or an S pole toward the N pole, the two unlike poles attract each other. In other words, *like poles repel each other,* and *unlike poles attract each other.* It can also be shown by experiment that these attractive or repulsive forces between magnetic poles vary inversely as the square of the distance between the poles.

Effects of an Electric Current

As a further experiment, connect a coil to a battery, as shown in Figure 1-2. The compass points to one end of the coil, but if the battery connections are reversed, the compass points away from that end. Thus, the direction of the current through the coil affects the compass in a manner similar to the permanent magnet in the previous experiment.

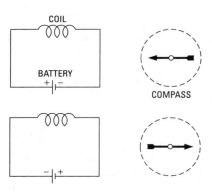

COIL

BATTERY

COMPASS

Figure 1-2   A demonstration showing that the direction of current flow through a coil affects a compass needle.

In the early part of the nineteenth century, Oersted discovered the relationship between magnetism and electricity. He observed that when a wire connecting the poles of a battery was held *over* a compass needle, the north pole of the needle was deflected as shown in Figure 1-3. A wire placed *under* the compass needle caused the north pole of the needle to be deflected in the opposite direction.

Magnetic Field of an Electric Current

Inasmuch as the compass needle indicates the direction of magnetic lines of force, it is evident from Oersted's experiment that an electric current sets up a magnetic field at right angles to the conductor. This can be shown by the experiment illustrated in Figure 1-4. If a strong current is sent through a vertical wire that passes through a horizontal piece of cardboard on which iron filings are placed, a gentle tap of the board causes the iron filings to arrange themselves

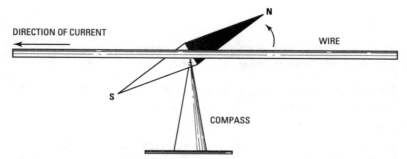

Figure 1-3   A demonstration showing that current flowing through a wire will cause a compass needle to deflect.

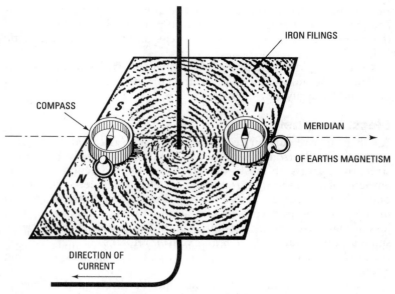

Figure 1-4   An experiment to show the direction of lines of force that surround a conductor carrying current.

in concentric rings about the wire. A compass placed at various positions on the board will indicate the direction of these lines of force, as shown in Figure 1-4.

A convenient rule for remembering the direction of the magnetic flux around a straight wire carrying current is the so-called *left-hand*

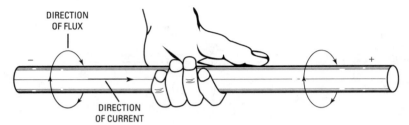

DIRECTION
OF FLUX

DIRECTION
OF CURRENT

Figure 1-5   Using the left-hand rule to determine the direction of magnetic lines of force (flux) around a conductor through which current is flowing.

rule. With reference to Figure 1-5, it can be seen that if the wire is held by the left hand, with the thumb pointing in the direction of the current, the fingers will point in the direction of the magnetic field.

Conversely, if the direction of the magnetic field around a conductor is known, the direction of the current in the conductor can be found by applying this rule.

Electromagnets

A soft-iron core surrounded by a coil of wire is called an *electromagnet.* The electromagnet owes its utility not so much to its great strength as to its ability to change its magnetic strength with the amount of the current through it. An electromagnet is a magnet only when the current flows through its coil. When the current is interrupted, the iron core returns almost to its natural state. This loss of magnetism is not absolutely complete, however, since a very small amount, called *residual magnetism,* remains.

An electromagnet is a part of many electrical devices, including electric bells, telephones, motors, and generators.

The polarity of an electromagnet may be determined by means of the left-hand rule used for a straight wire as follows: Grasp the coil with the left hand so that the fingers point in the direction of the current in the coil, and the thumb will point to the north pole of the coil (see Figure 1-6).

The strength of an electromagnet depends on the strength of the current (in amperes) multiplied by the number of loops of wire (turns)—that is, the *ampere-turns* of the coil (Figure 1-7). In practical electromagnets, it is customary to make use of both poles by

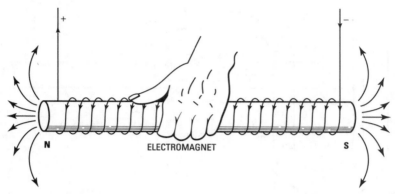

Figure 1-6 Using the left-hand rule to determine the polarity of an electromagnet.

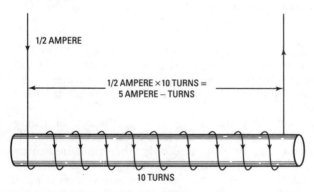

Figure 1-7 Ampere turns of a coil are equal to the product of the current (in amperes) flowing through the windings and the number of turns in the coil.

bending the iron core and the coil in the form of a horseshoe. It is from this form that the name *horseshoe magnet* is derived.

Induced Currents

If the ends of a coil of wire having many turns are connected to a sensitive galvanometer, as shown in Figure 1-8, and the coil is moved up and down over one pole of a horseshoe magnet, a deflection of the galvanometer pointer will be observed. It will also be noted that in lowering the coil, the deflection of the galvanometer

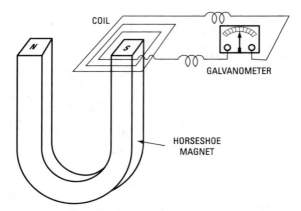

Figure 1-8 Demonstrating how the movement of a coil in a magnetic field generates an electric current.

pointer will be in a direction opposite that of the needle when the coil is raised. When the coil is lowered and held down, the galvanometer pointer returns to zero. This experiment shows that it is possible to produce a momentary electric current without an apparent electrical source.

The electrical current produced by moving the coil in a magnetic field is called an *induced current*. It is evident from the experiment that the current is induced only when the wire is moving, and that the direction of the current is reversed when the motion changes direction. Since an electric current is always made to flow by an electromotive force (*emf*), the motion of a coil in a magnetic field must generate and produce an induced electromotive force.

The direction of an induced current may be stated as follows: An induced current has such a direction that its magnetic action tends to resist the motion by which it is produced. This is known as Lenz's law.

The most useful application of induced currents is in the construction of electrical machinery of all sorts, the most common of which are the generator and motor.

A simple way to obtain a fundamental understanding of the generator is to think of the induced electromotive force produced in a single wire when it is moved across a magnetic field. Suppose wire *AB* in Figure 1-9 is pushed down through the magnetic field. An

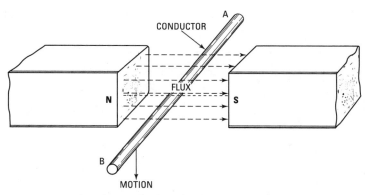

Figure 1-9 A voltage is induced when a wire cuts through magnetic lines of force.

induced emf is set up in *AB*, making point *B* at a higher potential than point *A*. This can be shown by connecting a voltmeter from *A* to *B*.

As long as the wire remains stationary, no current flows. In fact, even if the wire moves parallel to the lines of force, no current flows. Briefly, a wire must move so as to *cut lines of magnetic* force in order to have an emf induced in it.

Direct-Current (DC) Generators

A *generator* is a machine that converts mechanical energy into electrical energy. This is done by rotating an *armature,* which contains *conductors,* through a *magnetic field.* The movement of the conductors through a magnetic field produces an induced emf in the moving conductors. In any generator, a relative motion between the conductors and the magnetic field will always exist when the shaft is rotated.

Parts of a DC Generator

The principal parts of a DC generator are an armature, commutator, field poles, brushes and brush rigging, yoke or frame, and end bells or end frames (Figure 1-10). Figure 1-11 shows the parts of a DC generator.

Armature

The armature is the structure upon which the coils are mounted. These coils cut the magnetic lines of force. The armature is attached to a shaft. The shaft is suspended at each end of the machine by

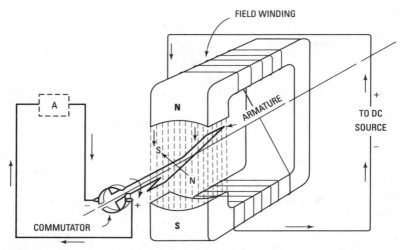

Figure 1-10 An elementary direct-current (DC) generator. A commutator keeps the current flowing in the same direction in the load circuit (A).

bearings set in the end bells, as shown in Figure 1-12. The armature core, which is circular in cross section, consists of many sheets of soft iron. The edge of the laminated core is slotted (Figure 1-13). Coil windings fit into these slots. The windings are held in place and in their slots by wooden or fiber wedges. Sometimes steel bands are also wrapped around the completed armature to provide extra support. On small generators, the laminations of the armature core are usually pressed onto the armature shaft.

Commutator

The commutator is that part of the generator that rectifies the generated alternating current to provide direct current output (Figure 1-12). It also connects the stationary output terminals to the rotating armature. A typical commutator consists of *commutator bars.* The bars are wedge-shaped segments of hard-drawn copper. These segments are insulated from each other by thin strips of mica. Commutator bars are held in place by steel V-rings or clamping flanges, as shown in Figure 1-14. These V-rings or clamping flanges are bolted to the commutator sleeve by hexagonal cap screws. The *commutator sleeve* is keyed to the shaft that rotates the armature. A mica collar or ring insulates the commutator bars from the commutator sleeve.

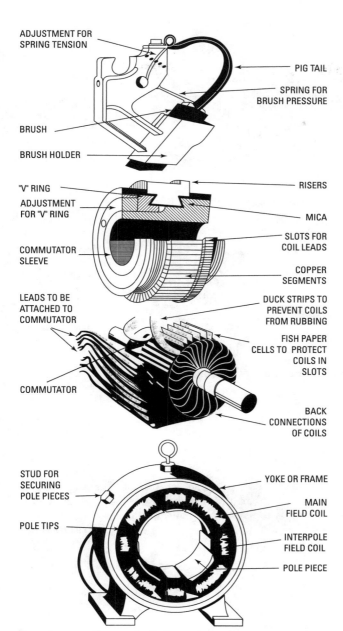

Figure 1-11   Parts of a DC generator, stationary type.

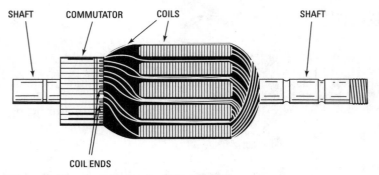

Figure 1-12   Armature of a DC generator.

Figure 1-13   Unwound armature core on a shaft.

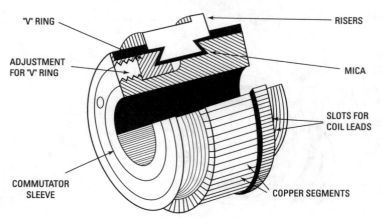

Figure 1-14   Commutator construction.

The commutator bars usually have risers or flanges to which the leads from the associated armature coils are soldered. These risers serve as a shield for the soldered connections when the commutator bars become worn. When risers are not provided, it is necessary to solder the leads from the armature coils to short slits in the ends of the commutator bars. The brushes contact the commutator bars.

The brushes collect the current generated by the armature coils. The brush holders transfer the current to the main terminals. The commutator bars are insulated from each other. Thus, each set of brushes, as it contacts the commutator bars, collects current of the same polarity. This results in a continuous flow of direct current. The finer the division of the commutator bars, the less the ripple that appears in the current, and therefore, the smoother the flow of the DC output.

Field Pole and Frame

The frame or yoke of a generator serves two purposes. It provides mechanical support for the machine. It is also a path for the completion of the magnetic circuit. The lines of force that pass from the north to the south pole through the armature are returned to the north pole through the frame. Frames are made of electrical-grade steel. The method of construction of field poles and frames varies with the manufacturer. Figure 1-15 shows the magnetic circuit of a two-pole generator.

Field Windings

The field windings are connected so that they produce alternate north and south poles, as shown in Figure 1-16. Connection is done that way to obtain the correct direction of emf in the armature conductors. The field windings form an electromagnet that establishes the generator field flux. These field windings may receive current from an external DC source, or they may be connected directly across the armature, which then becomes the source of voltage. When the windings are energized, they establish magnetic flux in the field yoke, pole pieces, air gap, and armature core (Figure 1-15).

Brushes and Brush Holders

The brushes carry the current from the commutator to an external circuit. Usually they are a mixture of graphite and metallic powder. Brushes are designed to slide freely in their holders because the commutator surface is usually uneven, and the brushes and commutator wear. The freedom thus allows the brushes to have good contact with the commutator despite wear or uneven surfaces.

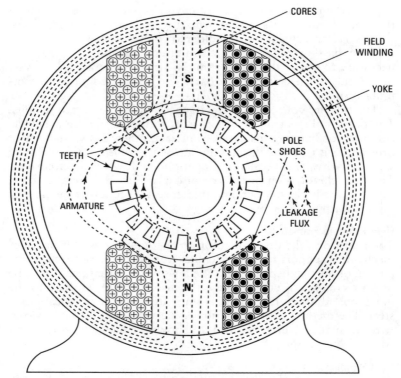

Figure 1-15 Magnetic circuit of a two-pole generator.

Proper pressure of the brushes against the commutator is maintained by means of springs. This pressure is usually about 1½ to 2 pounds per square inch of brush contact area. A low resistance connection—usually braided copper wire—is provided between the brushes and brush holders.

Armature Windings

The simplest generator armature winding is a *loop* or *single coil*. Rotating this loop in a magnetic field induces an emf. The strength of the magnetic field and the speed of rotation of the conductor determine the emf produced.

A *single-coil generator* is shown in Figure 1-17. Each coil terminal is connected to a bar of a two-segment metal ring. The two segments of the split rings are insulated from each other and the shaft. This forms a simple commutator. The commutator mechanically

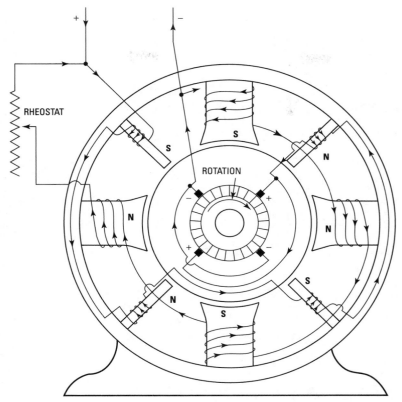

Figure 1-16   Schematic wiring diagram of a shunt generator.

reverses the armature coil connections to the external circuit at the same instant that the direction of the generated voltage reverses the armature coil. This process is known as *commutation*.

Figure 1-18 is a graph of a *pulsating current* (DC) for one rotation of a single-loop, two-pole armature. A pulsating current or *direct voltage* has ripple. In most cases, this current is not usable. More coils have to be added.

The heavy black line in Figure 1-19 shows the DC output of a two-loop (coil) armature. A great reduction in voltage ripple is obtained by using two coils instead of one. Since there are now four commutator segments in the commutator and only two brushes, the voltage cannot fall lower than point *A*. Therefore, the ripple is limited to the rise and fall between points *A* and *B*. Adding more armature coils will reduce the ripple even more.

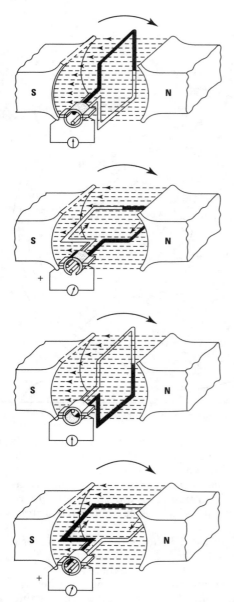

Figure 1-17 Single-coil generator with commutator.

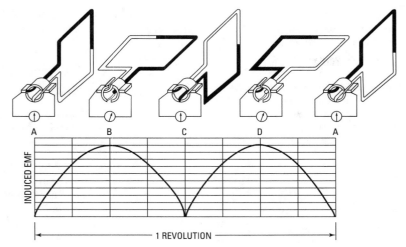

Figure 1-18   Output of a single-coil DC generator.

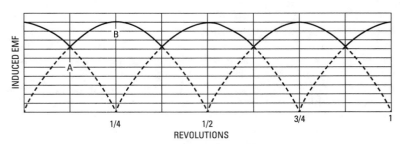

Figure 1-19   Output voltage from a two-coil armature.

Armature Losses

There are three losses in every DC generator armature. One is the *copper loss* in the winding. The second is the *eddy current loss* in the core. The third is the *hysteresis loss* caused by the friction of the revolving magnetic particles in the core.

Copper Losses

Copper loss is the power lost in heat in the windings due to the flow of current through the copper coils. This loss varies directly with the armature resistance and the square of the armature current. The armature resistance varies inversely with the cross-sectional area.

Armature copper loss varies mainly because of the variation of electrical load on the generator and not because of any loss occurring in the machine. This is because most generators are constant-potential machines supplying a current output that varies with the electrical load across the brushes. The limiting factor in load on a generator is the allowable current rating of the generator armature.

The armature circuit resistance includes the resistance of the windings between brushes of opposite polarity, the brush contact resistance, and the brush resistance.

Eddy Current Losses

If a DC generator armature core were made of solid iron and rotated rapidly in the field, as shown in Figure 1-20, part A, excessive heating would develop even with no-load current in the armature windings. This heat would be the result of a generated voltage in the core itself. As the core rotates, it cuts the lines of magnetic field flux at the same time the copper conductors of the armature cut them. Thus, induced currents alternate through the core, first in one direction and then in the other. These currents cause heat.

Such induced currents are called *eddy currents*. They can be minimized by sectionalizing *(laminating)* the armature core. For instance, a core is split into two equal parts, as shown in Figure 1-20, part B. These parts are insulated from each other. The voltage induced in each section of iron is thus one-half of what it would have been if it remained solid. The resistance of the eddy current paths is doubled. That is because resistance varies inversely with the cross-sectional area of the lamination.

If the armature core is subdivided into many sections or laminations, as in Figure 1-20, part C, the eddy current loss can be reduced to a negligible value. Reducing the thickness of the laminations reduces the magnitude of the induced emf in each section. It also increases the resistance of the eddy current paths. Laminations in small generator armatures are usually $\frac{1}{64}$ in. thick. Often the laminations are insulated from each other by a thin coat of lacquer. Sometimes they are insulated simply by the oxidation of the surfaces caused by contact with the air while the laminations are being annealed. The voltages induced in laminations are small; thus the insulation need not be great.

All electrical rotating machines and transformers are laminated to reduce eddy current losses.

Eddy current loss is also influenced by speed and flux density. The induced voltage, which causes the eddy currents to flow, varies

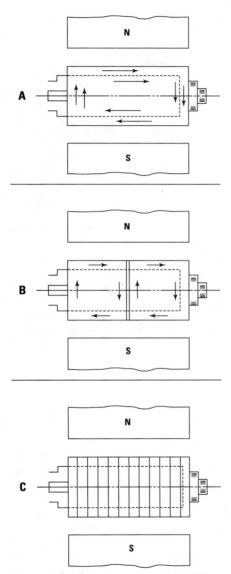

Figure 1-20 Eddy currents in a DC generator armature core.

with the speed and flux density. Therefore, the power loss,

$$P = \frac{E^2}{R}$$

varies as the square of the speed and the square of the flux density.

Hysteresis Losses

When an armature revolves in a stationary magnetic field, the number of magnetic particles of the armature that remain in alignment with the field depends on the strength of the field. If the field is that of a two-pole generator, these magnetic particles will rotate, with respect to the particles not held in alignment, one complete turn for each revolution of the armature. The rotating of the magnetic particles in the mass of iron produces friction and heat.

Heat produced this way is called *magnetic hysteresis loss.* The hysteresis loss varies with the speed of the armature and the volume of iron. The *flux density* varies from approximately 50,000 lines per square inch in the armature core to 130,000 lines per square inch in the iron between the bottom of adjacent armature slots (the *tooth root*). Heat-treated silicon steel having a low hysteresis loss is used in most DC generator armatures. The steel is formed to the proper shape. Then the laminations are heated to a dull red heat and allowed to cool. This annealing process reduces the hysteresis loss to a low value.

Armature Reaction

Armature reaction in a generator is the effect on the main field of the armature acting as an electromagnet. With no armature current, the field is undistorted (Figure 1-21, part A). This flux is produced entirely by the ampere-turns of the main field windings. The neutral plane *AB* is perpendicular to the direction of the main field flux. When an armature conductor moves through this plane, its path is parallel to the undistorted lines of force. Thus, the conductor does not cut through any flux, and no voltage is induced in the conductor. The brushes are placed on the commutator so that they short-circuit coils passing through the neutral plane. With no voltage generated in the coils, no current will flow through the local path formed momentarily between the coils and segments spanned by the brush. Therefore, no sparking at the brushes will result.

When a load is connected across the brushes, armature current flows through the armature conductors. The armature itself

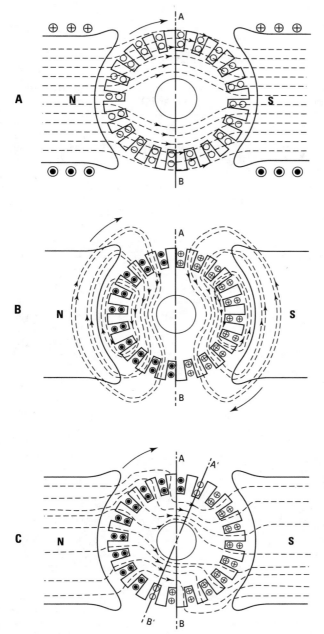

Figure 1-21 Flux distribution in a DC generator.

21

becomes a source of magnetomotive force. The effect of the armature acting as an electromagnet is shown in Figure 1-21, part B. The main field coils are de-energized, and full-load current is applied to the armature circuit from an external source. The conductors on the left of the neutral plane all carry current toward the observer. Those on the right carry current away from the observer. These directions are the same as those the current would follow under the influence of the normal emf generated in the armature with normal field excitation.

These armature-current-carrying conductors establish magnetomotive force that is perpendicular to the axis of the main field. In Figure 1-21, part B the force acts downward. This magnetizing action of the armature current is called *cross magnetization.* It is present only when current flows through the armature circuit. The amount of cross magnetization produced is proportional to the armature current.

When current flows in both the field and armature circuits, the two resulting magnetomotive forces distort each other (Figure 1-21, part C). They twist in the direction of rotation of the armature. The mechanical (no-load) neutral plane, *AB,* is now advanced to the electrical (load) neutral plane *A'B'.* When the armature conductors move through plane *A'B',* their paths are parallel to the distorted field. The conductors cut no flux. Thus, no voltage is induced in them. The brushes must, therefore, be moved on the commutator to the new neutral plane. They are moved in the direction of armature rotation. The absence of sparking at the commutator indicates the correct placement of the brushes. The amount that the neutral plane shifts is proportional to the load on the generator. That is because the amount of cross-magnetizing magnetomotive force is directly proportional to the armature current.

Effects of Brush Shift
When the brushes are shifted into the electrical neutral plane *A'B',* the direction of the armature magnetomotive force is downward and to the left, instead of vertically downward (Figure 1-22, part A). The armature magnetomotive force may now be resolved into two components (Figure 1-22, part B).

The conductors at the top and bottom of the armature within sectors *BB* produce a magnetomotive force that is directly in opposition to the main field and weakens it. This component is called the *armature-demagnetizing mmf.* The conductors on the right and left sides of the armature within sector *AA* produce a *cross-magnetizing*

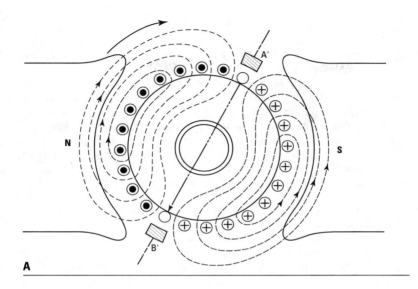

A

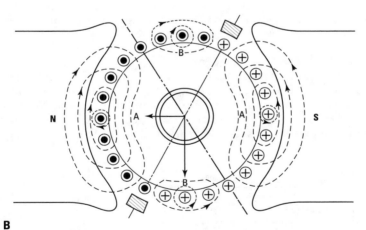

B

Figure 1-22   Effect of brush shift on armature reaction.

mmf at right angles to the main field axis. This cross-magnetizing force tends to distort the field in the direction of rotation. As mentioned, the distortion of the main field of the generator is the result of armature reaction. Armature reaction occurs in the same manner in multipolar generators.

Compensating for Armature Reaction

The effects of armature reaction are reduced in DC machines by the use of *high-flux density pole tips,* a *compensating winding,* and *commutation poles.*

The cross-sectional area of the pole tips is reduced by building field poles with laminations having only one tip. These laminations are alternately reversed when the pole core is stacked so that a space exists between alternate laminations at the pole tips. The reduced cross section of iron at the pole tips increases the flux density. Thus, they become saturated. Cross-magnetizing and demagnetizing forces of the armature do not affect the flux distribution in the pole face as much as they would reduced flux densities.

The compensating winding consists of conductors embedded in the pole faces parallel to the armature conductors. The winding is connected in series with the armature. It is arranged so that the ampere-turns are equal in magnitude and opposite in direction to those of the armature. The magnetomotive force of the compensating winding, therefore, neutralizes the armature magnetomotive force, and armature reaction is almost eliminated. Compensating windings are costly, so they are generally used only on high-speed and high-voltage large-capacity generators.

Motor Reaction in a Generator

When a generator supplies current to a load, the load current creates a force that opposes the rotation of the generator armature. An *armature conductor* is represented in Figure 1-23. When a conductor is moved downward and the circuit is completed through an external load, current flows through the conductor in

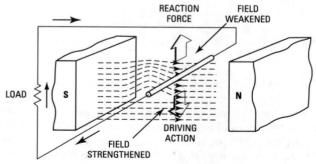

Figure I-23 Motor reaction in a generator.

the direction indicated. This causes lines of force around it in a clockwise direction.

The interaction of the conductor field and the main field of the generator weakens the field above the conductor and strengthens it below the conductor. The field consists of lines that act like stretched rubber bands. Thus an upward reaction force is produced that opposes the downward driving force applied to the generator armature. If the current in the conductor increases, the reaction force increases. More force must then be applied to the conductor to keep it from slowing down.

With no armature current, no magnetic reaction exists. Therefore, the generator input power is low. As armature current increases, the reaction of each armature conductor against rotation increases. The driving power to maintain the generator armature speed must be increased. If the prime mover driving the generator is a gasoline engine, this effect is accomplished by opening the throttle of the carburetor. If the prime mover is a steam turbine, the main steam-admission valve is opened wide so that more steam can flow through the turbine.

Types of DC Generators

DC generators are classified by how excitation current is supplied to the field coils. There are two major classifications:

- Separately excited
- Self-excited

Self-exciting generators are further classified by the method of connecting the field coils. These include series-connected, shunt-connected, and compound-connected generators.

Separately Excited Generators

A separately excited generator is one for which the field current is supplied by another generator, by batteries, or by some other outside source. Figure 1-24 shows a typical circuit.

Figure 1-25 shows the voltage characteristics of a separately excited generator. When operated at constant speed with constant field excitation but not supplying current, the terminal voltage of this type of generator equals the generated voltage. When the unit is delivering current, the terminal voltage is less than the generated voltage. The total amount of voltage drop equals the drop due to armature reaction plus the voltage drop due to the resistance of the armature and the brushes. Separately excited generators, however, are seldom used.

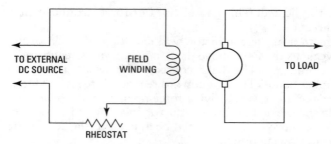

Figure 1-24 Connection of a separately excited DC generator.

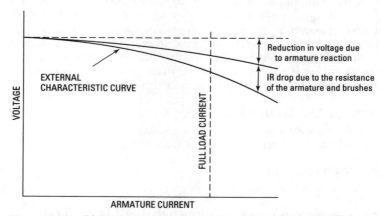

Figure 1-25 Voltage characteristics of a separately excited DC generator.

Self-Excited Generators
There are three types of self-excited generators:

- Series
- Shunt
- Compound

There are some variations of the compound type.

Series Generators
When all the windings are connected in series with the armature, a generator is series-connected. See Figure 1-26 for the typical series-connected circuit. Figure 1-27 shows the voltage characteristics of a

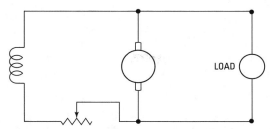

Figure 1-26   A typical series DC generator circuit.

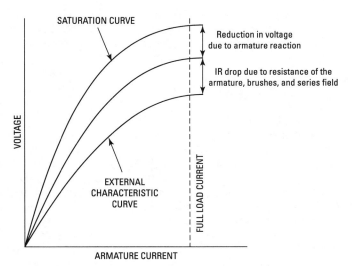

Figure 1-27   Voltage characteristics of a series DC generator.

series generator. With no load, the only voltage present is due to the cutting of the flux established by residual magnetism. (*Residual magnetism* is magnetism retained by the poles of a generator when it is not in operation.) However, when a load is applied or increased, the current through the field coil increases the flux. Therefore, the generated voltage increases. The voltage generated tends to increase directly as the current increases, but three factors lessen the voltage increase.

One factor is *saturation of the field core*. If field excitation is increased beyond the point at which the flux produced no longer increases directly as the exciting current, the core is said to be

28   Chapter I

saturated. The second factor is *armature reaction*. The effect of this reaction increases as the current load increases. The third factor is *loss in terminal voltage*. This loss is caused by ohmic resistance of the armature winding, brushes, and series field. This loss increases as the unit is loaded. Since the terminal voltage of a series generator varies under changing load conditions, it is generally connected in a circuit that demands constant current. When used that way, it is sometimes referred to as a *constant-current generator*, even though it does not tend to maintain a constant current itself. Constant current is achieved by connecting a variable resistance in parallel with the series field. The variable resistance can be manually or automatically controlled. Thus, as the load is increased, the resistance of the shunt path is decreased. This permits more of the current to pass through it and maintains a relatively constant field.

Shunt Generators

When the field windings are connected in parallel with the armature, the generator is shunt-connected. Figure 1-28 is a typical circuit of a shunt generator; Figure 1-29 shows the voltage characteristics of a shunt generator. A comparison of the voltage characteristics of a shunt generator shows they are similar to those of a separately excited generator. In both instances, the terminal voltage drops from the no-load value as the load is increased. But note that the terminal voltage of the shunt generator remains fairly constant until it approaches full load. This is true even though the graph of the shunt generator has an extra factor that causes the terminal voltage to decrease: the weakening of the field as the current approaches full load. It is, therefore, better to use a shunt generator, and not a separately excited or a series generator, when a constant voltage with a varying load is required. Shunt generators

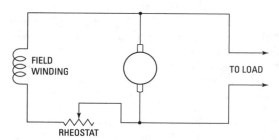

Figure 1-28   Connection of a shunt DC generator.

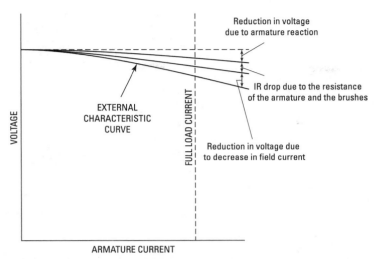

Figure 1-29   Voltage characteristics of a shunt DC generator.

are readily adaptable to applications where the speed of the prime mover cannot be held constant. Aircraft and automobile engines are typical examples of variable-speed prime movers that require a constant voltage. Constant voltage is obtained by controlling generator field current, which is accomplished by varying the shunt field resistance to compensate for changes in speed of the prime mover.

Compound Generators
If both a series and a shunt field are included in the same unit, it is possible to obtain a generator with a voltage-load characteristic somewhere between those of a series and a shunt generator. Figure 1-30 shows typical circuits of a compound-wound (series-shunt) DC generator. Figure 1-30, part A, is a cumulative compounded generator. Its series and shunt fields are wound to aid each other. Figure 1-31 shows the voltage characteristics of a compound-wound DC generator. By changing the number of turns in the series field, it is possible to obtain three distinct types of compound generators.

Overcompounded An *overcompounded generator* is one in which there are more turns in the series field than necessary to give about the same voltage at all loads. Thus, the terminal voltage at full load will be higher than the no-load voltage. This is desirable when

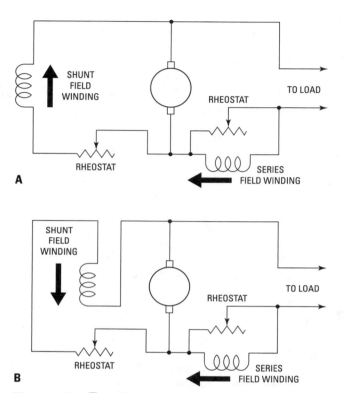

Figure 1-30 Typical compound-wound DC generator circuits.

power must be transmitted a long distance. The higher generated voltage compensates for the voltage loss in the transmission line.

Flat Compounded A *flat-compounded generator* is one in which the relationship of the turns in the series and shunt fields is such that their terminal voltage is about the same over the entire load range.

Undercompounded An *undercompounded generator* is one in which the series field does not have enough turns to compensate for the voltage drop of the shunt field. The voltage at full load is less than the no-load voltage. In an undercompounded generator, the series and shunt fields are connected so as to oppose rather than to aid one another. It is referred to as being *differentially compounded*. The terminal voltage of this type of generator decreases rapidly as the load increases. Undercompounded generators are used in applications where a short circuit might occur, as in an arc welder.

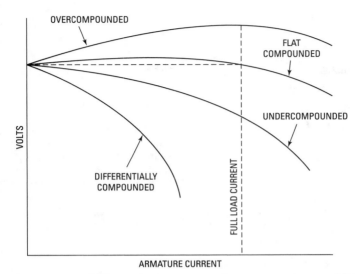

Figure 1-31 Voltage characteristics of a compound-wound DC generator.

Control of DC Generators

Generally, the DC generator is controlled by a resistor that produces variable resistance, called a *rheostat*. After the generator is brought up to speed by the prime mover, the rheostat is adjusted. The rheostat may be manually or automatically operated. The adjustment of the rheostat controls the amount of excited current fed to the field coils. Metering requires the use of a DC voltmeter and ammeter of appropriate ranges in the generator output circuit. Matched sets of shunt-wound or compound-wound generators with series-field equalizer connections are used for parallel operation. Precautions must be observed when connecting the machines to generator buses.

Series Generator

The series generator is classified as a *constant-speed generator*. It can be used to supply series motors, series arc-lighting systems, and voltage boosting on long DC feeder lines. The series generator is excited entirely by low-resistance field coils connected in series with the armature terminals and the load. The circuit of a DC series generator is shown in Figure 1-26. The voltage increases with load because the load current provides the necessary additional field excitation. Low-resistance shunts may be used across series field coils to obtain desired voltage characteristics. The series field of the generator is adjusted so the output voltage may be maintained at a constant value. Because series generators have poor voltage regulation, only a few are in use.

Shunt Generator

The shunt DC generator can be called a *constant-potential generator*. It is seldom used for lighting and power because of its poor voltage regulation. The field coils in this type of generator have a comparatively high resistance; they are connected across the armature terminals in series with the rheostat. A DC shunt generator circuit is shown in Figure 1-28. Shunt generators sometimes have separate excitation. This prevents reversal of the generator polarity and allows better voltage regulation. Shunt generators are frequently used with automatic voltage regulators as exciters for AC generators.

Compound Generator

The compound generator is the most widely used DC generator. The speed of a compound generator affects its generating characteristics. Therefore, the compounding can be varied. The engine governor can be adjusted for the proper no-load voltage. The range of the shunt-field rheostat and the engine characteristic limit the amount of speed variation that can be obtained. Compound generators can be connected either cumulatively or differentially.

Direct-Current Motors

A machine that converts electrical energy into mechanical energy is called a *motor*. The functions of a DC generator and a DC motor are interchangeable in that a generator may be operated as a motor, and vice versa. Structurally, the two machines are identical. The motor, like the generator, consists of an electromagnet, an armature, and a commutator with its brushes.

Figure 1-10 will serve to illustrate the operation of a direct-current motor as well as a generator. The magnetic field, as indicated, will be the same for a motor because of current flowing in the field windings. Now, let the outside current at *A* have a voltage applied that causes a current to flow in the armature loop, as indicated by the arrows.

It must be remembered that any current flowing in a loop or coil of wire produces a magnetic field. This is exactly what happens in the armature of this motor. In addition, a second magnetic field is produced, with poles N and S perpendicular to the armature loop. The north pole of the main magnetic field attracts the south pole of the armature, and since the loop is free, it will revolve. At the instant the north and south poles become exactly opposite, however, the commutator reverses the current in the armature, making the poles of the field and the armature opposite, and the loop is then repelled and forced to revolve further. Again the armature current is reversed when unlike poles approach, and the armature is free to revolve. This continues as long as there is current in the armature and field windings.

It should be observed that, in an actual motor, there is more than one loop (called an *armature coil*), each with its terminals connected to adjacent commutator segments (Figure 1-11). Hence, the attracting and repelling action is correspondingly more powerful and also more uniform than that of the weak and unstable action obtained with the single-loop armature described here.

The various types of direct-current motors as well as their operating characteristics and control methods are fully treated in a later chapter.

Alternating-Current Motors

When a coil of wire is rotated in a magnetic field, the current changes its direction every half turn. Thus, there are two alternations of current for each revolution of a bipolar machine. As previously noted, this alternating current is rectified by the use of a commutator in a direct-current generator. In an alternating-current generator, also termed an *alternator,* the current induced in the armature is led out through *slip rings* or *collector rings,* as shown in Figure 1-32.

A magnetic field is established between the north pole and the south pole by means of an exciting current flowing in winding W. A loop of wire, L, in this field is arranged so that it can be rotated on axis X, and the ends of this loop are brought out to slip rings SS, on which brushes BB can slide.

This circuit, of which the rotating loop is a part, is completed through the slip rings at A. When the loop is rotating, voltage is produced in conductors F and G, which will cause a current to flow out to A where the circuit is completed.

The simplified machine represented in Figure 1-32 is a two-pole, single-phase, revolving-armature, alternating-current generator. The magnetic field—the coils of wire and iron core—are called simply the field of the generator. The rotating loop in which the voltage is induced is called the armature.

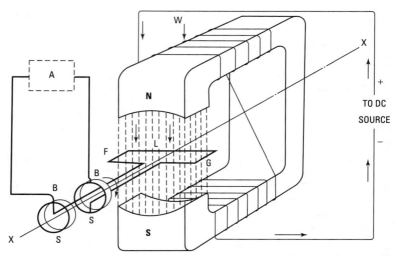

Figure 1-32 An elementary alternating-current (AC) generator of the rotating-armature type. The slip rings and brushes are used to collect the current from the armature.

The rotating-armature type of generator is generally used only on small machines, whereas large machines almost without exception are built with rotating fields.

If the voltage completes 60 cycles in one second, the generator is termed a *60-hertz* machine. The current that this voltage will cause to flow will be a 60-hertz current. The term *hertz* (Hz) indicates cycles per second.

Polyphase Machines

A two-phase generator is actually a combination of two single-phase generators, as shown in Figure 1-33. The armatures of these two machines are mounted on one shaft and must revolve together, always at right angles to each other. If the voltage waves or curves are plotted as in Figure 1-34, it will be found that when phase 1 is in such a position that the voltage is at a maximum, phase 2 will be in such a position that the voltage in it is zero. A quarter of a cycle later, phase 1 will be zero and phase 2 will have advanced to a position previously occupied by 1, and its voltage will be at a maximum. Thus, phase 2 follows phase 1 and the voltage is always exactly a quarter of a cycle behind because of the relatively mechanical positions of the armatures.

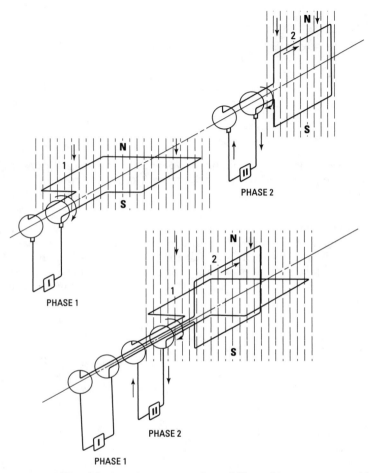

Figure 1-33 An elementary two-phase AC machine constructed by combining the two single-phase machines shown at the top of the illustration.

It has been found economical to have more than one coil for each pole of the field. Because of this, present-day AC generators are built as three-phase units in which there are three sets of coils on the armature. These three sets of armature coils may each be used separately to supply electricity to three separate lighting circuits.

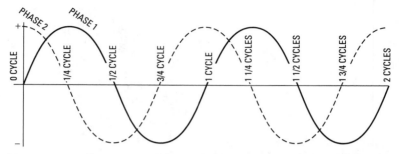

Figure 1-34 Curves representing the voltage in two separate loops of wire that are positioned at right angles to each other and rotating together.

In a three-phase generator, three single-phase coils (or windings) are combined on a single shaft and rotate in the same magnetic field, as shown in Figure 1-35. Each end of each coil is brought out through a slip ring to an external circuit. The voltage in each phase alternates exactly one-third of a cycle after the one ahead of it

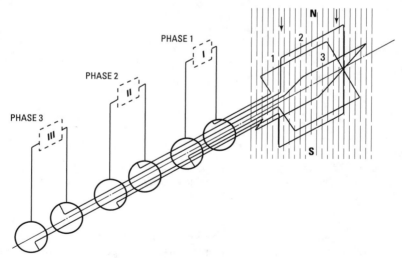

Figure 1-35 An elementary three-phase AC machine is one in which three separate loops of wire are displaced from one another at equal angles, the loops made to rotate in the same magnetic field, and each loop brought out to a separate pair of slip rings.

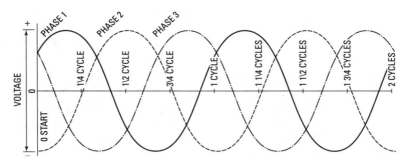

Figure 1-36 Curves illustrating the voltage variation in a three-phase machine. One cycle of rotation produces 1 Hz of alternating current.

because of the mechanical arrangement of the windings on the armature. Thus, when the voltage in phase 1 is approaching a maximum positive, as shown in Figure 1-36, the voltage in phase 2 is at a maximum negative, and the voltage in phase 3 is declining. The succeeding variations of these voltages are as indicated.

In practice, the ends of each phase winding are not brought out to separate slip rings, but are connected as shown in Figure 1-37.

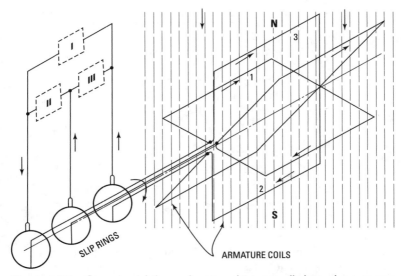

Figure 1-37 Commercial three-phase machines usually have the separate loops of wire connected as shown. This requires only three slip rings.

This arrangement makes only three leads necessary for a three-phase winding, each lead serving two phases. This allows each pair of wires to act like a single-phase circuit that is substantially independent of the other phases.

Revolving Magnetic Field

In the diagrams studied thus far, the poles producing the magnetic field have been stationary on the frame of the machine, and the armature in which the voltages are produced rotates. This arrangement is universally employed in direct-current machines, but alternating-current motors and generators generally have revolving fields because they need only two slip rings.

When the revolving-field construction is employed, the two slip rings need only carry the low-voltage exciting current to the field. For a three-phase machine with a rotating armature, at least three slip rings would be required for the armature current, which is often at a high voltage and therefore would require a large amount of insulation, adding to the cost of construction. A schematic of a single-phase AC generator with revolving field is shown in Figure 1-38.

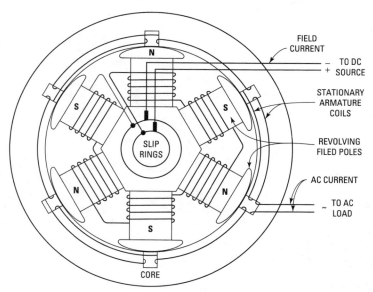

Figure 1-38 Construction details of an AC generator having six poles and a revolving field. If this generator is to deliver 60 Hz current, it must be driven at a speed of exactly 1200 rpm.

The operation of practically all polyphase alternating-current motors depends on a revolving magnetic field that pulls the rotating part of the motor around with it.

To produce a rotating field, assume that two alternating currents of the same frequency and potential, but differing in phase by 90°, are available. Connect them to two sets of coils wound on the inwardly projecting poles of a circular iron ring, as illustrated in Figure 1-39. It will be noted that when the current in phase 1 is at a maximum, the current in phase 2 is zero. Poles A and A_1 are magnetized, while poles B and B_1 are demagnetized. The magnetic flux is in a direction from N to S, as indicated by the arrow in the center of diagram I.

Referring to the voltage curves, it will be found that one-eighth of a cycle (45°) later, the current in phase 1 has decreased to the same value to which the current in phase 2 has increased. The four poles are now equally magnetized, and the magnetic flux takes the direction of the arrow shown in the center of diagram II.

One-eighth of a cycle later, the current in phase 1 has dropped to zero, while the current in phase 2 is at its maximum. With reference to diagram III, in Figure. 1-39, this condition indicates that poles A and A_1 are demagnetized, but that poles B and B_1 are magnetized, with the flux from N to S, as shown by the center arrow.

Continuing the analysis, notice that after an additional one-eighth of a cycle, the current in both phases 1 and 2 has decreased and that the four poles are again equally magnetized, with the magnetic flux in the direction as indicated by the center arrow in diagram IV. If this process is continued at successive intervals during a complete period or cycle of change in the alternating current, the magnetic flux represented by the arrow will make a complete revolution for each cycle of the current.

The action of the current is inducing a rotating magnetic field, which would cause a magnet to revolve on its axis according to the periodicity of the impressed alternating current. This analysis explains the action in a two-phase motor. The rotating magnetic field in a three-phase AC motor having any number of poles can be similarly obtained.

Synchronous Motors

Any AC generator can be employed as a motor, provided that it is first brought up to the exact speed of a similar generator supplying the current to it, and provided that it is then put in step with the alternations of the supplied current. Such a machine is called a *synchronous motor*. However, because of complications in starting,

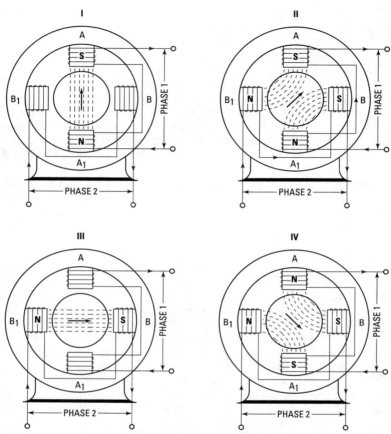

Figure 1-39 Illustrating how a rotating magnetic field is produced by two currents 90° apart.

most synchronous motors of late construction are equipped with a *damper* or *amortisseur* winding, which produces a starting torque, permitting them to be started as induction motors.

The speed of a synchronous motor depends on the frequency of the current supplied to it and the number of poles in the motor. The equation for the speed is

$$\text{revolutions per minute} = \frac{\text{frequency} \times 60}{\text{number of pairs of poles}}$$

Since a synchronous motor runs at exactly this speed, it is a relatively simple matter to calculate the speed of any motor provided that the number of poles and the frequency of the source are known. Thus, for example, an eight-pole synchronous motor operating from a 60-Hz source has a speed of

$$\text{rpm} = \frac{60 \times 60}{4} = 900 \text{ rpm}$$

Induction Motors

Although the synchronous motor is used commercially in certain applications, the induction motor is used more extensively because of its simplicity. There are two principal classes of polyphase motors, namely:

- Squirrel cage
- Wound rotor

By definition, an induction motor is one in which the magnetic field in the rotor is induced by currents flowing in the stator. The rotor has no connections whatever to the supply line.

Squirrel-Cage Motor

This type of induction motor consists of a *stator,* which is identical to the armature of a synchronous motor, with a "squirrel-cage" rotor with bearings to support it. Because the stator receives the power from the line, it is often called the *primary* and the rotor the *secondary.*

In an induction motor of this type, the squirrel-cage winding takes the place of the field in the synchronous motor. The squirrel cage consists of a number of metal bars connected at each end to supporting metal rings. As in the synchronous motor, a rotating field is set up by the currents in the armature.

As this field revolves, it cuts the squirrel-cage conductors, and voltages are set up in them exactly as though the conductors were cutting the field in any other motor. These voltages cause current to flow in the squirrel-cage circuit, through the bars under the north poles into the ring, back to the bars under the adjacent south poles, into the other ring, and back to the original bars under the north pole, completing the circuit.

The current flowing in the squirrel cage, down one group of bars and back in the adjacent group, makes a loop that establishes magnetic fields with north and south poles in the rotor core. This loop consists of one turn, but there are several conductors in parallel and the currents may be large. These poles in the rotor, attracted by the poles of the revolving field, set up the currents in the armature winding and follow them around in a manner similar to that in which the field poles follow the armature poles in a synchronous motor.

There is, however, one interesting and important difference between the synchronous motor and the induction motor: The rotor of the latter does not rotate as fast as the rotating field in the armature. If the squirrel cage were to go as fast as the rotating field, the conductors in it would be standing still with respect to the rotating field, rather than cutting across it. Thus, there could be no voltage induced in the squirrel cage, no currents in it, no magnetic poles set up in the rotor, and no attraction between it and the rotating field in the stator. The rotor revolves just enough slower than the rotating field in the stator to allow the rotor conductors to cut the rotating magnetic field as it slips by, and thus induces the necessary currents in the rotor windings.

This means that the motor can never rotate quite as fast as the revolving field, but is always slipping back. This difference in speed is called the *slip*. The greater the load, the greater the slip will be—that is, the slower the motor will run—but even at full load, the slip is not too great. In fact, this motor is commonly considered to be a constant-speed device. The various classes of squirrel-cage motors, and their operation and control, are given in a later chapter.

Wound-Rotor Motor

This type of induction motor differs from the squirrel-cage type in that it has wire-coil windings in it instead of a series of conducting bars in the rotor. These insulated coils are grouped to form definite polar areas having the same number of poles as the stator. The rotor windings are brought out to slip rings whose brushes are connected to variable external resistances (Figure 1-40).

By inserting external resistance in the rotor circuit when starting, a high torque can be developed with a comparatively low starting current. As the motor accelerates up to speed, the resistance is gradually reduced until, at full speed, the rotor is short-circuited. By varying the resistance at the rotor circuit, the motor speed can be regulated within practical limits.

This method of speed control is well-suited for the wound-rotor motor because it is already equipped with a starting resistance in each phase of the rotor circuit. By making these resistances of large-enough, current-carrying capacity to prevent dangerous heating in

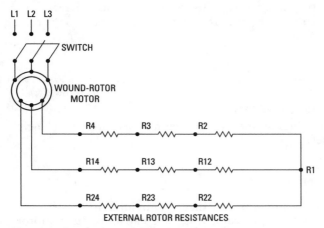

Figure 1-40 Wiring diagram showing the connections between the slip rings and external resistances for a wound-rotor motor. The resistances are connected to a drum controller, the drum rotation of which determines the amount of resistance in the circuit and thus the speed of the motor.

continuous service, the same resistances can also serve to regulate the speed. Although effective speed control is best secured by the use of direct-current motors, the wound-rotor motor, because of its adjustable rotor resistance, possesses one of the few methods of speed control available for alternating-current motors.

Slip
The speed of a synchronous motor is constant for any given frequency and number of poles in the motor. In an induction motor, however, this exact relationship does not exist, because the rotor slows down when the load is applied. The ratio of the speed of the field (relative to the rotor) to synchronous speed is termed the *slip*. It is usually written:

$$s = \frac{N_s - N}{N_s}$$

where s is the slip (usually expressed as a percentage of synchronous speed), N_s is the synchronous speed, N is the actual rotor speed.

For example, a six-pole, 60-Hz motor would have a synchronous speed of 1200 rpm. If its rotor speed were 1164 rpm, the slip would be:

$$s = \frac{1200 - 1164}{1200} = 0.03, \quad \text{or} \quad 3\%$$

Single-Phase Motors

Single-phase induction motors may be divided into two principal classes, namely

- Split-phase
- Commutator

Split-phase motors are further subdivided into resistance-start, reactor-start, split-capacitor, and capacitor-start motors. Commutator-type motors are subdivided into two groups, series and repulsion, and each of these is further subdivided into several types and combinations of types.

These two classes and subdivisions represent various electrical modifications of single-phase induction motors, where one modification must be used to produce the necessary starting torque. All methods serve to increase the phase angle between the main winding and the starting winding so as to produce a rotating magnetic field similar to that in a two-phase motor.

Resistance-Start Motors

A resistance-start motor is a form of split-phase motor having a resistance connected in series with the starting (sometimes called *auxiliary*) winding. A schematic diagram of this type of motor may be represented as in Figure 1-41. This shows a resistance connected in series with the starting winding to provide a two-phase rotating-field effect for starting. When the motor reaches approximately 75 percent of its rated speed, a centrifugal switch opens to disconnect the starting winding from the line. This motor is known as resistance-start, split-phase type and is commonly used on washing machines and similar appliances. It is not practical to build such motors for the heavier types of starting duty.

Split-Capacitor Motors

In a split-capacitor motor (Figure 1-42), two stationary windings are connected to a single-phase line. The capacitor has the peculiar characteristic of shifting the phase of the current in coil 2 with

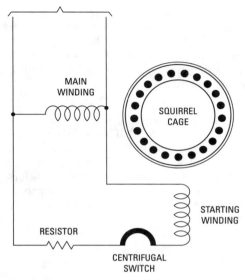

Figure 1-41 Schematic diagram showing the winding arrangement of a resistance-start, split-phase, induction motor.

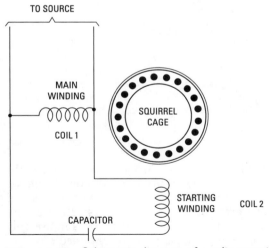

Figure 1-42 Schematic diagram of a split-capacitor motor.

respect to the current in coil 1. This provides the same action as in the two-phase motor discussed previously, producing the rotating field effect to rotate the squirrel-cage rotor. The capacitor is mounted permanently in the circuit. Because capacitors for continuous duty are expensive and somewhat bulky, it is not practical to make this motor for heavy-duty starting.

Capacitor-Start Motors

In applications where a high starting torque is required, a motor such as that shown in Figure 1-43 is employed. This is only another form of split-phase motor having a capacitor (or condenser, as it was once called) connected in series with the starting winding. The construction is similar to the split-capacitor motor, but differs mainly in that the starting winding is disconnected at approximately 75 percent of rated speed by a centrifugal switch, as in the case of the resistance-start motor.

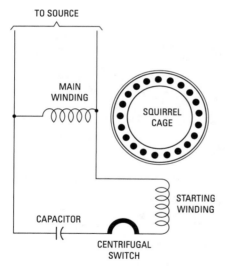

Figure 1-43 Winding connections in a capacitor motor.

The centrifugal switch is mounted on the motor shaft and, as the name implies, works on the centrifugal principle, disconnecting the starting winding when the speed at which the switch is set is reached. The capacitor-start motor has a greater starting ability than the resistance-start motor. Because the capacitor is in use only during the starting period, a high capacity can be obtained economically for this short-term duty.

Shaded-Pole Motors

Another type of single-phase induction motor is schematically represented in Figure 1-44. This motor consists principally of a squirrel-cage rotor and two or more coils with an iron core to increase the magnetic effect. Part of one end of this core is surrounded by a heavy copper loop known as the *shading ring*. This ring has the characteristic of delaying the flow of magnetism through it. With alternating current applied to the coil, the magnetism is strong first at *A,* and then slightly later at *B*. This gives a rotating-field effect that causes rotation of the squirrel cage in the direction in which the shading ring points. A motor thus constructed is known as a *shaded-pole motor.*

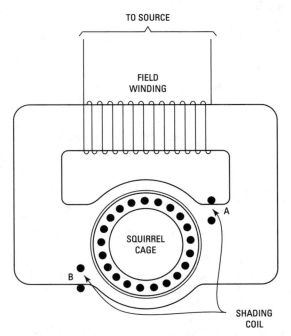

Figure 1-44 Arrangement of the windings in a skeleton-type, shaded-pole motor.

Because of the limitations of force and current possible in shading poles, it is not feasible to build efficient motors of this type larger than approximately ¹⁄₂₀ hp (or 37.3 W). These motors are

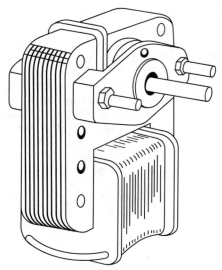

Figure 1-45   Small shaded-pole fan motor.

used principally on small fans, agitators, and timing devices (Figure 1-45).

Repulsion-Start Induction Motors

Repulsion-start induction motors are operated in various ways. In the running position, the brushes may or may not be raised. If the same rotor winding is used for both starting and running, the commutator is short-circuited at about 75 percent of rated speed to obtain a rotor winding approximating the squirrel cage in its functioning. Other designs have two rotor winding, that is, a squirrel cage and a wound winding for running and starting, respectively. In this type, no rotor mechanism is required because the magnetic conditions automatically transfer the burden from one winding to the other as the motor comes up to speed.

Repulsion starting may best be explained by the action of a wire connected to a battery and moved across the face of a magnet. Here, there is a force on the wire that, for example, tends to move it upward or downward, depending on the direction in which the current is flowing. It can thus be demonstrated that a current-carrying wire in a magnetic field has a force acting on it that tends to move it in a certain direction. Also, if the direction of the current flowing through the wire is reversed, the force and motion are also reversed.

Repulsion starting operates on this principle. Current is caused to flow in the wires of the rotor winding, and these wires are affected by a magnetic field.

Figure 1-46 shows a stationary C-shaped iron core on which is mounted a coil connected to a single-phase supply line. In the opening of the C is a ring of iron on which is wound a continuous and uniform coil. The path of the magnetism produced by the coil wound on the C-shaped core is around through the C-shaped core, and, dividing equally, half of the magnetism passes through each half of the iron ring.

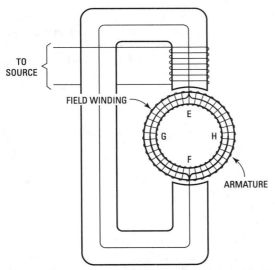

Figure 1-46 Illustration to show the operating principles of a repulsion-start induction motor.

The winding and the magnetism are identical in both halves of the ring. Thus, any effect that the magnetism may have on the winding between E and G will be the same as that produced in windings E and H. This can be proven by connecting an ammeter between points G and H. It will be found that no current is flowing between these two points. By further tests it can be shown that maximum current will flow when a wire is connected between E

and *F*. Thus, the first requirement of our principle has been satisfied: With a wire connecting *E* and *F*, there is a current flowing in the rotor winding.

Assume that the current flows upward in this wire from *F* to *E*. At point *E*, it divides equally, half going to the winding to the left of *E*, and the other half to the right. Referring to the wires on the outer surface of the ring, those on the right have the current flowing toward the observer, while those on the left have the current flowing away. Thus the magnetic field from the C-shaped core tends to force the wires on the right in one direction and those on the left in the other direction. The forces are equal and opposite so that they neutralize each other and no motion takes place.

In order for the rotor to rotate, it is necessary to add a magnetic field that can effectively react with the current in the rotor winding. This is done readily by adding another C-shaped core with its own coil, as shown in Figure 1-47. The rotor-winding current under each tip of this C-shaped core is all in the same direction, and rotation is obtained. The wire from *E* to *F* in Figure 1-46 has been replaced with stationary brushes so that a connection is maintained as the rotor turns.

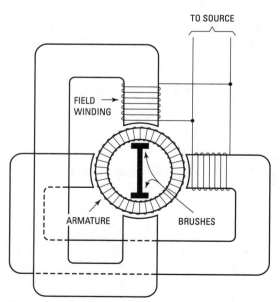

Figure I-47 Schematic diagram of a repulsion-start induction motor.

Motor Control

Although the function of motor control is fully covered later in this book, a brief outline of its essentials will be of aid in further study of the various types of motors and their associated control circuits.

The elementary functions of control are starting, stopping, and reversing of the motor. These, however, are only a few of the many contributions that the control renders to efficient operation of industrial motors.

The most common control functions of industrial motors are

- To limit torque on the motor and machine
- To limit motor starting current
- To protect the motor from overheating
- To stop the motor quickly
- To regulate speed
- Miscellaneous functions

Limiting Torque

One example of the need for limiting torque is that of a belt-driven, motor-operated machine throwing the belt when the motor is started. The pulleys may be correctly lined up and the belt tension may be correct, and yet the belt is thrown off in starting. This is the result of applying torque too quickly at standstill, and can be avoided by limiting the torque on the motor in starting. As another example, the blades on centrifugal fans can be sheared off if too much torque is applied to the fan in starting.

Limiting Starting Current

It is common to see DC motors flash over at the commutator when too much current is applied to the motor in bringing it up to speed. Also, it is common to see lights blink when a motor on the same power circuit is started. True, this blinking of lights can be reduced by selecting a motor with the right characteristics, but usually the real solution is the selection of a control that limits the starting current, either by inserting resistance in the circuit or by using a reduced-voltage source of power.

Protection from Overheating

Motors are designed to produce full-load torque for a definite period without overheating. While the motor is capable of exceeding its normal output for limited periods, there is nothing inherent in the motor to keep its temperature within safe limits. It is therefore the function of the control to prevent the motor from overheating excessively without shutting it down unnecessarily.

Quick Stopping

Where a driven machine has high inertia, it will continue to run for a considerable time after the power has been disconnected. There are several types of controls, such as electric brakes, to stop a motor quickly. The one most generally used on AC motors is the plugging switch. To plug a motor, it is necessary only to disconnect it from the line, and then reconnect it so that the power applied to the motor tends to drive it in the opposite direction. This brakes the motor rapidly to a standstill, at which time the plugging switch cuts off the reverse power.

Speed Regulation

Fans are sometimes run at various speeds, depending on the ventilation requirements. For some applications, it is advisable to use a two-speed motor, but where a greater variety of speeds is required, a motor with variable speed control may be the best solution to the problem.

Miscellaneous Control Functions

Adequate control equipment covers various other protective functions, which are not as common as those enumerated previously. Among these are *reverse-phase protection,* which prevents a motor from running in the wrong direction if a phase is inadvertently reversed; *open-phase protection,* which prevents the motor from running on single phase in case a fuse blows, and *undervoltage protection,* which prevents a motor from starting after a power failure unless started by the operator.

Summary

If the south-seeking (S) pole of a magnet is brought near the S pole of a suspended magnet, the poles repel each other. Likewise, if the two north-seeking (N) poles are brought together, they repel each other. However, if an N pole is brought near the S pole or if an S pole is brought near the N pole, the two unlike poles attract each other. In other words, *like poles repel each other, and unlike poles attract each other.* Experiments have shown these attracting or repelling forces between magnetic poles vary inversely as the square of the distance between the poles.

Oersted discovered the relation between magnetism and electricity in the early nineteenth century. He observed that when a wire connecting the poles of a battery was held *over* a compass needle, the N pole of the needle was deflected in one direction when the current flowed, and a wire placed *under* the compass needle caused the N pole of the needle to be deflected in the opposite direction. The compass needle indicates the direction of the magnetic lines of force, and an electric current sets up a magnetic field at right angles to the conductor. The so-called *left-hand rule* is a convenient method for determining the direction of the magnetic flux around a straight wire carrying a current: *If the wire is held in the left hand, with the thumb pointed in the direction of the current, the fingers will point in the direction of the magnetic field.* Conversely, if the direction of the magnetic field around a conductor is known, the direction of the current in the conductor can be determined by applying the rule.

An *electromagnet* is a soft-iron core surrounded by a coil of wire. The magnetic strength of an electromagnet can be changed by changing the strength of its applied current. When the current is interrupted, the iron core returns to its natural state. This loss of magnetism is not complete, however, because a small amount of magnetism, or *residual magnetism,* remains. The electromagnet is used in many electrical devices, including electric bells, telephones, motors, and generators. The polarity of an electromagnet can be determined by means of the left-hand rule, as follows: *Grasp the coil with the left hand; with the fingers pointing in the direction of the current in the coil, the thumb will point to the north pole of the coil.*

If a coil of wire having many turns is moved up and down over one pole of a horseshoe magnet, a momentary electric current without an apparent electrical source is produced. This current produced by moving the coil of wire in a magnetic field is called an *induced current.* Lenz's law states that an induced current has such a direction that its magnetic action tends to resist the motion by which it is produced. The generator and the motor are examples of useful applications of induced currents.

A *generator* converts mechanical energy into electrical energy. Its essential parts are a magnetic field, usually produced by permanent magnets, and a moving coil or coils called the *armature.*

The DC generator is classified either as a separately excited or a self-excited type. The separately excited generator has very little practical use; the self-excited generator is the one most often used. The self-excited DC generator can be broken down into a number of classificatons: the series, shunt, compound, overcompound, flat compound, and undercompound. Each type has particular advantages and disadvantages according to its load and speed of rotation.

Some of these generator types cannot be regulated in terms of a constant voltage output; they are used for other purposes where voltage regulation is not so important.

A *motor* converts electrical energy into mechanical energy. The motor, like the generator, consists of an electromagnet, an armature, and a commutator with its brushes.

The two principal classes of polyphase induction motors are the *squirrel-cage* motor and the *wound-rotor* motor. By definition, an induction motor is one in which the magnetic field in the rotor is induced by currents flowing in the stator. The rotor has no connection whatever to the supply line.

Single-phase motors can be divided into two principal classes as follows:

1. Split-phase
   a. Resistance-start
   b. Split-capacitor
   c. Capacitor-start
   d. Repulsion-start
2. Commutator
   a. Series
   b. Repulsion

Review Questions

1. How can the direction of the magnetic field around a straight wire carrying a current be determined?
2. Describe the basic construction of an electromagnet.
3. How can the polarity of an electromagnet be determined?
4. How is an "induced current" produced?
5. What is the basic difference between a generator and a motor?
6. What is the function of a DC generator?
7. How are the commutator segments of a DC generator insulated?
8. What materials are used to make brushes for a DC generator and/or motor?
9. How is the neutral plane of a generator shifted?
10. How are DC generators classified?

11. What is the only type of compound generator commonly used?

12. What is the name of the mechanical power source used to drive generators?

13. How are the effects of armature reaction overcome or reduced permanently in a generator?

14. Why will a shunt generator build up to full terminal voltage with no external load connected?

15. What is the name given to power lost in heat in the windings of a generator due to the flow of current through the copper?

16. As armature current of a generator increases, what happens to the motor reaction force?

17. How can compound generators be connected?

18. What is the name given to the part of a DC generator into which the working voltage is induced?

19. What are the two principal types of induction motors?

20. What are the two principal types of single-phase motors?

Chapter 2

Small DC and Universal Motors

DC motors are made in series, shunt, and compound configurations. The series motor is also used on AC. When used on AC or made to be able to use both AC and DC, it is called a *universal motor*.

The latest development in DC motors is the PM, or *permanent-magnet*, type. Instead of a field coil, this motor uses a permanent magnet and a wound armature.

Shunt DC Motors

The shunt DC motor (Figure 2-1) is one of the most versatile of DC types. It is a relatively constant-speed motor. Solid-state circuitry can be used to control its speed over a wide range. It has a wound field coil, or coils, and a wound rotor (armature). It does, however, use brushes and a commutator. The armature is connected across (in shunt or in parallel with) the field windings.

These motors are available in base speeds of 1140, 1725, 2500, and 3450 rpm. They can be wound for almost any speed or for special applications.

The shunt motor is reversible at rest or during operation. Reversing is done by simply reversing the polarity of the armature or the field voltage. Usually it is the armature that is reversed, since the field winding has high inductance. High inductance can cause excessive arcing at the switch contacts as the voltage is reversed.

The shunt DC motor is used for many purposes. It can be used to run windshield wipers on cars, the fans in car heaters, and special types of printing equipment. Printing presses with DC power available have better control of the printing process with DC shunt motors. They run at almost constant speed when the voltage and load are kept constant.

Plugging (reversing the motor when it is running) can be harmful to the motor. This subjects the armature to approximately twice the rated voltage. Dynamic braking (placing a short across the armature when the power is removed) should be handled with caution.

Brush life is good on a shunt motor. This and the armature commutator segments are the main concern for maintenance. Plugging and dynamic braking can severely limit the life of the brushes. Keep

Figure 2-1 Shunt-wound DC motor. Note the snap-together case.
(Courtesy Voorias)

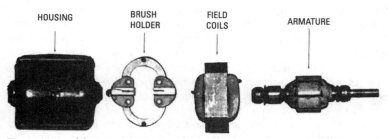

Figure 2-2 Notice the brushholder of the small shunt DC motor.

in mind that one brush will wear faster than the other. This is normal operation for a shunt DC motor (Figure 2-2).

Compound Motors

Since the DC compound motor is usually larger than one horsepower, it will be considered just briefly here. The compound motor has a series and a shunt coil and can be connected in various combinations. The SCR speed control and the permanent-magnet motor have replaced the compound motor in fractional-horse-power sizes.

The compound motor has constant speed and high starting torque. It has the qualities of both series and shunt motors and can be used for such things as elevators, grain mill operations, and any-where there is a need for good starting torque and fairly constant speed.

Compound motors are designed in two configurations: *cumulative compound* and *differential compound*. Cumulative compound DC motors have the two fields wound in the same direction. Speed depends on the sum of the two fields. Torque is higher than that for a shunt motor, but speed regulation is less than that for a shunt motor.

Differential compound DC motors have the coils wound in the opposite direction; that is, the shunt coil is wound in one direction and the series coil is wound in another. Speed depends on the difference of the two fields. Torque is lower than that for a shunt motor for the same amount of armature current. As a load is applied, the speed *increases*.

Occasionally the differential compound motor has been specially made to perform at a particular point. The term *flat-compounded* is used in this case. It means that a relatively flat speed-torque curve results from the design.

Permanent-Magnet Motors

The permanent-magnet (PM) motor has a permanent magnet that replaces the field coil. That means this type of DC motor will have only a wound armature (Figure 2-3). The magnet is interesting since it takes advantage of the latest technology in oriented strontium ferrites (ceramic magnet). In some cases, rare earth metals are used for the magnet, but this results in a rather expensive type of motor.

The best reason for using a PM motor is its size. It is physically smaller than a comparable shunt-type DC motor (Figure 2-4). This type of motor produces relatively high torques at low speeds. The PM motors are often used as substitutes for gearmotors. Permanent-magnet motors cannot be continuously operated at the high torques they are able to generate because they will seriously overheat.

Permanent-magnet motors draw less current from the battery than the shunt type and are therefore more efficient. The permanent magnet also produces some braking. The coasting of the armature

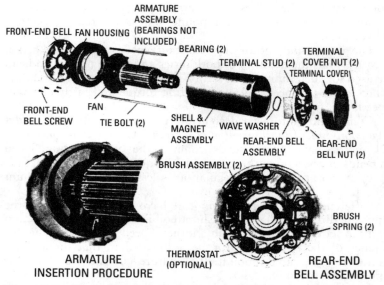

ARMATURE
ASSEMBLY
FRONT-END BELL FAN HOUSING (BEARINGS NOT
INCLUDED)
BEARING (2)
TERMINAL
TERMINAL STUD (2) COVER NUT (2)
TERMINAL COVER

FRONT-END FAN
BELL SCREW TIE BOLT (2)
SHELL & WAVE WASHER
MAGNET REAR-END
ASSEMBLY REAR-END BELL ASSEMBLY BELL NUT (2)

BRUSH ASSEMBLY (2)

BRUSH
SPRING (2)

ARMATURE THERMOSTAT REAR-END
INSERTION PROCEDURE (OPTIONAL) BELL ASSEMBLY

Figure 2-3 Exploded view of a permanent-magnet motor. When disassembling for maintenance, use a clean bench free of steep parts or chips. When replacing brushes, check the commutator for wear. If commutator is worn down more than $1/32$ inch on the diameter, turning and undercutting is recommended. Usually three sets of brushes can be used for one commutator turning. *(Courtesy Doerr)*

after power is removed can be stopped by using dynamic braking. This is done by simply shorting across the armature. This type of motor can be reversed by reversing the polarity of the armature leads.

This type of motor has some limits. It can lose some of its magnetism at temperatures below 0°F, and there is some change in the working flux when exposed to high temperatures. However, there are many applications for PM motors. They are used in the automobile industry for window lifts, heaters, blowers, defrosters, seat adjusters, windshield wipers, and rear deck defrosters. In marine equipment, you will find them used in variable-speed pumps, water pumps, fishing reels, trollers, winches, and blowers. Other applications include office machines, door-latch mechanisms, fans, motorized valves, and cordless appliances. Armatures can be wound to

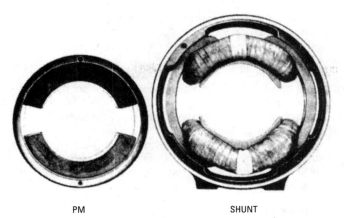

PM   SHUNT

Figure 2-4   Note the size difference, especially the outside diameter, of the PM motor as compared to the shunt-wound motor with the same diameter for the armature. *(Courtesy Bodine)*

handle from 60 to 230 volts (V). The brushes are about the only part of the motor that requires maintenance, but the commutator may need some attention after prolonged use. Remember, these are generally designed for intermittent duty. The bearings in most are bronze with a felt reservoir for holding lubrication. Some are available with sealed ball bearings.

Series or Universal Motors

As already stated, the series DC motor is used as a universal motor since it can be designed to run on either AC or DC. The series wound motor is among the most popular of fractional and subfractional motors. It can deliver high speed and high torque at starting, and provides a wide variety of speeds when used with speed controllers. At first glance, the series motor may look just like the shunt-type of DC motor. However, upon looking closely, you will see that the windings are different. The difference between operation on 50 Hz as compared to 60 Hz is negligible.

All series motors are not necessarily universal. Some may have been designed for operation on DC only; they may not perform well on AC or they may fail completely. Check the nameplate to be sure. Usually the universal motor will run more slowly on AC than on DC. In some cases it is necessary to place a resistor in series with the power source to slow down the DC-operated motor to where it would operate normally on AC. This is the case only when speed is critical. Placing a resistor in series with a universal motor does dramatically reduce its starting torque. A good example of this is the sewing machine. When the foot-pedal speed control is used (it is nothing more than a resistor being varied by the foot pedal), its resistance causes the sewing machine to have little or no torque. Most sewing machines have a hand-operated wheel so the motor can be started manually.

Series motors are the least expensive in terms of dollar per horsepower of any type of motor, which is one reason why you find them in almost all household appliances. They are the only small motors capable of doing more than 3600 rpm. In fact, they will operate at 10,000 to 20,000 rpm. *Caution: They should not be operated without a load.* Once they are loaded, you can hear the speed decrease.

Direction of Rotation

One thing you should keep in mind when working with universal motors is that the direction of rotation is important. Universal motors are usually supplied with one direction specified. This is primarily to improve operating efficiency, since the brushes sit at a point where the armature is rotating in a given direction. If the direction is changed, the nature of the wear on the brush will likewise change. Therefore, the brush life is affected by the direction of rotation.

Some Disadvantages

The series motor does not have good speed regulation. It will vary in speed with the load applied. Bearings and brushes wear faster when the armature turns fast. Most home appliance motors have from 200 to 1200 hours of operation on brushes before they have to be replaced. Brush replacement is the most common maintenance procedure needed with this type of motor. Compare the characteristics of the series DC motor with those of other motors (Table 2-1).

Table 2-1 Motor Characteristics

Type	Shunt Wound	Series Wound	Compound Wound
Duty	Continuous	Intermittent	Continuous
Power Supply	DC	DC	DC
Reversibility	At rest or during rotation	Usually unidirectional	At rest or during rotation

(continued)

Table 2-1 *(continued)*

Type	Shunt Wound	Series Wound	Compound Wound
Speed	Relatively constant and adjustable	Varying with load	Relatively constant and adjustable
Starting Torque	125 to 200% of rated torque	175% and up of rated torque	125% and up of rated torque
Starting Current	Normal	Normal	Normal

Types of Universal Motors

Series motors used for AC and DC are made in a number of different configurations. Figure 2-5 shows the simplest of the universal motors. It has a C-frame type of construction.

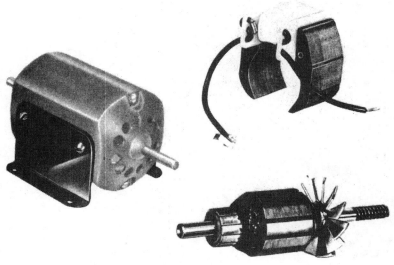

Figure 2-5 Universal motor. *(Courtesy Voorlas)*

Figure 2-6 shows an example of the A-frame series of universal motor. It has two coils connected in series with each other and also in series with the armature through brushes.

Figure 2-7 shows the E-frame universal motor. Check its characteristics in Table 2-2. This one does not have a fan mounted on the shaft; it may be totally enclosed without a fan. Note the shape of the armature.

Figure 2-6 A-frame universal motor.
(Courtesy Voorlas)

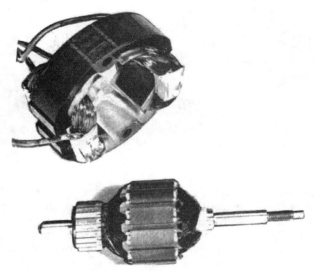

Figure 2-7 E-frame universal motor. *(Courtesy Voorlas)*

Table 2-2   Series Motor Types and Characteristics

	Maximum Wattage Output	RPM	Normal Wattage Output	RPM
C-Frame				
C-16	50	10000	30	20000
C-20	65	9000	40	17500
C-24	85	10000	55	15000
A-Frame				
224-18	125	9500	90	14000
224-24	140	8500	100	13000
224-28	165	8000	110	12500
224-32	230	8000	125	14000
B-Frame				
243-16	130	8000	105	12500
243-20	180	7500	135	11500
243-24	220	8000	160	12500
243-28	260	7500	190	11500
262-16	130	8000	105	12500
262-20	180	7500	135	11500
262-24	220	8000	160	12500
262-28	260	7500	190	11500
E-Frame				
287-16	180	7000	120	13000
287-20	250	8000	150	13000
287-24	285	7500	175	12500
287-28	320	6000	200	11000
287-30	335	6000	215	10000
287-34	370	6000	240	10000
K-Frame				
318-14	350	6000	170	12000
318-17	460	6000	220	12000
318-19	570	6000	280	12000
318-22	740	6000	340	12000
318-33	880	6000	430	12000

(continued)

Table 2-2 *(continued)*

Oversize B–Maximum Frame	Maximum Wattage Output	RPM	Normal Wattage Output	RPM
P.Q.-Frame				
368-16	330	7500	200	13000
368-20	450	6000	230	11000
368-24	550	6000	280	11000
368-28	675	5500	325	11000
368-32	800	5500	400	10000
368-36	1000	4100	500	12000
368-40	1100	4100	700	9500

Courtesy Voorlas.

In Figure 2-8 you will find the oversized B-frame type of universal motor with a fan. Note the number of leads from the coil. This also has a brushholder mounted on the frame. The PQ-frame motor is shown in Figure 2-9, while the K-frame universal motor is shown in Figure 2-10. Note, in Table 2-2, that the notation "C-16" means that there are 16 segments to the commutator on the C-frame type of motor; the C-20 has 20 segments; and so on. The number 224-18 for the A-frame means 2.24 in. in diameter for the outside measurement and 18 segments in the commutator. Follow the same interpretation for corresponding numbers in the rest of the table. Note the variation in rpm as the maximum and normal wattages are compared. The larger the load, the slower the motor. This, again, is typical of the series motor.

Armatures

In order to keep down vibration and noise, it is necessary to balance the armature since it rotates at about 20,000 rpm. Figure 2-11 shows how a typical armature (on the left) has been balanced by drilling holes in the laminated core. The number of holes and the depth determine the balance. The fan blades are press-fitted onto the shaft. Note the white silicon used to hold the wires from the coils in place near the connection to the commutator segments. The centrifugal force is so great at 20,000 rpm that it can cause the windings to separate, since the force of the rotating armature will be exerted on anything not held to the shaft. If the connections to the commutator are broken, the circuit is broken at this particular coil and will cause some rough operation and excessive arcing. The

Figure 2-8   B-frame universal motor. *(Courtesy Voorlas)*

armature on the right uses the conventional method of connection: more insulation (varnish). The silicon is supposed to be an improved method. Most connections are now made to the commutator segments by welding and not soldering.

The Hand Drill

One of the most popular tools around the house is the hand drill (Figure 2-12). It is a prime example of the use of a universal motor. This series motor is loaded by the gear reduction located near the Jacob's chuck or the end where the drill bit is attached to the tool. Note how the armature is inserted into the coil mounted on the drill housing (Figure 2-13).

Most of the maintenance of this machine is caused by the misuse of the tool. If the brushes are sparking too much, it is time to

Figure 2-9 PQ-frame universal motor. *(Courtesy Voorlas)*

remove them and the armature, smooth the commutator segments with sandpaper (*not* emery paper) and blow them clean, and then replace the brushes. The grease in the gearbox is usually sufficient for the life of the drill. It can be damaged by extremes in heat, however, and may need to be replaced.

Speed Control

Most hand-held electric drills are variable-speed. The speed control unit is nothing more than a variable resistor being controlled as the switch is pulled back toward the handle (Figure 2-14). The gear-operated resistor changes the electronic circuit so that the SCR operates properly and increases the speed as you pull harder on the trigger. The SCR controls the amount of current allowed through the coils. A heat sink is used to dissipate the heat for cooler operation of the SCR. Keep in mind that the SCR rectifies the AC and produces DC; this means that the drill is now operating on DC. In most instances this also means a greater efficiency since the series motor is more efficient on DC.

Slower speeds are used for starting holes without skipping or center punching. A slower speed can also be used for mixing paints and

Figure 2-10 K-frame universal motor. *(Courtesy Voorlas)*

drilling ceramic tile. Medium speed is used for drilling plastics and metals. Faster speeds are used for drilling wood and for powering accessories such as buffers and grinders. Once the trigger has been squeezed all the way back to the handle, it clicks a switch and takes the SCR out of the circuit and the drill operates on the full 120 V.

Figure 2-11 Armature for a hand-held electric drill. Note the holes drilled in the laminations for purposes of balance.

BEARING

Figure 2-12 Disassembled hand drill. Note that the bearing is the rounded piece of bronze near the gasket.

Figure 2-13 Coils of the series motor used on a hand drill. Farther back you can see the brushholder.

Portable Saws

Portable saws are powered by a universal motor. The arrow in Figure 2-15 shows where to look on the portable saw to check for excess sparking at the brushes. Too much arcing will cause serious damage to the commutator. If the motor is stalled, it is subject to overheating and excess current flow through the windings of the armature and the coils.

Proper maintenance requires keeping the ventilation area free from sawdust and other types of dirt that may cause damage to the bearings and get caught between the armature and the stationary part of the motor.

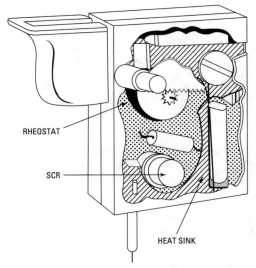

RHEOSTAT

SCR

HEAT SINK

Figure 2-14   Variable-speed control for a series motor used as a hand drill. *(Courtesy Black & Decker)*

Repairing a Series (Universal) Motor

Upright vacuum cleaners have been in use for many years. Many of them are out of service simply because the motor was not properly maintained. In this section we are going to disassemble the motor of an upright vacuum cleaner to point out some of the problems related to the operation of such a machine.

The upright we are going to disassemble is shown in Figures 2-16 and 2-17. The motor is located under the hood, which means that the hood must be removed before the motor can be reached for maintenance or repair. Note the location of the brushes. They are easily located by following the two wires that come from the motor housing. Figure 2-18 shows a screwdriver being inserted into the screw that holds the brush cover in place. Once the brush cover is removed, a small piece of brass holds the spring-loaded brush in its channel. Remove the brush (Figure 2-19). Move over to the other side of the motor and remove the other brush.

Next, remove the top plate from the motor. Remove the four screws. While you have the cover plate off, place a drop of machine oil in the bearing (Figure 2-20). You must hold the other end of the motor armature if you want to remove the small fan on top of the armature (Figure 2-21).

Figure 2-15   The arrow points to the area to check for excess arcing of the brushes.

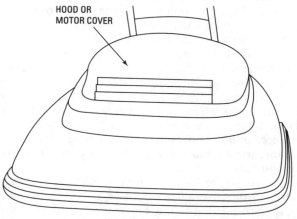

Figure 2-16   Upright vacuum cleaner using a universal motor.

VACCUM CLEANER

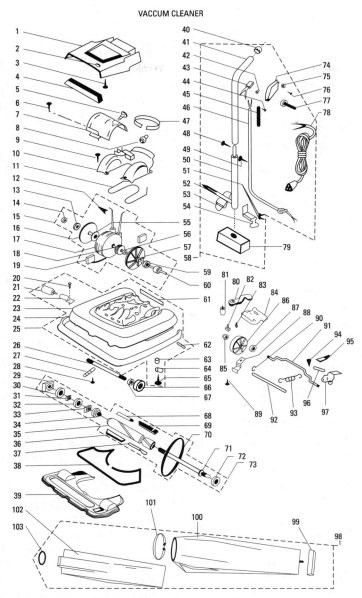

Figure 2-17   Exploded view of vacuum cleaner.

Key	Description	Key	Description
1	Top Cover w/#2	31	Felt Washer (2)
2	Trade Label ("Floor Genie")	32	Ball Beaing (2)
		33	Loading Spring (2)
3	Screw (#8–18 ω ¾ in. Lg. Rd Hd Thd Form, C.R.)(3)	34	Bearing Retainer (2)
		35	Brush Holder
4	Lens	36	Bristle Assembly (Right Side)
5	Bushing	37	Beater Bar (Belt Side)
6	Screw (#10–16 ω ⅝ in. Lg.)(3)	38	Gasket Access Plate
		39	Access Plate Assembly w/#38
7	Motor Hood Complete	40	Plug Button
8	Solderless Connector (2)	41	Tubular Handle (Upper)
9	Screw (#10–16 ω ⅝ in. Lg.)(11)	42	Solderless Connector (3)
		43	Switch
10	Blower Cover w/#11	44	Screw (Tension Spring) (#6–20 ω ¼ in. Lg.)
11	Gasket		
12	Solderless Connector (3)	45	Tension Spring
13	Washer—Fan (R.H.)	46	2-Conductor Cord
14	Nut, Suction Fan	47	Gasket, Bag Assembly
15	Motor Complete, Universal	48	Screw (Upper Handle)
16	Suction Fan (R.H.) w/#13, #55	49	"T" Nut (Curved Flange)
		50	Tubular Handle (Lower)
17	Field Core w/#19	51	"T" Nut (Curved Flange) (2)
18	Armature w/Bearings (Threaded Ends)	52	Screw (2)
		53	Bifurcated Handle Support (Right)
19	Carbon Brush Assembly (Set of 2)	54	Bifurcated Handle Support (Left)
		55	Shoulder Washer, Fan (R.H.)
20	Rivet—Light Socket	56	Shoulder Washer, Fan (L.H.)
21	Light Bulb	57	Suction Fan (Left Side) w/#56, #59
22	Light Socket		
23	Chassis	58	Handle Assembly Complete
24	Bumper	59	Washer—Fan (L.H.)
25	Chassis Assembly Complete	60	Motor Shaft Pulley
26	Axle—Front Wheel	61	Nozzle Adjustment Decal
27	Retaining Plate—Front Axle (2)	62	Rivet Bumper (6)
		63	Spacer (3)
28	Screw (#10–16 ω ⁷⁄₁₆ in. Lg.)(2)	64	Retainer Cup—Access Plate (3)
29	End Cap (2)		
30	Cup (L.H. Threads)		

Figure 2-17 Legend for exploded view of upright vacuum cleaner.

Key	Description	Key	Description
65	Screw (#10–16 ω ⅝ in. Lg.)(3)	84	Wheel & Cam Retaining Plate
66	Wheel—Front (2)	85	Wheel Cap (2)
67	Washer (4)	86	Rear Wheel (2)
68	Beater Bar (Right Side)	87	Washer (2)
69	Bristle Assembly (Belt Side)	88	Retaining Plate—Rear
70	Belt	89	Screw (2)
71	Shaft, Floor Brush	90	Retaining Ring
72	Cup (Belt Side)—(Marked w/"B")	91	Shaft—Rear Wheel
		92	Trunion Arm Shaft
73	Floor Brush Complete (Universal)	93	Tension Spring Pedal—Pedal
		94	Screw
74	Switch Box	95	Retainer—Pedal Shaft
75	Switch Lock Nut	96	Shaft—Pedal
76	Screw (Switch Box Mounting)	97	Pedal—Handle Release
77	Cord Strain Relief	98	Dust Bag Assembly
78	Cord	99	Bag Top
79	Handle Support Top	100	Dust Bag Assembly w/#47
80	Elevating Cam Lever	101	Dust Bag Strap Assembly
81	Spacer	102	Disposable Inner Bag
82	Screw (#10–16 ω ⅝ in. Lg.)	103	Garter Spring
83	Screw (#10–16 ω ⁷⁄₁₆ in. Lg.)		

Figure 2-17   (continued)

SCREWDRIVER

Figure 2-18   Screwdriver is inserted into the screw head that holds the brush cover in place.

Figure 2-19 Removing the brush. Note that it is spring-loaded.

Figure 2-20 Oiling the top bearing on the motor.

Figure 2-21   Unscrew the top fan from the armature to be able to clean the armature. You have to hold the bottom of the armature with a screwdriver while doing this.

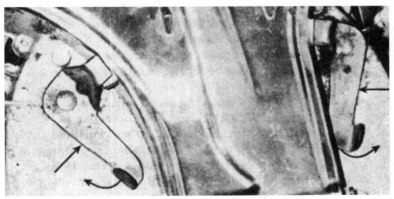

Figure 2-22   In order to get at the bottom of the armature you have to remove this cover plate. Pull outward on the two tabs shown by arrows.

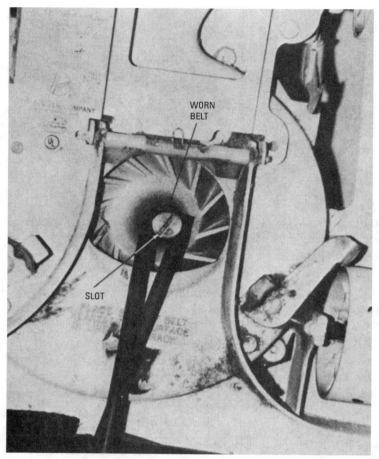

Figure 2-23 The fan or blower for the vacuum is shown. Note the condition of the brush drive belt.

In order to get at the armature from the bottom of the sweeper it is necessary to remove the cover plate shown in Figure 2-22 by sliding the two handles shown here by arrows. The ends are hooked under the spring at the back of the sweeper, and the whole plate comes off easily. Once it is off, you can see the slot in the motor armature. In Figure 2-23, you can see that the belt also needs replacing. Some fuzz from carpeting is also evident around parts of the sweeper. This is a good time to clean up the surface areas.

Note, in Figure 2-24, that the motor used here has a C-frame. Only one coil is used to furnish the power for the machine. The armature has slots in the commutator that need attention. Clean the armature and reassemble in reverse order. Make sure the brushes are reinserted or new ones are seated properly. Clean the area around the armature. Make sure the entire area around the motor is clean before you reassemble. Brushes will need attention in 500 to 2000 hours of operation. The heavier the carpet and the more frequently the vacuum is used, the more attention it will need.

Figure 2-24   The armature is now visible and can be cleaned or checked for wear. Note: One coil is all that is used for this type of vacuum cleaner.

Tank Vacuum Cleaners

The tank-type vacuum cleaner has a slightly different motor design. Tank-type motors are made in a number of designs for the household cleaner. Replacement brushes are available at electrical supply stores that sell motor parts. Commercial and industrial vacuum cleaners use a slightly different type of motor housing, but all use universal motors. They are different only in the shape and design needed for a specific purpose.

Motor Installation and Mounting

There are many different methods of mounting vacuum cleaner motors. The ideal method involves clamping the fan case between two sponge rubber gaskets (Figure 2-25). The gaskets should be compressed to ensure an airtight seal for the vacuum chamber. This method also acts as a shock mounting for the motor.

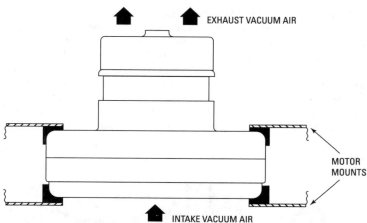

Figure 2-25   Motor installation and mounting for a vacuum.
(Courtesy Northland)

In clamping the fan case between the two mounting gaskets, care should be taken to allow sufficient pressure to resist start-up torque of the motor. Too much compression of the gaskets will decrease the shock protection provided by the sponge rubber gaskets.

Another method of mounting uses only one gasket between the motor fan case and the mounting flange (Figure 2-26). Screws are assembled through the mounting flange into the fan case, compressing the gasket and providing a vacuum seal. This method does not provide as good a shock mounting as the preceding method. Because of the vacuum created by the motor opposing the seal, the clamping must be tighter in order to ensure a good seal.

Bypass vacuum motors should be mounted in a manner that keeps the working air separated from the cooling air in order to prevent overheating of the motor due to recirculation. Figure 2-27 outlines the preferred method of mounting, which separates the intake and exhaust of both the working and cooling air systems. Caution should be taken in mounting so that cooling air is not restricted from flowing around the upper bearing, since this may cause overheating and severely reduce bearing life.

Miscellaneous Information

It may be helpful, in some instances, to know that vacuum motors may be combined to achieve additional performance. Two or more

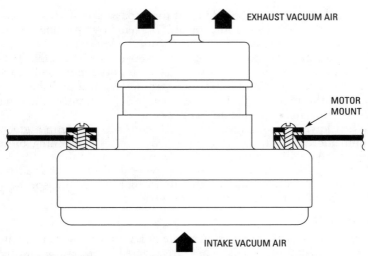

EXHAUST VACUUM AIR

MOTOR
MOUNT

INTAKE VACUUM AIR

Figure 2-26 Another method of mounting a vacuum motor. *(Courtesy Northland)*

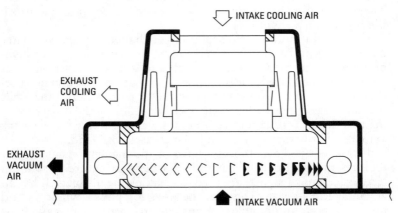

INTAKE COOLING AIR

EXHAUST
COOLING
AIR

EXHAUST
VACUUM
AIR

INTAKE VACUUM AIR

Figure 2-27 Operation of a properly mounted vacuum motor.
(Courtesy Northland)

motors may be connected in (air) series. This arrangement significantly increases vacuum performance with little increase in the
cubic feet per minute (CFM) of air moved. Connecting two motors
in (air) series is normally accomplished by mounting two or more
motors in a long tube. The lower motor can be of the flow-through

type and the upper units should be of the bypass type. This arrangement of a flow-through and a bypass prevents overheating of the upper unit due to the passage of heat from the lower unit.

Vacuum motors may also be connected in (air) parallel. This arrangement significantly increases CFM with little increase in vacuum. Connecting motors in (air) parallel is accomplished by mounting the motors on the cover of a large container, allowing them to work in the same airspace.

When lower performance is wanted than is available in a given vacuum motor, it can operate on a lower voltage. Running on a lower voltage will also increase the life expectancy of the motor.

Reducing the speed of a motor will also increase the life expectancy. A rough rule of thumb is that when the speed is reduced by 50 percent, the brush life is tripled.

Summary

DC shunt motors are the most versatile of DC types. They are available in a variety of speeds. They are reversible at rest or during operation. Plugging is the reversing of a motor when it is running and can be harmful to the motor. DC shunt motors have many uses. They can be used in printing presses for better color printing. Brush life is good on a shunt motor.

Compound motors are usually larger than 1 horsepower. They have a series and a shunt coil and can be connected in various combinations. The SCR speed control and the permanent-magnet motor have replaced the compound motor in fractional-horsepower sizes. Differential compound DC motors have the coils wound in the opposite direction. Speed depends on the difference of the two fields. Torque is lower than for a shunt motor for the same amount of armature current. Speed increases with load.

Permanent-magnet motors have a permanent magnet that replaces the field coil. It has only a wound armature. Strontium ferrites (ceramic magnet) are used for making the magnets. It has an advantage in its size. They can be used for power windows and power seats in automobiles. Permanent-magnet motors cannot be continuously operated at the high torques for which they are capable. To do so causes them to generate high heat and they will seriously overheat. This type of motor is also limited in its size. This type of motor has many applications.

The series or universal motor is so labeled because it can be used on AC or DC. Some, of course, have been especially designed for DC operation and may not perform very well on AC. The sewing machine uses a series motor with a resistor inserted to aid in speed regulation. This type of motor has high torque on starting, but stalls easily when overloaded. It slows rapidly when the load is increased. A large number of household appliances have series motors, such as vacuum cleaners. They are capable of doing more than 3600 rpm. They can be used with speeds up to 20,000 rpm. Brush life can be influenced by changing the direction of rotation.

Series motors can be obtained in C-Frame, B-Frame, E-Frame, and K-Frame configurations, as well as the oversize B-Frame and P-Q Frame. Keep in mind that the larger the load placed on the motor the slower the speed.

The armatures on series motors are wire-wound and their laminations are made so that the core is balanced, which enables them to handle high speeds. If no load is applied to the motor it can run away or keep gaining speed until the wire on the armature is thrown off and jams the motor.

One of the most commonly used series motor around the house is the electric hand drill. It is a prime example of a universal motor. A variable resistor is used in the circuit to control the speed of the drill. A silicon-controlled rectifier (SCR) is also used in the control circuit of today's models of hand drills. Portable saws are also powered by universal motors that have brushes. As the saw blade is engaged in cutting wood the motor slows down. If the load is applied too rapidly the saw blade will stall.

Upright vacuum motors are of the universal type. They also have brushes that can wear down and malfunction. The CFM (cubic feet per minute) of air moved by the vacuum determines its ability to do the job. Reducing the speed of the motor will also increase the motor's life expectancy. When the brush speed is reduced by 50 percent, the brush life is tripled.

Review Questions

1. How is a shunt DC motor different from a series?
2. Why does a DC motor have brushes and a commutator?
3. What makes a series motor a universal motor?
4. Where are shunt DC motors used?
5. Where are series DC motors used?
6. What type of motor does the hand drill use?
7. Which electric motor has high starting torque?
8. Which electric motor type tends to run away in speed if left unloaded?

9. What is strontium ferrite used for?

10. What are some uses for the permanent magnet motor?

11. What is the main advantage of the PM motor?

12. Why is CFM important in a vacuum cleaner?

13. How is speed regulation accomplished with a series motor?

14. Why would you use a DC motor instead of an AC motor?

15. What is the purpose of an SCR in a motor circuit?

Chapter 3

Commutators and Brushes

An electric motor consists of a number of parts. The shape and function of each part are determined by the type of motor and its practical application. Armatures are found in DC motors and universal motors (see Chapter 2). The wound rotor or armature is used in both DC and AC motors. If the rotor is wound and the motor operates on AC, it is most likely a universal motor and its application will be that of a small hand drill, sander, or similar device. The universal motor is also used in vacuum cleaners. Some types of heavy-duty industrial motors use a wound rotor. However, they are not included here since they have special applications and are made in sizes larger than 1 horsepower.

Fractional-horsepower motors that burn out are replaced rather than repaired. However, the price of new motors is becoming such that people are considering learning more about rebuilding rather than replacing small motors. Most of the armatures used in hand drills, saws, and sanders are machine wound, and are almost impossible to rewind. However, some maintenance can be performed on them to make sure the commutator segments are in proper shape and the brushes are making good contact with the commutator surface. Dirt and lint removal is also a necessary part of the maintenance program for small motors.

Before you throw out or replace a small motor, check it over to learn just what is wrong with the armatures. There may be broken wire that can be soldered, or some other problem may be visible and easily correctable. Remember, the armature is connected in series with the field coils, and if there is an open circuit anywhere within the motor wiring, the circuit is not completed and the motor will not run.

The Armature

The armature is made up of a shaft, commutator, armature poles or core, laminations (called teeth), and copper wire that is wound around the pole pieces.

In Figure 3-1 the armature is stripped and ready for rewinding. Note that in Figure 3-2 the bottom lead of the coil of wire that has been wound around the armature poles is brought out and soldered to the commutator segment.

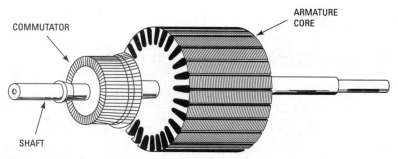

Figure 3-1   A stripped commutator ready to be rewound.

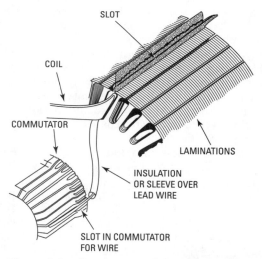

Figure 3-2   Connection of the coil lead to the commutator.

In Figure 3-3 a completely wound armature is shown. The entire procedure of winding an armature will be discussed and illustrated later in this chapter.

Magnet Wire Characteristics
Today, magnet wire is used to rewind stators and armatures. It is a quality product with a coating that can be trusted to bend (with the wire) without creating damaging cracks that can later develop into trouble spots.

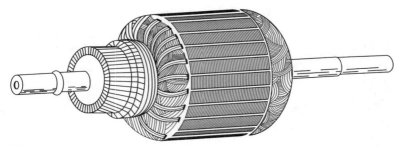

Figure 3.3 A wound armature.

At one time, the wire available for motor winding was coated with cotton fabric. In some cases it was double-coated cotton (dcc). This was soaked in an insulation solution after winding to produce a winding that would take the wear. Then came a clear varnish and then a black varnish with asphalt as one of its ingredients. It worked very well with round wire.

In 1938, polyvinyl formal (FORMVAR) was introduced. It is the primary type of insulation used on today's magnet wire employed in the winding of motors in fractional-horsepower sizes. There are about 300 types of coatings available for magnet wire. The Japanese use polyethylene for a coating on magnet wire used for submersible pump motors. The magnet wire is first coated with polyethylene and then topped with an overcoat of nylon.

Table 3-1 shows the turns per square inch of magnet wire per size. It also indicates the diameter of the magnet with FORMVAR insulation.

Table 3-1   Turns per Square Inch of Magnet Wire and Diameter of the Wire with FORMVAR Insulation

Size AWG	Turns per Square Inch	Diameter of Magnet Wire
16	361	52.6
17	453	47.0
18	570	42.0
19	711	37.5
20	893	33.5
21	1110	30.0

Table 3-1 *(continued)*

Size AWG	Turns per Square Inch	Diameter of Magnet Wire
22	1400	26.8
23	1760	23.9
24	2200	21.3
25	2740	19.1
26	3440	17.0
27	4270	15.3
28	5370	13.6
29	6760	12.2
30	8530	10.8
31	10,600	9.7
32	13,100	8.8
33	16,500	7.8
34	20,400	7.0
35	25,900	6.2
36	31,900	5.6
37	40,800	5.0
38	50,200	4.5
39	64,700	4.0
40	79,600	3.6

Diameter is given in mils, or thousandths of an inch.
Single FORMVAR insulation is shown in the diameter column.

The insulation is no more than 1 to 2 mils (0.001 to 0.002 in.) in thickness. However, in some cases it may reach 15 to 20 mils (0.015 to 0.020 in.). It is best to remember, when checking the size of wire with a wire gage, to allow one size for the thickness of the insulation. For instance, if you measure the wire and find it is No. 17 on the wire gage, take into consideration the thickness of the insulation and consider the actual size of the wire as No. 18. That is, the copper wire diameter was No. 18 before the insulation was added.

Today's insulations are capable of long-term operation at higher temperatures than those with double-coated cotton insulation. It does not mean, however, that the motor can be operated for long periods when overheated. The heat buildup in the armature or the stator coils will cause the insulation to break down and damage the winding permanently, requiring a rewind of the coils or armature.

Table 3-2 show the wire size in a number of insulation thicknesses, bare wire diameter, weight for 1000 feet, area in circular mils, and resistance per 1000 feet at 77°F(25°C). This should be sufficient for checking any type of insulation found on old and new motors of fractional horsepower size.

Figure 3-4 shows how magnet wire is wound one turn on top of the other inside the slot of a motor's stationary pole piece. Note how the wire is insulated from touching the metal wall by a piece of insulating material.

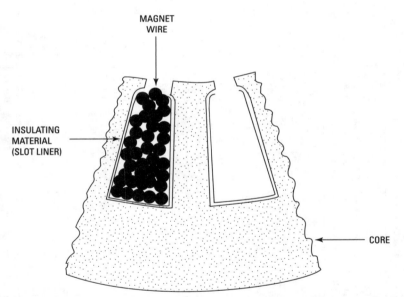

Figure 3-4 Placing the magnet wire coil into a slot in the stator of a motor.

Types of motors other than DC and universal also need wound armatures. Figure 3-5 shows the stripped repulsion-induction motor armature. Keep in mind that the explanation of this type of motor is included here to illustrate its type and not because it is a motor that you will be encountering. This is an industrial-type motor and is made in large-horsepower sizes.

Table 3-2 Wire Table—Characteristics of Magnet Wire for Motor Winding

		Diameter Over Insulation								Weight (pounds per 1000 ft)					
AW Gage No.	Diameter, Bare	Enam. or Single FORM-VAR	Double FORM-VAR	Quadruple FORM-VAR	Single Paper Enam.	Single Cotton Enam.	Single Silk (or Nylon) Enam.	Double Glass Bare	Single Glass Heavy FORM-VAR	Bare	Double FORM-VAR	Single Cotton Enam.	Double Glass Bare	Area in Circular Mils	Resistance in Ohms per 1000 ft at 77°F (25°C)
13	0.072	0.0738	0.0753	0.0779	0.077	0.0786	—	0.0795	0.0803	15.68	15.9	16.0	16.3	5178	2.043
14	0.064	0.659	0.0673	0.0699	0.0691	0.0707	—	0.0716	0.0723	12.43	12.6	12.7	13.0	4107	2.575
15	0.057	0.0588	0.0602	0.0628	0.062	0.0637	0.0604	0.0646	0.0652	9.858	10.05	10.1	10.3	3257	3.247
16	0.051	0.0525	0.0539	0.0563	0.0557	0.0571	0.0541	0.0583	0.0589	7.818	7.96	8.06	8.24	2583	4.094
17	0.045	0.0469	0.0482	0.0506	0.0501	0.0515	0.0485	0.0528	0.0532	6.20	6.34	6.41	6.56	2048	5.163
18	0.040	0.0418	0.0432	0.0456	0.045	0.0466	0.0434	0.0478	0.0482	4.917	5.02	5.10	5.23	1624	6.51
19	0.036	0.0374	0.0386	0.0409	0.0406	0.0421	0.039	0.0434	0.0436	3.899	4.00	4.06	4.18	1288	8.21
20	0.032	0.0334	0.0346	0.0368	0.0366	0.038	0.035	0.0395	0.0396	3.092	3.17	3.24	3.34	1022	10.35
21	0.0285	0.0299	0.031	0.0331	0.0331	0.0345	0.0315	0.036	0.0360	2.452	2.51	2.57	2.67	810.1	13.05
22	0.0254	0.0267	0.0278	0.0298	0.0299	0.0313	0.0283	0.0328	0.0328	1.945	1.99	2.05	2.14	642.4	16.46
23	0.0226	0.0239	0.0249	0.0269	0.027	0.0284	0.0254	0.0301	0.0299	1.542	1.58	1.63	1.72	509.5	20.76
24	0.0201	0.0213	0.0224	0.0243	0.0245	0.0259	0.0229	0.0276	0.0274	1.223	1.26	1.31	1.39	404.0	26.17
25	0.0179	0.0191	0.0201	0.022	0.0215	0.0233	0.0207	0.0232	0.0233	0.9699	0.998	1.04	1.12	320.4	33.00
26	0.0159	0.0170	0.018	0.0198	0.0194	0.021	0.0186	0.0212	0.0213	0.7692	0.793	0.837	0.900	254.1	41.62
27	0.0142	0.0153	0.0161	0.0178	0.0177	0.0193	0.0169	0.0195	0.0194	0.610	0.630	0.662	0.727	201.5	52.48
28	0.0126	0.0136	0.0145	0.016	0.016	0.0178	0.0152	0.0179	0.0177	0.4837	0.501	0.532	0.0588	159.8	66.17
29	0.0113	0.0122	0.013	0.0145	0.0146	0.0164	0.0138	0.0166	0.0162	0.3836	0.396	0.427	0.477	126.7	83.44
30	0.0100	0.0109	0.0116	0.0131	0.0133	0.0151	0.0125	0.0153	0.0149	0.3042	0.316	0.344	0.387	100.5	105.2
31	0.0089	0.0098	0.0105	—	—	0.0139	0.0113	0.0142	0.0137	0.2413	0.251	0.278	0.314	79.7	132.7
32	0.0080	0.0088	0.0094	—	—	0.013	0.0104	0.0133	0.0126	0.1913	0.198	0.224	0.255	63.71	167.3
33	0.0071	0.0079	0.0085	—	—	0.0119	0.0094	0.0124	0.0117	0.1517	0.158	0.182	0.208	50.13	211.0
34	0.0063	0.0071	0.0075	—	—	0.0111	0.0086	0.0116	0.0108	0.1203	0.126	0.148	0.169	39.75	266.0
35	0.0056	0.0063	0.0067	—	—	0.0103	0.0078	0.0109	0.0100	0.0954	0.0996	0.120	0.138	31.52	335.5
36	0.0050	0.0057	0.0060	—	—	0.0093	0.0072	0.0103	0.0093	0.0757	0.0791	0.100	0.113	25.00	423.0
37	0.0045	0.0051	0.0055	—	—	0.0087	0.0066	0.0098	0.0087	0.060	0.0628	0.080	—	19.83	533.4
38	0.0040	0.0045	0.0049	—	—	0.0082	0.0061	0.0093	0.0081	0.0476	0.0498	0.068	—	15.72	672.6
39	0.0035	0.0040	0.0043	—	—	—	0.0055	0.0088	0.0075	0.0377	0.0397	0.060	—	12.47	848.1

SQUIRREL-CAGE
SHORTING RING

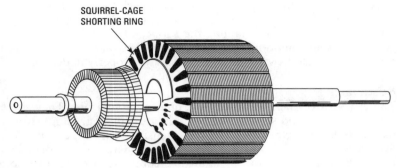

Figure 3-5   Squirrel-cage shorting ring on a rotor designed for a repulsion-induction motor.

Notice the location of the squirrel-cage winding on the armature. Little current flows in the windings of this type of armature during starting moments. The heavy-gage shorting bar pointed out in Figure 3-5 has very low inductance; the coil has very high inductance. Inductive reactance keeps the current low in the winding but very high in the shorting squirrel cage. This means that the shorting ring has high current and dominates the effect of the total armature in the circuit. The motor will start the same as any simple repulsion motor. Once it is started, however, and normal speed is attained, the inductance of the squirrel-cage winding decreases. The current flowing in the shorting ring decreases, but not to such an extent that it has no effect on the operation of the motor. It still dominates over the wound coil. That accounts for the characteristics of this type of motor. No short circuiting device is needed as is the case in the repulsion-start motor. The characteristics of this type of motor are high starting torque and constant speed. It is not the type of motor found in appliances around the home, but is primarily a special-application industrial motor. It does require maintenance since it has brushes and a wound rotor.

Commutator Maintenance

The most important factor, and the one on which the success or failure of a DC motor and commutator-type AC motor depends, is commutation. Satisfactory commutation means operating under reasonable conditions without excessive sparking, burning of commutator bars or brushes, or other conditions requiring excessive maintenance.

Assuming that the design of a machine is such that good commutation is to be expected, one of the best means of securing satisfactory operation is through maintaining the surface of the commutator in good operating condition. Generally, this means that the commutator surface should be smooth, concentric, and properly undercut.

Resurfacing

There are three methods used in truing commutators. They are sandpapering, hand stoning, and grinding or turning. Hand stoning is used, or was used, more extensively in large motors. Sandpapering is used on small commutators. Grinding or turning can be used on small and medium-size commutators that can be mounted between the tail stock and chuck on a lathe. The method used depends on the degree of damage and how you are equipped to handle the commutator.

Sandpapering

This is a satisfactory method of removing deposits from a commutator surface. It can be used to correct roughness or reduce high mica provided the accurate contour of the surface has not been disturbed. However, sandpaper cannot be depended on to remove flat spots even of small size.

One of the principal objections to the use of sandpaper for cleaning commutators is that it very rarely leaves the bars properly round. Oil should not be used when sanding a slotted commutator. It will cause the copper dust and sand to collect in the slots, while if no oil is used, centrifugal force will throw the dust from the slots.

Figure 3-6 shows how a piece of sandpaper can be attached to a curved block to aid in the proper sanding of a commutator.

Caution

Emery paper or emery cloth should never be used on a commutator. Emery is an electrical conductor and particles are likely to become embedded between the segments and cause short circuits.

Hand Stoning

The tendency of sandpaper to broaden rather than remove flat spots and to distort the contour of the commutator segments results from the flexibility of the paper. The use of a commutator stone is, therefore, recommended in preference to sandpaper. This is, of course, for larger DC motors and generators designed for industrial use. In small fractional-horsepower motors, it is unnecessary to

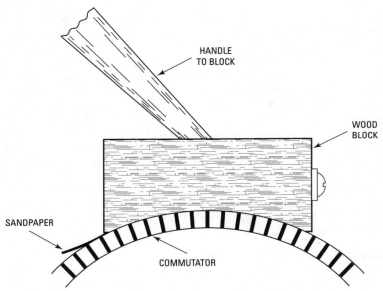

Figure 3-6 Making a sanding block for smoothing the commutator surface. Used on very large DC motors.

stone the commutator. Various sizes of commutator hand stones are made for use on larger commutators (Figure 3-7). The stone is pressed firmly against the commutator and moved slowly from side to side.

Grinding or Turning

While sandpapering offers a satisfactory method of resurfacing a commutator that is dirty or on which the mica is just beginning to build up, hand stoning will do a better job of removing high mica and even flat spots of considerable size. The most satisfactory method of resurfacing a commutator that is badly out of round is by grinding or turning with a tool that has a rigid support. All cases of eccentricity come under this head, for no hand method of finishing can entirely eliminate eccentricity in a commutator.

Figure 3-7 Hand-held stones for turning down a commutator. Used on very large DC motors.

When truing the commutator in a lathe, you can support the armature on the lathe centers or on sleeve bearings. The former is simpler but the latter method has some advantage in point of accuracy.

Before starting to turn or grind a commutator, the windings of the armature (adjacent to the commutator) should be wrapped in cloth or plastic to protect them from copper chips and dust. In turning a commutator with a steel tool, use what is known as a diamond point tool or one with a very sharp point. The point of the tool should be rounded sufficiently so that the cuts will overlap and not leave a rough thread on the commutator surface. Use a light cut, taking several cuts if necessary, to remove bad flat spots or a considerable degree of eccentricity.

After turning the armature's commutator, it is a good idea to use very fine sandpaper and hold it on the surface while the commutator is turning. Then clean the dust from the slots between the commutator segments.

Undercutting Commutator Mica

The object of undercutting commutator mica is to remove the mica between the copper segments so that the segments will wear evenly. The removal of the mica is necessary.

Figure 3-8 shows how a hacksaw blade has been used to undercut the mica insulation between the commutator segments. The blade may have to be ground to fit the width of the slot. Figure 3-9 shows techniques for working on commutators.

Figure 3-10 shows two correct ways and two incorrect ways to undercut the mica. Keep in mind the main reason for doing this is to make the motor operate smoothly. If the mica is sticking up above the surface of the copper, the brushes cannot make contact with the commutator segments.

Be sure to clean the commutator before replacing it in the motor, and make sure that under no conditions are grease and oil allowed to become part of the commutator surface. Small electrical tools have pitted commutator segments and may need a slight sanding to bring them back into normal operation. The quick way to check is to look at the motor while it is operating and see if there are many sparks where the brushes touch the commutator. There is supposed to be some sparking or arcing; the trick is to learn what is the correct amount for normal operation. If the commutator is pitted and black, it probably needs to be resurfaced or sanded. It is a good idea to check every time you change brushes in the motor. Most small motor manufacturers suggest that you

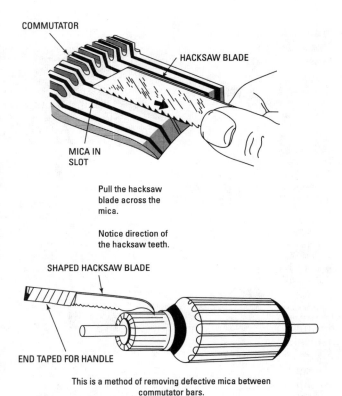

COMMUTATOR

HACKSAW BLADE

MICA IN
SLOT

Pull the hacksaw
blade across the
mica.

Notice direction of
the hacksaw teeth.

SHAPED HACKSAW BLADE

END TAPED FOR HANDLE

This is a method of removing defective mica between
commutator bars.

Figure 3-8   Using a hacksaw blade to cut the high mica on a
commutator.

inspect the commutator for wear each time you replace the brushes.
If the commutator is worn down more than $\frac{1}{32}$ in. (0.80 mm) in
diameter, turning and undercutting is recommended. Usually three
sets of brushes can be used for one commutator turning.

Brushes

Brushes are necessary to complete the path for current flow
through the armature, which puts some very heavy demands on
them. They must be smooth enough in their contact with the copper conductor, yet not have oil or lubricant on them, to glide over
without causing undue damage; they have to make good electrical

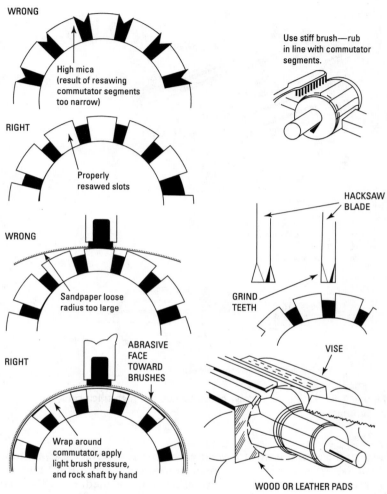

WRONG

High mica
(result of resawing
commutator segments
too narrow)

RIGHT

Properly
resawed slots

WRONG

Sandpaper loose
radius too large

RIGHT

ABRASIVE
FACE
TOWARD
BRUSHES

Wrap around
commutator, apply
light brush pressure,
and rock shaft by hand

Use stiff brush—rub
in line with commutator
segments.

HACKSAW
BLADE

GRIND
TEETH

VISE

WOOD OR LEATHER PADS

Figure 3-9 Working on commutators.

contact; and they have to be able to handle the current needed in the specific situation where they are used. Current density is the current-carrying capacity of brushes expressed in amperes per square inch of useful brush area in normal contact with the commutator surface. The current density varies with the application.

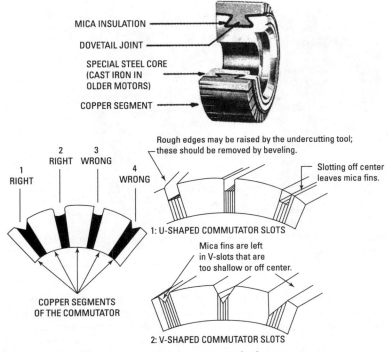

MICA INSULATION

DOVETAIL JOINT

SPECIAL STEEL CORE
(CAST IRON IN
OLDER MOTORS)

COPPER SEGMENT

Rough edges may be raised by the undercutting tool;
these should be removed by beveling.

Slotting off center
leaves mica fins.

1: U-SHAPED COMMUTATOR SLOTS

Mica fins are left
in V-slots that are
too shallow or off center.

2   3
RIGHT WRONG

1
RIGHT

4
WRONG

COPPER SEGMENTS
OF THE COMMUTATOR

2: V-SHAPED COMMUTATOR SLOTS

Figure 3-10   Right and wrong ways to cut high mica.

The highest density is about 150 amperes per square inch for certain super-baked grades.

Brushholders
Brushes must be held securely to the surface of the commutator, which usually requires spring action. The spring tension is supplied by a screw-in type of cap or by a coiled spring mounted on the outside of the motor (Figure 3-11).

Brush Types
There are four popular types of brush materials. The *carbongraphite brush* is made of hard carbon graphite and is particularly well suited for use with motors operating with flush mica commutators, where appreciable polishing action is required. The density is 35 to 45 amperes per square inch, which means it is used only on low-current fractional-horsepower motors.

BRUSH   BRUSH

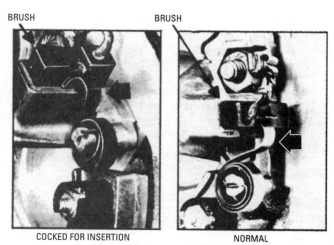

COCKED FOR INSERTION   NORMAL

Figure 3-11   Brush cocking procedure for a permanent-magnet DC motor. *(Courtesy Doerr)*

The *electro-graphite brush* is made by subjecting carbon to intense heat. That means its conversion to graphite is a physical, not a chemical, change. It is useful in higher-speed motors. It is less abrasive than carbon graphite, and is tougher and has greater current density—75 amperes per square inch.

Natural graphite is a mined product, just like coal. *Artificial graphite* is made in an electric furnace. It is the artificial graphite that is most often used for brushes in motors. This type of brush has a more polishing effect than electro-graphite grades. This type of brush gives good riding qualities and can be used on extremely high-speed devices, such as series motors used in routers and drills. The current density of this type of brush is in the range of 50 to 65 amperes per square inch. Artificial graphite brushes are found in a large variety of small appliances where sparking may be a problem. They are quieter than other types of brushes.

Metal-graphite brushes are made of copper and graphite mixed according to the demand for certain characteristics. This type of brush can carry extremely high currents. They have a current density of 150 amperes per square inch. That is, of course, if the copper content of the brush is 50 percent.

Selecting the Right Brush

There are a number of factors that work together to make a special brush necessary for a particular job. For example, series motors often gain higher efficiency with a low ampere-turns ratio, but with the normally used grade of brush, sparking becomes more noticeable and the commutator is blackened or burned. A proper substitute brush choice for that use would be a hard grade of brush with a slight cleaning action.

Frequent starting and stopping of a motor causes some heavy loads for brushes. When you select a brush, there is no easy method to use to make sure you are getting the right one. Recommendations from the brush and motor manufacturer will help narrow the choice of brush selection. Starting and stopping frequently, reversing, overload capacity, vibration, and brush noise are all factors that enter into the selection of the right brush for a given motor. Manufacturers of motors have most likely been busy testing their motors under various load conditions and will recommend the proper type of brush for their machines. Even humidity and temperature have a bearing on brush life and operation. Refer to the data supplied by the manufacturer of the motor for the best possible brush to use in it.

Figure 3-12 shows the location of the brushes in a portable saw used in general carpentry work. In Figure 3-13 you can see the location of brushes in a popular hand drill. Look through these holes to observe the degree of arcing. This arcing will indicate whether or not the commutator needs to be refinished.

Brush Seating

For lowest possible friction between the commutator and brushes, and for optimum commutation of armature current, an adherent and uniform film must be developed on the commutator surface (Figure 3-14). To accomplish this, brushes must first be "run in" or seated. During this running period, brush material is transferred to the commutator by the direct wiping motion of the brushes and by electrolytic action. As this occurs, the initial contact points are gradually worn away and the brushes begin to conform to the curvature of the commutator surface. Softer-grade brush materials can be seated faster since this initial process is mostly mechanical.

Electrical Brush Wear

After the brushes have been fully seated, electrical wear surpasses mechanical wear as the most important factor in brush operation.

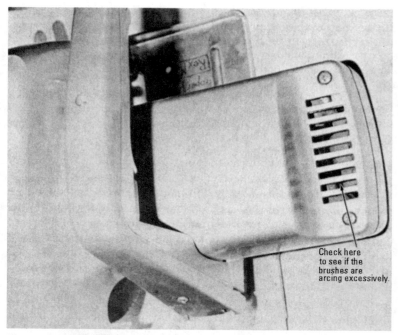

Check here
to see if the
brushes are
arcing excessively.

Figure 3-12   Brushes are located at this end of the saw.

The passage of current through the brush and intervening film causes
a high electrical wear rate. Very high temperatures at the minute conduction points cause copper melting and even carbon vaporization
when the arc voltage is high enough. Electrical wear is most evident
at the negative, or cathode, brush. Here, the wear rate can be two or
more times greater than at the anode (+) brush (Figure 3-15).

Overall Wear Rate

Mechanical and electrical wear contribute to the overall wear rate
of brushes. This rate may increase or decrease by a number of additional factors such as humidity. With a dew point below approximately −10°C, or an absolute humidity of less than 0.13 grams of
water vapor per cubic foot, the commutator film will dry out. This
can occur independently of operating altitude. Loss of the normal
water vapor component of the film gradually leads to the breakdown and eventual disintegration of the film itself.

Figure 3-13　Check for arcing on the electric drill near the brushes.

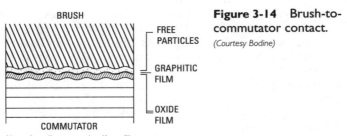

Figure 3-14　Brush-to-commutator contact.

(Courtesy Bodine)

BRUSH

FREE PARTICLES

GRAPHITIC FILM

OXIDE FILM

COMMUTATOR

Note: An adherent and uniform film must be developed on the commutator surface.

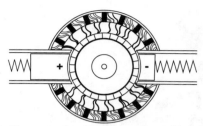

Wear rate at the cathode (−) brush can be two or more times that at the anode (+).

Without this protective film, friction can increase from two to five times, leading to the most rapid brush wear rate, known as "dusting." Dusting at high altitudes or in dry ambient air can be corrected by rehumidifying the atmosphere, or by using brushes with special additives.

Vibration from a connected load or other external source can also cause rapid brush wear. Excessive vibration will affect the way the brush rides on the commutator surface, causing it to conduct intermittently or *spark*. Intermittent contact can greatly accelerate electrical wear.

Fit, or the combination of clearance in the brushholder and brush spring pressure, is also critical to brush performance. If the brush is too loose in the brushholder or the brush spring pressure is too light, premature electrical wear can result. However, when brush pressure is too strong, heavy mechanical wear can be expected.

Commutator Insulation

If not properly undercut, the mica insulation between commutator bar segments can break off in small abrasive particles (Figure 3-16). These particles will rapidly wear down most brush grades now in use. If they become embedded in the brush faces, the mica particles can also scrape across the commutator. This makes grooves in its surface.

Oil on the brushes or commutator, sometimes picked up in the film from the air, acts as an insulator. It can also act as an adhesive to hold mica particles of dirt and cause abrasive wear to the brushes and commutator. For this reason, special care must be taken when lubricating sleeve-bearing motors. Overoiling will often adversely affect brush life. Figures 3-17 and 3-18 show examples of commutator wear from various sources.

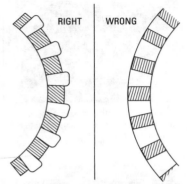

RIGHT   WRONG

Figure 3-16 Proper and improperly cut mica on a commutator can affect brush life. *(Courtesy Bodine)*

Mica insulation between commutator bars must be undercut (Mica—cross hatched in the above diagram—is shown greatly exaggerated in width for purposes of illustration).

Airborne Contaminants

Airborne contaminants, such as the decomposition products of silicones, can also cause poor commutation and excessive brush wear rates. When adhered to the commutator, they act like an abrasive insulating film. Other gaseous substances such as methyl alcohol can reduce the copper oxide in the commutator film and increase brush friction.

Contaminants such as turpentine, paint fumes, and acetone can increase brush friction and accelerate brush wear. Other harmful vapors come from chlorinated hydrocarbon solvents and acid-forming gases such as sulfur dioxide and hydrogen chloride. Even tobacco smoke can raise friction enough to double the brush wear rate.

There is no magic formula for selecting a suitable brush to combat these effects. Motor and brush manufacturers work to select optimum brushes for a given application and environment.

Brush Maintenance

To get the best performance and longest life from motor brushes, the user must develop a regular maintenance procedure. This should include

1. Periodic inspection of brushes and replacement when they have worn down to ¼ in. (7 mm).

2. Inspection of brush springs and replacement when damaged or collapsed.

(1) PITCH BAR-MARKING produces low or burned spots on the commutator surface that equals half or all the number of poles on the motor.

(2) STREAKING on the commutator surface denotes the beginning of serious metal transfer to the carbon brush.

(3) HEAVY SLOT BAR-MARKING involves etching of the trailing edge of commutator bar in relation to the numbered conductors per slot.

(4) THREADING of commutator with fine lines is a result of excessive metal transfer leading to resurfacing and excessive brush wear.

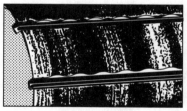

(5) COPPER DRAG is an abnormal amount of excessive commutator material at the trailing edge of bar. Even though rare, flashover may occur if not corrected.

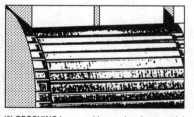

(6) GROOVING is caused by an abrasive material in the brush or atmosphere.

Figure 3-17 Commutator wear and markings: Types of problems.
(Courtesy of Reliance Electric)

3. Periodic cleaning and removal of built-up dust inside the motor. The decomposing or worn brush will deposit its dust.

4. The motor should always be disconnected from the power source before inspecting or replacing the brushes.

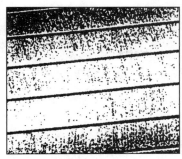

(1) LIGHT TAN FILM over entire commutator surface is a normal condition.

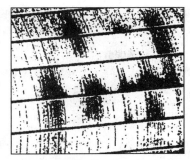

(2) MOTTLED SURFACE with random film patterns is satisfactory.

(3) SLOT BAR MARKINGS appearing on bars in a definite pattern depicts normal wear.

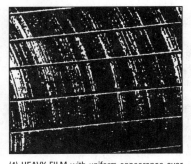

(4) HEAVY FILM with uniform appearance over entire commutator surface is acceptable.

Figure 3-18   Commutator wear and markings: Examples of normal wear. *(Courtesy of Reliance Electric)*

Summary

If the rotor of a motor is wound and it operates on AC, it is most likely to be a universal motor. It is used for small hand drills, sanders, or similar devices. The universal motor is also used in vacuum cleaners. Some types of heavy-duty industrial motors use a wound rotor. However, they are not included in our discussion. They are usually made in sizes over 1 horsepower.

In most cases fractional-horsepower motors that burn out or malfunction in some way are usually replaced rather than repaired. It is possible to maintain small motors and in some cases, repair them inexpensively.

The armature is made up of a shaft, commutator, armature poles, laminations, and lots of copper wire wound around the pole pieces and on the motor's field coils.

Today, magnet wire, insulated by a coating of Formvar-type of varnish is used to rewind stators and armatures. Early motors were wound with wire covered with cotton thread and dipped in shellac for insulation purposes. Formvar is polyvinyl formal and is used in insulating the magnet wire that makes a motor. There are over 300 types of coatings available for magnet wire. The Japanese use polyethylene to coat magnet wire and then overcoat it with nylon.

Operating an overheated motor for a long period of time can cause it to become permanently damaged and need to be rewound. The squirrel-cage motor has high current in the shorting ring. The squirrel-cage motor will start the same as any simple repulsion motor. Once it is started, however, and normal speed is attained, the inductance of the squirrel-cage winding decreases. The current flowing in the shorting ring then decreases.

The squirrel-cage motor has high starting torque and constant speed. It is not, however, found in appliances around the home. It is a special purpose industrial motor. Since it has a wound rotor it does require maintenance.

A hand stone is preferable to sandpaper for removing flat spots of a motor's commutator segments. Various sizes of commutator hand stones are made for use on larger commutators.

Before starting to turn or grind a commutator, the windings of the armature (adjacent to the commutator) should be wrapped in cloth or plastic to protect them from copper chips and dust.

The object or purpose of undercutting commutator mica is to remove the mica between the copper segments so that the segments will wear evenly. The removal of the mica is necessary. Defective mica can be removed by using a hacksaw blade. Pitted and black commutators indicate there has been sparking and the brushes are not properly riding on the surface of the commutator.

Brushes are needed to complete the electrical path for current flow through the armature, which puts some very heavy demands on them. Brushes must be held securely to the surface of the com-mutator, which usually requires spring action. There are four popu-lar types of brush materials: carbon graphite brush, electro-graphite brush, natural graphite brush, and metal-graphite brush. The latter one is made of copper and graphite mixed according to the demand for certain characteristics. The metal graphite brush can carry extremely high currents.

Series motors often gain higher efficiency with a low ampere-ratio, but with the normally used grade of brush, sparking becomes more noticeable and the commutator is blackened or burned.

Frequent starting and stopping, reversing, capacity overload, vibration, and brush noise are all factors that enter into the selection of a brush for a particular motor.

Brushes must be run in or seated to make sure they operate with greatest efficiency. Softer-grade brush materials can be seated faster since this initial process is mostly mechanical. Electrical wear is most evident at the negative, or cathode, brush. Wear at the – brush is twice that of the + brush. Mechanical and electrical wear contribute to the overall wear rate of brushes. Excessive vibration will affect the way the brush rides on the commutator surface, causing it to conduct intermittently or spark. Intermittent contact can greatly accelerate electrical wear. If not properly undercut, the mica insulation between commutator bar segments can break off in small abrasive particles that will rapidly wear down most brush grades now in use. Oil on brushes or commutator sometimes picks up the film from the air and the mica particles act as an insulator. The oil can also hold mica particles or dirt and cause abrasive wear to the brushes and commutator. Overoiling will adversely affect brush life. Methyl alcohol can reduce the copper oxide in the commutator film and increase brush friction. Contaminants such as turpentine, paint fumes, and acetone can increase brush friction and accelerate brush wear. Other solvents and acid-forming gases can also affect brush wear rate. There is no magic formula for selecting a suitable brush to combat these effects. Motor and brush manufacturers work to select optimum brushes for a given application and environment. Regular maintenance is needed to make sure maximum efficiency is obtained from a motor.

Review Questions

1. What type of motor has a wound rotor?
2. Where are wound-rotor motors used?
3. What is done with most fractional-horsepower motors when they malfunction?
4. What is an armature? Where is it found?
5. Describe how the armature is constructed.
6. What kind of material is used today to insulate magnet wire?
7. What are the characteristics of a squirrel-cage type motor?
8. For what would you use a hand stone?
9. What are motor brushes made of?
10. What is meant by the run-in of a motor's brushes?

11. Which brush of a DC motor wears down the fastest?
12. What are the two types of brush wear?
13. What is mica used for?
14. How is a commutator constructed?
15. What are some contaminants that may affect brush wear on a motor?

Part II

Single-Phase and Three-Phase Motors

Chapter 4

Split-Phase Motors

One form of the fractional-horsepower motor is the split-phase induction motor. This motor has two sets of windings: the *run* (main) winding and the *start* (auxiliary) winding. The run winding is the main workhorse of this motor. The start winding is used only when the motor is started; therefore, it is called an auxiliary winding. For a two-pole motor, the start winding is placed 90° electrically from the main winding (Figure 4-1). For a four-pole vector, the angle would be 45°. It is connected in parallel to the run winding (Figure 4-2). Without the auxiliary start winding, this motor would have no starting torque. However, with the start winding, the rotor will reach between 67 to 75 percent of the top or

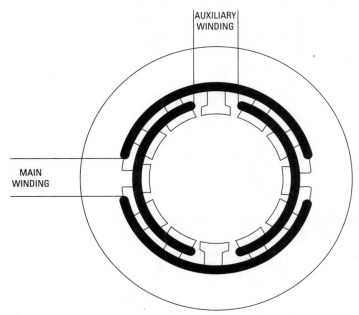

Figure 4-1   Two-pole, single-phase stator with auxiliary (start) winding and main (run) windings. Note the 90° displacement of the centers of the windings.

Figure 4-2 Four-pole, single-phase stator with start and run windings mounted on a frame. *(Courtesy Bodine)*

synchronous speed. At this speed, the motor develops a good running torque and the start winding is no longer needed. The start winding is disconnected from the motor circuit by an automatic starting switch within the motor case. This is usually a centrifugal switch.

Starting a Single-Phase Motor

In a single-phase AC motor (Table 4-1), the field pulsates instead of rotating, as does a two- or three-phase motor. No rotation of the rotor occurs. However, a single-phase pulsating field may be seen as two rotating fields revolving at the same speed but in opposite directions. The rotor will revolve in either of these directions at nearly synchronous speed. But it must be given an initial push in one direction or the other. The precise speed of the initial rotational push velocity varies widely with different motors, but a velocity of more than 15 percent of the synchronous speed is usually enough to

Table 4-1 Typical Characteristics of AC Motors

	HP Ratings	Full Load Speeds, rpm (60 Hz)	Starting Torque	Breakdown Torque	Starting Current	Comparative Cost (100=Lowest)	Guidelines
Shaded-pole	1/65 to 1/20	1650	very low	low	low	100	Low-cost motor for light-duty applications. Compact, rugged, easy to maintain
Permanent-split capacitor	1/50 to 1/3 1/60 to 1/6	3250 1625	low	moderate	low	140	Very compact, easy to maintain. High efficiency, high power factor. Can operate at several speeds with simple control devices.
Split-phase	1/40 to 1/3 1/50 to 1/6	3450 1725	moderate	moderate	high	120	For constant speed operation, varying loads. Where moderate torques are desirable, may be preferable to more expensive capacitor start.

(continued)

Table 4-1 (continued)

	HP Ratings	Full Load Speeds, rpm (60 Hz)	Starting Torque	Breakdown Torque	Starting Current	Comparative Cost (100=Lowest)	Guidelines
Capacitor-start	1/40 to 1/3 1/50 to 1/6	3450 1725	high	high	moderate	150	Suitable for constant speed under varying load, high torques, high overload capacity.
Polyphase	1/30 to 1/3 1/75 to 1/6	3450 1725	high	high	moderate	150	Generally suited to same applications as capacitor start motors if polyphase power is available. Gets to operating speed smoothly and quickly

Courtesy of Robbins & Myers

cause the rotor to accelerate to rated speed. A single-phase motor, of which the split-phase is but one type, can be made self-starting if means can be provided to give the effect of a rotating field.

Starting a Split-Phase Motor

The split-phase motor has a stator composed of slotted laminations that contain an auxiliary (start) winding and a run (main) winding (Figure 4-3). The axes of these two winding are displaced by an angle of 90° electrically. The start winding has fewer turns and smaller wire than the run winding. It also has higher resistance and less reactance. The main winding occupies the lower half of the slots and the start winding the upper half. The two windings are connected in parallel across the single-phase line that supplies power to the motor.

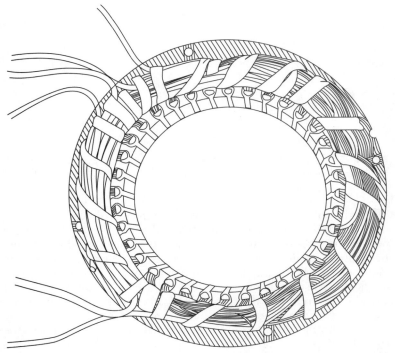

Figure 4-3   Split-phase motor windings. Note that the start and run windings have leads that can be connected in a number of ways to affect the direction of rotation.

The motor takes its name from the action of the stator during the start period. The single-phase motor is split into two windings, or phases, which are separated by 90°. These phases contain currents displaced in time phase by an angle of approximately 15°. The current in the start winding lags the line voltage by about 30°. It is less than the current in the main windings because of the higher impedance (or opposition) of the start winding. The current in the main winding lags the applied voltage by about 45° (Figure 4-4). The total current (I_{line}) during the starting period is the vector sum of the start and main winding currents.

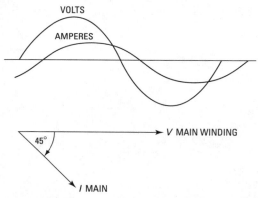

Figure 4-4 Phase relationship between current and voltage in the split-phase motor.

At the motor's start, these two windings produce a magnetic revolving field. The field rotates around the stator air gap (space between motor and stator) at synchronous speed. As the rotating field moves around the air gap, it cuts across the rotor conductors and induces a voltage. The maximum voltage is in the area of highest field intensity. Therefore, it is "in phase" with the stator field. The rotor current lags the rotor voltage at the start by an angle that approaches 90° because of the high rotor reactance.

As the rotor currents and the stator field interact, they cause the rotor to accelerate in the direction in which the stator field is rotating. During acceleration, the rotor voltage, current, and reactance are reduced and the rotor currents come closer to an in-phase relationship with the stator field.

When the rotor reaches about 75 percent of synchronous speed, a centrifugally operated switch disconnects the start winding from the line supply, and the motor continues to run on the main winding alone. Thereafter, the rotating field is maintained by the interaction of the rotor magnetomotive force and the stator magnetomotive force. These two magnetomotive forces are pictured as the vertical and horizontal vectors in Figure 4-5, part C.

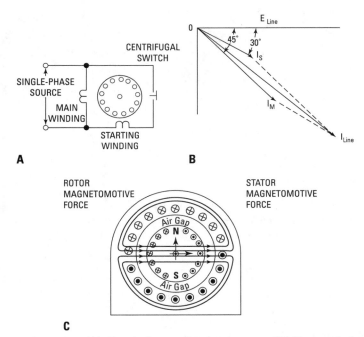

Figure 4-5 (A) Circuit for a split-phase motor. (B) Phase relationships between current and voltage in a split-phase motor. (C) Magnetomotive force angles in a split-phase motor.

The stator field is assumed to be rotating at synchronous speed in a clockwise direction, and the stator currents correspond to the instant that the field is horizontal and extending from left to right across the gap. The left-hand rule for magnetic polarity of the stator indicates that the stator currents will provide a north pole on the left side of the stator and a south pole on the right side (see Figure 4-6). The motor indicated in the figure is wound for two poles.

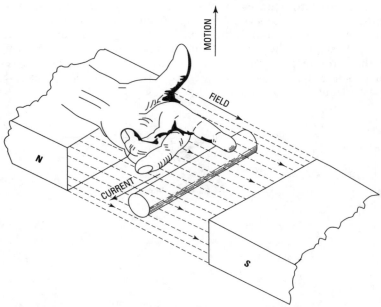

Figure 4-6 Using the left-hand rule to explain the direction of motion of the conductor, direction of the field, and the direction of electron flow in a piece of wire under the influence of a magnetic field.

By applying the left-hand rule (see Chapter 1) to find induced voltage in the rotor (the thumb points in the direction of motion of the conductor with respect to the field), we see that the induced voltage is less than the rotor voltage by an angle whose tangent is the ratio of rotor reactance to rotor resistance. This is a relatively small angle because the slip is small. When we apply the left-hand rule for magnetic polarity to the rotor winding, the vertical vector pointing upward represents the direction and magnitude of the rotor magnetomotive force. This direction indicates the tendency to establish a north pole on the upper side of the rotor and a south pole on the lower side, as indicated in Figure 4-5, part C. Thus, the magnetomotive forces of the rotor and the stator are displaced in space by 90° and in time by an angle that is considerably less than 90° but sufficient to maintain the magnetic revolving field and the rotor speed.

Characteristics of the Split-Phase Motor

The split-phase motor has the constant-speed, variable-torque characteristic of the shunt DC motor. Most of these motors are designed to operate on 120 or 240 V. For the lower voltage, stator coils are divided into two equal groups that are connected in parallel. For the higher voltage, the groups are connected in series. The starting torque is 150 to 200 percent of the full-load torque and the starting current is six to eight times the full-load current. The direction of rotation of the split-phase motor can be reversed by interchanging the start winding leads. Fractional-horsepower, split-phase motors are used in various machines, such as washers, oil burners, and ventilating fans.

After the start winding has been removed from the line, there is no rotating field. The rotation cannot be changed until the motor has come to rest or at least has slowed to the speed at which the automatic switch closes. Special starting switches and special reversing switches are available that can be used to shunt the open contacts of the automatic switch while the motor is running. Thus, the split-phase motor can be reversed while rotating.

Table 4-2 shows the characteristics of a split-phase AC motor (nonsynchronous).

Table 4-2   Motor Characteristics

Split-Phase AC Motor (nonsynchronous)

Duty:	Continuous
Power Supply:	AC
Reversibility:	At rest (with a special switch, can be reversed while rotating)
Speed:	Relatively constant
Starting Torque:	100 to 150 percent of full load
Starting Current:	High

Dual-Frequency Operation

The split-phase motor is not very suitable for operation on dual frequencies. In some cases the necessary compromise may be permissible. If the 60-Hz motor is designed to operate close to the flux saturation point, then the no-lead watts may be more than doubled, producing more heat, when operated on 50 Hz. An additional problem is the operating speed. This means trying to obtain the proper operating speed so that the centrifugal switch will operate and remove the start winding from the circuit. Also, finding a start relay suitable for both 50- and 60-Hz operation is difficult.

If you decide to rewind a 25-Hz motor to operate on 60-Hz current, the start-winding switch will have to be changed. The number of turns and the whole design of the motor must be considered. In most cases conversion is possible, but problems should be expected later in motor operation. Indeed, the performance of the motor may never be satisfactory.

If the change in frequency is small, say, from 50 to 60 Hz, no change in the winding of general-purpose motors is ordinarily needed unless the motor is severely overloaded. A change from 60 to 50 Hz normally requires no winding change either. If the motor has high torque, it may be necessary to increase the number of turns of the 60-Hz motor by 10 percent. This is done to obtain satisfactory operation on 50 Hz without overheating. However, when changing either a general-purpose or high-torque motor from 60 to 50 Hz, the rotating part of the starting switch should be changed. The switching torque is affected by the change in frequency. If the change is from 50 to 60 Hz and the torque requirements are not too severe, there may not be any reason for changing the windings or the switch.

The main concern is the inductive reactance (X_L) of the run winding. It changes as the frequency changes, because $X_L = 2\pi\, FL$, where F = frequency and L = inductance. However, if the motor was designed to operate on 50 *or* 60 Hz, then the operating frequency and either the 50- or 60-Hz frequency would result in little significant heat generation that was not designed in the motor in the first place.

Motor Noise and Vibration

There are two types of motor noise: *mechanical* and *electrical*. Mechanical noise can be caused by dynamic imbalance; this means that the rotor is unbalanced. When a rotor is unbalanced, small holes are drilled into it (Figure 4-7). The loss of the metal removed from these holes helps to balance the rotor. To check for rotor imbalance, turn off the motor and listen for a vibrating noise as the motor coasts. If there is vibration, the dynamic balance of the rotor is probably causing problems. In a split-phase motor, the centrifugal switch can also cause problems and should be checked to see if it is operating properly (Figure 4-8).

Noise may also be created by bearings. Usually, the trouble is closely tied to bearing preload. Preload is the axial thrust on the outer bearing race to eliminate rattling of unloaded balls. The

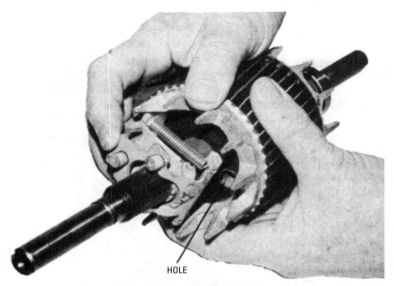

HOLE

Figure 4-7   Note the hole behind the centrifugal switch mechanism on this rotor. It is put there to aid the balance of the rotor.

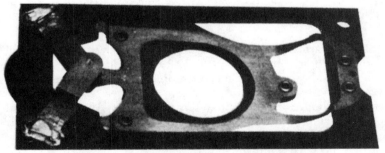

Figure 4-8   Start or centrifugal switch for a split-phase motor.

amount of preload that causes rattling noises is very low—below 2 lb for most motors. Noise levels for ball-bearing motors will not fall below 40 dB in any event. Also, a slight variation in noise level between identical motors is normal; thus, do not expect to find a very quiet ball-bearing motor (Figure 4-9).

Figure 4-9 To prevent ball-bearing movement, a special spring washer is placed in the bell housing.

Sleeve bearings produce much lower noise levels than do ball bearings. If noise must be kept low, sleeve bearings can be used in place of ball bearings. The biggest problem with sleeve bearings is to control noise from the thrust washers. These washers have a tendency to produce an intermittent scraping sound. Another problem is that bearing knocking or pounding occurs after the motor has been operated for a long time. This happens when the oil for the bearings has been thinned by heat.

Fans can be a major source of noise. Even fans with low-speed motors can produce noticeable levels of noise. The swish or rumble from an air exhaust opening can be annoying to some people. In electric motors, fans are used to keep air flowing over the rotor and stator windings during operation. High-speed motors can produce a siren-like noise if the fan housing is not properly designed. In reassembling a motor, make sure the fan blades do not touch the motor housing. If they do, the noise will be heard right away. Spacing washers are used to prevent end play that causes the fan to hit the motor housing; properly installed, they prevent noise from the fan. Of course, the proper size of washer is important.

Finally, gear trains may or may not be noisy. Worm-gear-type trains are almost noiseless. Helical gearing is also quiet. Spur gearing makes the most noise. Gear-related noise shows up when the load increases. It becomes more intense and evident as the load is increased. Backlash noise occurs in gear trains when the motor is operated at little or no load. Some noise at low speeds and low load is normal for all gear-type trains.

Getting Rid of Mechanical Noises

The first step in getting rid of mechanical noise is to find the source of the noise. To help eliminate noise, (1) enclose the motor to reduce the airborne noise that may come from the inside of the motor housing; (2) use a resilient mounting to help dampen motor vibrations that reach the mounting structure (Figure 4-10); (3) use flexible couplings to reduce noise; and (4) incorporate rubber, cork, and felt in the mounting design to absorb sound.

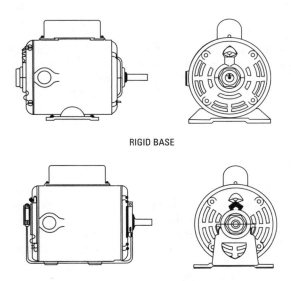

RIGID BASE

RESILIENT BASE

Figure 4-10 Rigid-base and resilient-base mountings for electric motors.

Electrical Motor Noise

The hum of a running motor may be music to some people, but it is very annoying to others. The noise itself may be caused by the electrical characteristics of the motor. Such noise is made up of loudness and frequency.

Slip in a motor can cause annoying noise. Slip is the difference between the speed of the rotating field and actual speed of the rotor. Induction motors, including the split-phase, have a lot of slip. They are quieter during operation than reluctance-synchronous motors, which operate with no slip. The amount of slip has a direct bearing on the level of noise generated by the electrical characteristics of a motor.

The *air gap* can also cause noise. The gap between the rotor and the stationary poles is important in noise production. When the air gap is too small, the stator teeth can become oversaturated and the motor noise will increase.

The *distribution of the field flux* makes a great deal of difference in the quietness of an electric motor. The quietness of motor operation depends on the strength of the magnetic flux and how it is distributed. For example, the permanent-split capacitor motor has two windings and thus has a more even flux distribution. It is less noisy than the split-phase motor. Troubleshooting and maintenance of the split-phase motor is found in detail in Chapter 21.

Bearings can also be a source of electrical noise. Sleeve bearings often come installed in small motors. They provide bearings that are, in a general-purpose system, smooth operating. The sleeve bearing is quieter than the ball bearing. Ball bearings have less friction, but are more susceptible to damage when bumped on the end. Larger motors use roller bearings while very large motors utilize a forced-oil bearing.

Summary

The split-phase motor has two sets of windings: the run winding and the start winding. Without the auxiliary start winding, this motor would have no starting torque. It reaches about 75 percent of its rated speed before a switch disconnects the start-winding from the circuit. This switch is referred to as a centrifugal switch.

In a single-phase AC motor, the field pulsates instead of rotating as it does in a two- or three-phase motor. No rotation of the rotor occurs. The rotor needs a push to get started. A single-phase motor can be made self-starting if means can be provided to give the effect of a rotating field. A split-phase motor has a stator with slotted laminations that contain both an auxiliary (start) winding and a run (main) winding. The axes of these windings are displaced by an angle of 90° electrically. The start winding has fewer turns and smaller wire than the run winding. The current in the start winding lags the line voltage by about 30°. It is less than the current in the main windings because of the higher impedance or opposition of the starting winding. When the motor starts the two windings produce a magnetic revolving field. The field rotates around the stator air gap at synchronous speed. As the rotor currents and stator field interact, they cause the rotor to accelerate in the direction in which the stator field is rotating. When the rotor reaches about 75 percent of the synchronous speed, a centrifugally operated switch disconnects the start-winding from the line supply and the motor continues to run on the main winding alone.

By applying the left-hand rule to find induced voltage in the rotor we see that the induced voltage is less than the rotor voltage by an angle whose tangent is the ratio of rotor reactance to rotor resistance. This is a relatively small angle because the slip is small.

The split-phase motor has the constant-speed, variable-torque characteristic of the shunt DC motor. Most of these motors are designed to operate on 120 and 240 V. The 120-volt motor has the stator coils divided into two equal groups that are connected in parallel. For the higher voltage, the groups are connected in series. Starting torque is 150 to 200 percent of the full-load torque and the starting current is six to eight times the full-load current. The direction of rotation of the split-phase motor can be reversed by interchanging the start winding leads. These motors are used in various machines, such as washers, oil burners, and ventilating fans. With the use of special switches the split-phase motor can be reversed while running. The split-phase motor is not suitable for operation on dual frequencies.

There are two types of motor noise: mechanical and electrical. Mechanical noise can be caused by dynamic imbalance and bearings. A slight difference in noise in any two motors is normal. Sleeve bearings produce much lower noise levels than do ball bearings. Fans in motors can also be a source of noise. Gear trains may or may not be noisy. Worm gear-type trains are almost noiseless. Spur gearing makes the most noise. Backlash noise occurs in gear trains when the motor is operated at little or no load. Some noise at low speeds and low load is normal for all gear-type trains.

The first step in getting rid of noise is to find the source. The hum of a running motor may be music to some people, but it is very annoying to others. The noise itself may be caused by the electrical characteristics of the motor. Such noise is made up of loudness and frequency. Slip can cause motor noise as can air gap. The gap between the rotor and the stationary poles is important in noise production. The distribution of the field flux makes a great deal of difference in the quietness of an electric motor. The quietness of motor operation depends on the strength of the magnetic flux and how it is distributed.

Review Questions

1. How many sets of windings does the split-phase motor have?
2. How is the split-phase motor made to self-start?
3. Where is the split-phase motor used?
4. Does the split-phase motor draw high, medium, or low amounts of current at start up?
5. What is the synchronous speed of a split-phase motor?
6. What does the left-hand rule tell you about a motor?
7. What is slip in a motor?
8. What type of torque does the split-phase motor have?
9. In what voltages are split-phase motors made?
10. What are the two types of noise made by motors?
11. What causes mechanical noise in a motor?
12. What causes electrical noise in a motor?
13. In reference to magnetic flux, on what does the quietness of the motor depend?
14. What is backlash in a motor?
15. How can a split-phase motor be reversed when it is running?

Chapter 5

Capacitor-Start/ Capacitor-Run Motors

The capacitor-start motor is a modified form of the split-phase motor. It has a capacitor in series with the start winding. An external view shows the capacitor sitting on top or on the side of the motor frame (Figures 5-1 and 5-2). The capacitor produces a greater phase displacement of currents in the start and run windings than is produced in the split-phase motor.

Figure 5-1 Capacitor-start motor. Note the location of the start capacitor on the top of the motor. *(Courtesy Westinghouse)*

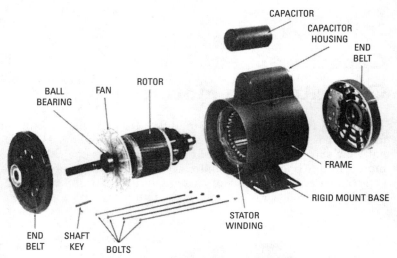

Figure 5-2 Exploded view of the capacitor-start motor.

(Courtesy Westinghouse)

The start winding is made of many more turns of larger wire than in the split-phase motor, and the winding is in series with the capacitor. The start-winding current is displaced 90° from the run-winding current. Since the axes of the two windings are also displaced by an angle of 90°, these conditions produce a greater starting torque than that of the split-phase motor. The start torque of the capacitor may be as much as 350 percent of the full-load torque.

If the start winding is cut out after the motor has increased in speed, the motor is called a capacitor-start motor. If the start winding and capacitor are designed to be left in the circuit continuously, the motor is called a *capacitor-run motor*. Electrolytic capacitors for capacitor-start motors vary in size from about 80 microfarads (μF) for ⅛-horsepower motors to 400 microfarads for 1-horsepower motors. Capacitor motors of both types are made in sizes ranging from fractional horsepower to about 10 horsepower. They are used to drive grinders, drill presses, refrigerator compressors, and other loads that require relatively high starting torque. The direction of rotation of the capacitor motor may be reversed by interchanging the start-winding leads. Some table saws are powered by capacitor motors. This provides high pull-out torque that is needed when sawing hard wood or if a knot is hit when ripping through a board.

Starting a Capacitor-Start Motor

The capacitor-start motor is similar to the split-phase motor in starting, with one exception: the insertion of a capacitor in series with the winding. The windings of the motor are displaced 90° electrically. In order to develop a rotating field, it is necessary to develop locked-rotor torque. The currents in the start and run windings must be displaced by 90°. In both the capacitor and split-phase motors, the windings are connected in parallel to the line or power supply. In the split-phase motor, resistance is deliberately built into the auxiliary winding to bring the current more nearly in phase with the line voltage. In a capacitor-start motor, the capacitor causes the start-winding phase current to *lead* the main-phase voltage, obtaining a large angle of displacement between the currents in the two windings. The line current in this motor is only two-thirds of the line current of the split-phase motor. This motor, however, has more than twice the locked-rotor torque of the split-phase motor. The capacitor is more effective as a starting device than the resistance in the split-phase motor. However, it must be remembered that the start current is rather high and must have a fuse or a circuit breaker that will allow the higher current draw from the line for a short period before disconnecting the circuit.

The Electrolytic Capacitor

Most modern-day capacitors used on AC are encased in a black Bakelite container (Figure 5-3). Many capacitors have two terminals that allow for a slip-on type of solderless wire connector. In other types a screw-type terminal is mounted on the electrolytic capacitor. This type of starting arrangement has been in use since 1892. However, it was not until 1930 that the capacitor-start motor was generally accepted as a standard power source for devices that start under load conditions.

The dry electrolytic-type capacitor is made by winding two sheets of aluminum foil into a cylindrical shape. The two sheets of aluminum foil are separated by an insulator, which is usually gauze. Two layers of paper are also used in combination with an electrolyte. The electrolyte is usually ethylene or some similar chemical composition. On each of the two layers of foil, an anodic film is produced by electromechanical means; that is, the finished capacitor has a DC voltage applied to its terminals. This means it has a positive (+) and a negative (−) marking. The capacitor is sealed in an aluminum case (Figure 5-4). There are other types of cases, more widely used than aluminum, for the outside of the capacitor.

DRY ELECTROLYTIC
CAPACITOR

Figure 5-3 Capacitor-start motor. Note how the Bakelite case of the electrolytic capacitor shows after the metal cover has been removed. A piece of corrugated cardboard is inserted underneath the capacitor to insulate it from the case and reduce noise.

Figure 5-4 Electrolytic capacitors used for capacitor-start motors. The round ones are usually used for capacitor-start operations. The other shapes are usually used for capacitor-run operations.

If AC is applied to an electrolytic that has been formed with a positive (+) and a negative (−) orientation on its terminals, the capacitor will *explode* in about 5 seconds. That is why the AC capacitor is actually two capacitors in one. The two capacitors are connected back to back in series; that is, the positive (+) of one capacitor is connected to the positive (+) of the other capacitor. That produces a *nonpolarized* electrolytic capacitor for use on AC (Figure 5-5).

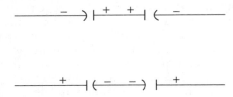

Figure 5-5   Series connections of capacitors (polarized types) to make them nonpolarized for use on AC motors.

Characteristics of the Electrolytic Capacitor

Electrolytic capacitors used on motors are usually of a larger size than those used in electronic circuits for several reasons. One reason is that they have to be able to take the vibration and mechanical abuse of a motor. Another is that they are located in a space that is not too confining and they can be larger to dissipate the heat buildup that occurs in the capacitor during its brief stay in the circuit. Some capacitors are designed to remain in the circuit in the capacitor-run type of electric motors (Figure 5-6).

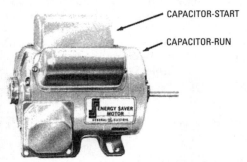

Figure 5-6   Small motor with two capacitors. The capacitor-start and capacitor-run motor is shown mounted on the outside of the motor frame. *(Courtesy General Electric)*

The voltage rating is stamped on the outside of the capacitor case. This is very important. It tells you the safe operating voltage for that particular capacitor. It becomes very important when you switch from 120-volt operation of the motor to 240 V. The capacitor must be able to take the higher voltage, as should the motor windings. However, since most of these motors are made to run on either 120 or 240 V, the engineers see to it that the capacitor will operate on both. The voltage rating tells what voltage it takes for the capacitor to break down.

About the only problem with normal operation of the capacitor-start motor is with the capacitor. After a few years of use, or even after no use, the capacitor has a tendency to dry out. This causes the capacitor to open. It no longer acts as a capacitor and the motor will not start; it will merely hum. However, if the motor is started manually, it will start in either direction. It will come up to speed and run properly. Replacement of the capacitor should always be with a unit of the same capacitance and voltage rating.

A *farad* is the unit of measurement for capacitance. It is described as the capacitance of a capacitor in which a charge of 1 coulomb produces a change of 1 volt in the potential difference between its plates or terminals. This means that there are a lot of electrons and the one-farad capacitor is huge, especially if the working voltage is 120 or 240. Therefore, it is necessary for most electronic and electrical applications to use only very small amounts of capacitance to get the job done. The term *micro* means one-millionth (0.000001); hence the microfarad is one-millionth of a farad. Eighty microfarads equal eighty-millionths of a farad.

Most electrolytic capacitors will operate in temperatures up to 176°F (80°C). That means they are useful for most motor applications. They are useful for shorter periods of time if they operate at elevated temperatures.

Duty cycle for motor-starting capacitors is rated on the basis of twenty 3-second periods per hour. Sixty 1-second periods per hour would be one equivalent duty cycle. Table 5-1 lists the ratings and test limits for AC electrolytic capacitors.

Caution
When you replace a defective capacitor, it is imperative that the new capacitor be of the same voltage and microfarad rating.

Characteristics of Capacitor Motors
There are many types of motors that use capacitors in their start circuit, in their start and run circuit, and in combinations of start

Table 5-1 Ratings and Test Limits for AC Electrolytic Capacitors

Capacity Rating, Microfarads			110-Volt Ratings		125-Volt Ratings		220-Volt Ratings	
Nominal	Limits	Average	Amps. at Rated Voltage, 60 Hz	Approx. Max. Watts	Amps. at Rated Voltage, 60 Hz	Approx. Max. Watts	Amps. at Rated Voltage, 60 Hz	Approx. Max. Watts
	2.5–30	27.5	1.04–1.24	10.9	1.18–1.41	14.1	2.07–2.49	43.8
	32–36	34	1.33–1.49	15.1	1.51–1.70	17	2.65–2.99	52.6
	38–42	40	1.56–1.74	15.3	1.79–1.98	19.8	3.15–3.48	61.2
	43–48	45.5	1.78–1.99	17.5	2.03–2.26	22.6	3.57–3.98	70
50	53–60	56.5	2.20–2.49	21.9	2.50–2.83	28.3	4.40–4.98	87.6
60	64–72	68	2.65–2.99	26.3	3.02–3.39	33.9	5.31–5.97	118.2
65	70–78	74	2.90–3.23	28.4	3.30–3.68	36.8	5.81–6.47	128.1
70	75–84	79.5	3.11–3.48	30.6	3.53–3.96	39.6	6.22–6.97	138
80	86–96	91	3.57–3.98	35	4.05–4.52	45.2	7.13–7.96	157.6
90	97–107	102	4.02–4.44	39.1	4.57–5.04	50.4	8.05–8.87	175.6
100	108–120	114	4.48–4.98	43.8	5.09–5.65	56.5	8.96–9.95	197
115	124–138	131	5.14–5.72	50.3	5.84–6.50	65		
135	145–162	154	6.01–6.72	62.8	6.83–7.63	85.8		
150	161–180	170	6.68–7.46	69.8	7.59–8.48	95.4		
175	189–210	200	7.84–8.71	81.4	8.91–9.90	111.4		
180	194–216	205	8.05–8.96	83.8	9.14–10.18	114.5		
200	216–240	228	8.96–9.95	93	10.18–11.31	127.2		
215	233–260	247	9.66–10.78	106.7	10.98–12.25	145.5		
225	243–270	257	10.08–11.20	110.9	11.45–12.72	151		
250	270–300	285	11.20–12.44	123.2	12.72–14.14	167.9		
300	324–360	342	13.44–14.93	147.8	15.27–16.96	201.4		
315	340–380	360	14.10–15.76	156				
350	378–420	399	15.68–17.42	172.5				
400	430–480	455	17.83–19.91	197.1				

and run. These motors are single-phase induction motors with main winding arranged for direct connection to the power source. The start winding has a capacitor connected in series with it. There are three types of capacitor motors: the capacitor-start, the permanent-split capacitor, and the two-value capacitor motor. The capacitor-start motor has the capacitor in the circuit only during starting. The permanent-split capacitor motor has the same capacitor and capacitor phase in the circuit for both starting and running. The two-value capacitor motor has capacitors of different capacitance values for starting and for running.

The *capacitor-start motor* is suitable for operation only on AC. Its windings are designed for continuous duty. It may be reversed at standstill or when rotating at a speed low enough to ensure that the start winding is in the circuit. The motor can be reversed at full speed by the use of a special type of centrifugal switch or by a relay.

The *permanent-split capacitor motor* is suitable for operation only on AC. Its standard windings are designed for continuous duty. This type of motor may be reversed at standstill or, under favorable conditions, when running. High-inertia loads retard the reversing action. Friction loads present no problems with reversing.

This permanent-split capacitor motor has fairly constant speed, as does the split-phase motor. It does not have low starting torque, though, and pulls very little current on starting. It is the quietest-operating and smoothest-performing of all the classes of single-phase fractional-horsepower motors.

The *two-capacitor motor* (Figure 5-6) is suitable for operation only on AC. The windings are also designed for continuous duty. Its starting and reversing characteristics are similar to those of the capacitor-start motor. It has less breakdown torque than the polyphase motors, but has similar run characteristics.

Table 5-2 shows the characteristics of capacitor motors.

Table 5-2   Motor Characteristics

Capacitor Motor

Duty:	Continuous
Power Supply:	AC
Reversibility:	At rest or during rotation
Speed:	Relatively constant
Starting Torque:	75–150 percent of rated torque
Starting Current:	Normal

Reversing the Motor

An induction motor will not always reverse while running. It may continue to run in the same direction, but at a reduced efficiency. An inertia-type load is difficult to reverse. Most motors that are classified as reversible while running will reverse with a noninertia-type load. They may not reverse if they are under no-load conditions or if they have a light load or an inertia load. A permanent-split capacitor motor that has insufficient torque to reverse a given load may just continue to run in the same direction.

One of the problems related to the reversing of a motor while it is still running is the damage done to the transmission system connected to the load. In some cases it is possible to damage a load. One of the ways to avoid this is to make sure the right motor is connected to a load.

Dual-Frequency Operation

Capacitor-start motors are least suitable for the dual-frequency operation. However, it may be possible in some cases.

The permanent-split capacitor motor is best suited for 50/60-Hz operation. This is due primarily to the two different effects on the two windings with a change in frequency. When the frequency is changed from 60 to 50 Hz, the current in the main winding will decrease. This means that the total current may remain the same. It is possible to wind a permanent-split capacitor motor so that it will not draw more power from the line at one frequency than the other. However, other operating characteristics may change. It is best to know under what conditions the motor will operate and then select the proper motor for doing the job.

Mechanical and Electrical Noises

The capacitor motors have about the same types of mechanical noises as the split-phase motor (see Chapter 4). They are essentially built the same mechanically, with the addition of a start capacitor mounted to the case in most applications. In the case of a refrigerator motor, the capacitor is mounted on the inside wall of the refrigerator cabinet or under the main unit. In the case of a washing machine, it is usually mounted on a removable panel in the rear of the machine. This means that it does little to add to the mechanical noise level.

Electrical noises are almost identical to those generated by the split-phase motor (see Chapter 4). However, most of the capacitor motors are smoother-running than the split-phase motors, and they have a lower noise level than other types of single-phase motors.

Most of the noise can be eliminated by using a resilient type of mounting for the motor and by checking closely the location of the motor in reference to the device it is powering.

The Capacitor-Run Motor

The capacitor-run motor is, for all practical purposes, a two-phase motor that is powered by a single-phase source. It has two windings, one of which is directly connected to the source. The other winding is also connected to the source, but in a series with a paper capacitor. The capacitor-fed winding has a large number of turns of relatively small wire in comparison to the directly connected winding.

This motor runs very quietly and is often used in hospitals and sound studios where a low level of motor noise is desired. It has a high power factor due to the capacitor being in the circuit continuously even while the motor is running. No centrifugal switch is needed inasmuch as the capacitor stays in the circuit. It does, however, have a low starting torque.

This type of motor acts, for all practical purposes, as a two-phase motor only when it operates at full load. Under full load, the fluxes, ϕ, in the two windings are equal and out of phase by 90°. This eliminates vibration and the motor is able to operate very quietly. This type of motor is available in sizes under 500 W or under ¾ of a horsepower.

Reversing the Capacitor-Run Motor

In order to reverse motors, it is usual practice to reverse the leads of either the auxiliary winding or the main winding. However, if the single-phase motor has a centrifugal switch, its direction of rotation cannot be reversed while it is running. Since there is only one lead in the circuit when running, the reversal of the leads allows the motor to continue rotating in the same direction. But in the case of the capacitor-run motor, it can be reversed since both windings are in the circuit at all times. It can be done with a simple double-throw switch that changes which winding has the capacitor in series with it (see Figure 5-7). Since the two windings are identical, it simply switches the capacitor arrangement in reference to the coils. When switched, the running motor will come to a complete stop and then start rotating in the opposite direction.

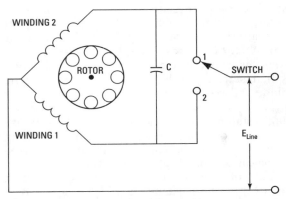

Figure 5-7   Reversible single-phase motor: capacitor-run type.

Summary

The capacitor-start motor is a modified form of the split-phase motor. It has a capacitor in series with the start winding. The capacitor produces a greater phase displacement of currents in the start and run windings than is produced in the split-phase type. The start torque may be as much as 350 percent of the full-load torque. If the capacitor is removed after the rotor comes up to speed then it is called a capacitor-start motor. If the capacitor stays in the circuit after the motor is started then it is called a capacitor-start, capacitor-run motor. Electrolytic capacitors used in these type of motors are 80 microfarads (μF) for $\frac{1}{8}$-horsepower motors to 400 microfarads for 1-horsepower motors.

Capacitor motors of both types are made in sizes ranging from fractional horsepower to about 10 horsepower. They are used in drill presses, refrigerator compressors, and other loads that require relatively high starting torque. The direction of rotation of the capacitor motor may be reversed by interchanging the start-winding leads. In some instances where high pull-out torque (such as in table saws that hit a knot when cutting relatively soft wood) is needed, the capacitor motor is preferred. One advantage of this motor over the split-phase is its starting torque and its ability to develop pull-out torque when needed. Another advantage is that the line current is only two-thirds that of the line current of the split-phase motor. However, the starting current is much higher than the split-phase motor.

Most of the electrolytic capacitors are encased in a metal cover and mounted on the outside of the motor frame. They should be of the AC type, which means they do not have polarity. The capacitor motor dates back to 1892, but it wasn't until the 1930s that the capacitor-start motor was generally accepted as a standard power source for devices that start under load conditions. Electrolytic capacitors can explode and you need to be careful when working on them or nearby.

The voltage rating is stamped on the capacitor as well as its size in microfarads (μF). Capacitors can dry up when not in use for long periods of time. Replacement of the capacitor should always be with a unit of the same capacitance and voltage rating.

A *farad* is the unit of measurement for capacitance. It is described as the capacitance of a capacitor in which a charge of 1 coulomb produces a change of 1 volt in the potential difference between its plates or terminals. This means the 1 microfarad capacitor is very large, so it is usually made in smaller sizes that are in micro (0.000001) farads. Micro means *one millionth*.

Types of capacitor motors include capacitor-start; capacitor-start, capacitor run; and permanent-split capacitor, as well as the two-value capacitor motor. The two-value capacitor motor has two capacitors with different ratings as far as size is concerned. The permanent split-capacitor motor is suitable for operation only on AC. Its standard windings are designed for continuous duty and may be reversed at standstill. High-inertia loads retard the reversing action. Friction loads present no problems with reversing. It has fairly constant speed, as does the split-phase motor. It is the quietest operating and smoothest-performing of all the classes of single-phase motors in fractional-horsepower sizes.

Capacitor-start motors are least suitable for dual-frequency operation. It is possible in some cases, though. The permanent-split motor is best suited for 50/60-Hz operation.

These motors have about the same types of mechanical noises as the split-phase motor. They are essentially the same mechanically. In the case of a refrigerator motor, the capacitor is mounted on the inside wall of the refrigerator cabinet or under the main unit.

Electrical noises are almost identical to those generated by the split-phase motor. However, most of the capacitor motors are smoother-running than other types of single-phase motors. Location and mounting for the motor will keep the noise level down.

Review Questions

1. What other type of motor is the capacitor-start motor similar to?
2. What is the purpose of the capacitor in the capacitor-start motor?
3. What type of capacitor is used in this type motor?
4. Where is the capacitor-start used today?
5. What is one of the most valuable characteristics of the capacitor-start motor?
6. How is the capacitor-start motor reversed? Under what conditions may it be reversed?
7. What is the advantage of having high pull-out torque in a motor design?
8. Why does the capacitor-start motor have high starting current requirements?
9. How are motor electrolytic capacitors packaged?
10. What is a farad? How big is a one-farad capacitor?
11. Which is the quietest running of the capacitor motors?
12. Which motors are the least suitable for dual-frequency operation? Which is the best suited?
13. What is meant by a high-inertia load?
14. Why does a high-inertia load make a motor hard to reverse?
15. Why can't you operate a capacitor-start motor on DC?

Chapter 6

Shaded-Pole Motors

The shaded-pole motor is a single-phase induction motor with an auxiliary short-circuited winding or windings. The shorted winding is displaced in magnetic position from the main winding (Figure 6-1). Shaded-pole motors are suitable for operation on AC only. The standard windings are designed for continuous duty when used to power fans that will draw air over the windings. The shaded-pole motor is not normally reversible. It takes a stator with two windings wound differently to make it reversible (Figure 6-2). The speed of a shaded-pole motor is fairly constant, but it will vary once a load is placed on it. It has very low starting torque—less than 100 percent in most cases. Starting current, however, is also very low—less than twice full-load current (Table 6-1).

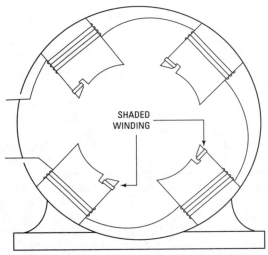

Figure 6-1   Four-pole motor stator showing the shaded windings.

Figure 6-2   Dual stator for reversibility. *(Courtesy Brevel)*

Table 6-1   Characteristics of the Shaded-Pole Motor

Characteristic	Rating	Comment
Efficiency	Low	20–40%
Power factor	Low	50–60%
Starting torque	Low	Plus 3rd Harmonic Dip
Noise and vibration	High	120 Hz plus winding harmonics
Cost	Low	

The shaded-pole motor has one coil that is connected across the line at all times. One of its distinguishing characteristics is its shaded pole, or the short circuited winding it has around one end of the stator pole piece. The shorted pole creates a varying flux field that gives the effect of having two coils out of phase with one another, and thereby provides the rotating field for starting the motor. The starting torque is very low, usually 30 to 50 percent of the rated motor torque. This means it is good only for mechanical loads, at starting. Such loads are fans that start out with a low torque requirement and then require more to cause them to move when they reach full-load at rated speed.

The shaded-pole motor is inefficient due to its shaded pole losses. That means it can be used where the need for efficiency is less important than initial cost. The efficiencies of the larger motor (up to 0.1 hp) are up to 30 percent while the very small motors have efficiencies as low as 5 percent. It has a variety of uses, but is made in 0.0025 to 0.1 horsepower ratings. In most cases, the motors are rated in terms of watts. It takes 746 W to equal 1 horsepower, so a 0.0025-horsepower motor would be about 2 W. The 0.1 horsepower motor would be about 75 W. As you can see, they may be left on the line without drawing too much current and, in some cases, the lower wattage ones will not even cause the watt-hour meter to turn, that is, if nothing else is drawing current on that line. This makes it ideal for a device used to time recharge cycles for a soft-water conditioner or similar devices. It is often used as a timer for automatic irrigation systems for home lawns.

Starting the Shaded-Pole Motor

Just as with any other motor, when the main field coils in a shaded-pole motor are energized, a magnetic field is set up between the pole pieces and the rotor. A portion of the magnetic field is cut by a *shading* coil (Figures 6-3 and 6-4). This coil, made of a large-diameter piece of wire shorted together in the form of a ring, shades the pole piece of the main windings. This shorted winding makes the magnetic field between the pole piece and rotor slightly out of phase with the main portion of the pole piece. This has the effect of producing a two-phase voltage situation at every pole piece. There is very little torque generated by this phase difference, but it is enough to get the motor started if it is not overloaded. The rotor is a squirrel-cage configuration, which has a high resistance. The net effect of the shading of the pole piece is to produce a displacement that causes a shifting flux in the air gap, always shifting toward the shading coil. The direction of rotation of a shaded-pole motor is always from the unshaded to the shaded portion of the pole.

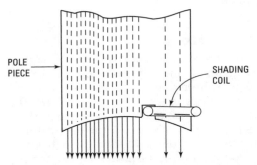

Figure 6-3   Shading ring on shaded-pole motor.

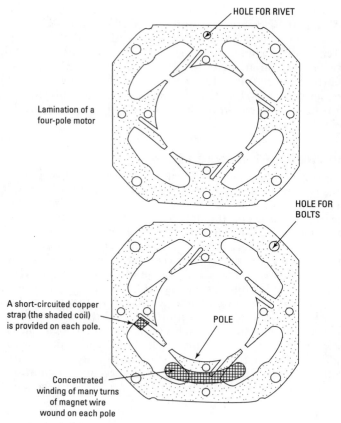

HOLE FOR RIVET

Lamination of a
four-pole motor

HOLE FOR
BOLTS

A short-circuited copper
strap (the shaded coil)
is provided on each pole.

POLE

Concentrated
winding of many turns
of magnet wire
wound on each pole

Figure 6-4 Shaded-pole motor laminations.

Types of Shaded-Pole Motors

Shaded-pole motors are of three basic types and are built in a
wide variety of ratings, with any number of performance characteristics. They are made in subfractional and fractional horsepower only. Horsepower ratings vary from 0.0025 to 0.1
horsepower. Some serious design characteristics prevent them
from being made in larger horsepower sizes. In the upper horsepower range, most of the motors are of the four-pole construction
method (Figure 6-5), while the lower-horsepower types use only
two poles.

Figure 6-5 Four-pole, shaded-pole motor. *(Courtesy General Electric)*

Speeds for the shaded-pole motor are stated with no-load conditions, and range from 2700 to 3500 rpm. Full-load speeds are from 1600 to 3000 rpm. The most common 60-Hz types with four poles have a speed of 1750 to 1770 rpm. Full-load speeds are then 1450 to 1620 rpm. A good many of these motors are designed with a reduction gearbox attached. They can be reduced to as much as 450:1, which means that a wide variety of speeds is available for use in many places. General Electric makes four-pole, shaded-pole motors in 1000, 1100, and 1500 rpm.

Construction of Shaded-Pole Motors

The three types of shaded-pole motors are salient-pole, skeleton, and distributed-winding. *Salient-pole* construction has many main-winding coils. The number of poles is the same as the number of coils. In most cases there are four coils. Some of the older refrigerator fans are salient-pole, shaded-pole motors. They used a felt soaked in oil for lubricating the rotary part.

The *skeleton* type is used for horsepowers from 0.00025 to 0.03 (Figures 6-6 and 6-7). This type of motor uses triple shading. Three shading coils of different throw for each pole are used. In this type the bearings are self-aligning oilite. Wick-type oilers are used to spread the lubrication so that it covers the whole rotating shaft. A squirrel-cage rotor is used in the skeleton design. A spring causes the rotor to be slightly offset from the pole, and it is easily compressed as

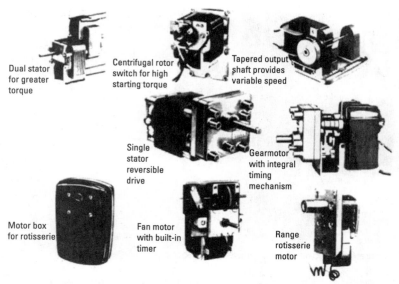

Dual stator for greater torque

Centrifugal rotor switch for high starting torque

Tapered output shaft provides variable speed

Single stator reversible drive

Gearmotor with integral timing mechanism

Motor box for rotisserie

Fan motor with built-in timer

Range rotisserie motor

Figure 6-6   Different types of special-design, shaded-pole motors.
(Courtesy Brevel)

Figure 6-7   Typical bearing brackets for skeleton-type, shaded-pole motors. *(Courtesy Brevel)*

the coil is energized. The sucking effect of the coil causes the rotor to be pulled back inside the hole in the laminated core. This type of motor is used on can openers, small knife sharpeners, clocks, and timers. It has an advantage over the synchronous type originally used in clocks, as it is self-starting and will start again if the power fails. The old clock-type synchronous motors had to be started by hand.

The third type of construction of shaded-pole motors is the *distributed-winding*. The stator laminations are similar to those used for single-phase or polyphase induction motors. The main winding is also similar to the single-phase motor. The short-circuited auxiliary winding is similar to the start winding except that it is short-circuited upon itself. It is displaced from the main winding by less than the 90° usually found in induction motors.

Performance of Shaded-Pole Motors

If this type of motor is properly lubricated—in most cases the skeleton type is sealed—it will last (continuously operated) for over 25 years. Most clock motors and fan motors are limited only by the physical abuse they receive. If they are kept plugged in (in the case of a clock) and operating continuously as timers or similar devices, without overloading, there is no reason why they cannot last indefinitely. There is only one coil in the skeleton-type motor. If the voltage is stable and the temperature is normal, there is no reason why it will not continue to operate without maintenance of any kind. Various types of physical and electrical abuse can cause them to fail, however.

If the timer motor has been used to power the timing mechanism of an oven, then it will be only a matter of eight to ten years of sporadic use that will cause it to lose its lubrication. The heat from the oven can cause it to lose its oil and the seals to break down.

Applications of Shaded-Pole Motors

The efficiency of a shaded-pole motor suffers due to the presence of winding *harmonics*. A dip can be observed in the speed-torque curve. It is caused by the third harmonic at approximately one-third synchronous speed. The shaded-pole motor is the least efficient and the noisiest of the single-phase motors. It is used mostly in air-moving applications where its low starting torque and third harmonic dip can be tolerated. Extra main windings can be added to provide additional speeds in a manner similar to that used on permanent split-capacitor motors. This is the case in the 1000-, 1100-, and 1500-rpm three-speed designs. This type of motor is used in portable fans, space heaters, forced-draft types of heaters, ventilators, room air conditioners, recorders, humidifiers, and range hoods.

Figure 6-8 illustrates a four-pole, shaded-pole motor with a double shaft. In some cases you may want to buy this type and cut off

Figure 6-8   Double-shaft, four-pole, shaded-pole motor for replacement of damaged fan motors. *(Courtesy General Electric)*

CONNECTION DIAGRAMS

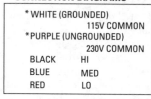

```
* WHITE (GROUNDED)
                    115V COMMON
* PURPLE (UNGROUNDED)
                    230V COMMON
   BLACK      HI
   BLUE       MED
   RED        LO
```

* Caution: Using any other col-
  or than white or purple as
  common will burn out motor.

KSM59 SHADED POLE [GENERAL ELECTRIC]

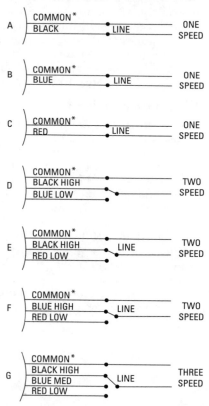

Figure 6-9 Connection diagrams for General Electric shaded-pole replacement motors. *(Courtesy General Electric)*

one end to make it fit the needed application. The mounting screws are long so they too can be cut to fit any type of replacement requirement.

Diagram connections for the General Electric shaded-pole motors are shown in Figure 6-9. Note the two-speed and three-speed color codes. Most companies have a standard color code for their wiring or hookups.

Figure 6-10 shows another manufacturer's type of four-pole, shaded-pole motor. This is a standard four-pole 1550-rpm, 115-V, 60-Hz design. It is used in the usual air-moving jobs and in the shoe polisher and the tape recorder. It is designed for face-mount with two #8-32 case screws. It has sleeve bearings and an oil hole to keep it properly lubricated. This type of motor needs little maintenance except for a few drops of oil at regular intervals.

There are a number of types and shapes attached to the shaded-pole motor. In Figures 6-11, 6-12, and 6-13, you can see some of the wide variety of shapes and designs. Note which are the two-pole, which are the four-pole, and which are the six-pole; it is possible to identify them without taking them apart and checking

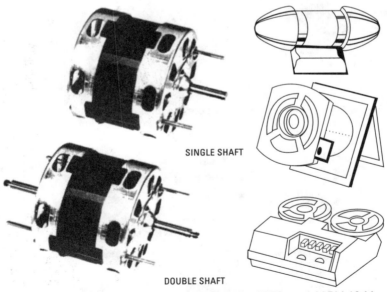

SINGLE SHAFT

DOUBLE SHAFT

Figure 6-10   Shaded-pole motors—four-pole, 1550 rpm, 115 V, 60 Hz.
(Courtesy Voorias)

Figure 6-11   Two-pole, shaded-pole motors. *(Courtesy Fasco)*

Figure 6-12   Four-pole, shaded-pole motors. *(Courtesy Fasco)*

Figure 6-13   Six-pole, shaded-pole motors. *(Courtesy Fasco)*

the coil. All of these motors use sleeve bearings. The only maintenance needed is oiling and keeping the surface area around the
coils free of dust and dirt that may cause the rotor to jam.

The skeleton shaded-pole motor has a very limited number of
parts. It has only one coil, unless it is used in a special application,
as shown in Figure 6-6. This motor may be used to drive inexpensive turn-tables or for clocks. The horsepower rating will vary with

Figure 6-14   Skeleton-type, shaded-pole motor.

the size of the laminations, coil, and rotor.

Figure 6-14 shows a skeleton-type, shaded-pole motor. This one was used to power an inexpensive record player—that is why the shaft has varying diameters.

In Figure 6-15 you can see the rotor inside the motor lamination. The top bracket and bearing have been removed so that you can see the rotor in place.

Figure 6-16 shows the entire motor disassembled. The coil is still on the frame. There are two spacing washers and a spring in addition to the frame. The end caps or bearings have an oil-retaining wick to keep the shaft lubricated while it rotates. About the only necessary maintenance is oiling once in many years of operation. If it has operated in high temperatures or has a humidity problem surrounding it, as when it is used for a timer clock on an electric oven, you may have to lubricate it more often or check for corrosion in some cases.

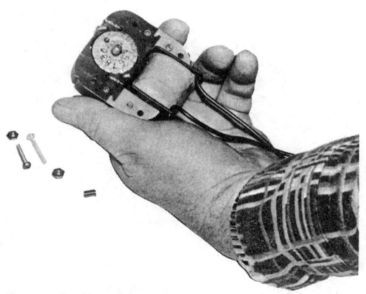

Figure 6-15   Skeleton-type with one bearing and bracket removed.

Figure 6-16   All the parts of a skeleton-type, shaded-pole motor.

In some cases the shaded-pole motor has been designed to do specific jobs. One of those special jobs is the inexpensive grinder, buffer, or a similar type of tool. K-Mart, Black & Decker, and some other 6-in. grinders are powered with a shaded-pole motor. Note the shading coil on the motor in Figure 6-17. It has only two windings; they

Figure 6-17   Close-up view of a shaded-pole motor used in a grinder-buffer. Note the arrows, which show the shading coil.

are connected in series and to a switch and continue out to a plug. This type of motor is sufficient for low-usage or home-workshop applications. It is not sufficient for use in production. Production machines have a split-phase motor to power the grinders, wire brushes, and buffers.

As the price of shaded-pole motors continues to decrease in comparison with other types, you will probably find more of this size of shaded-pole motor in machines that once used split-phase or similar motors. This one is inexpensive and has no starting mechanism.

Summary

The shaded pole motor is a single-phase motor with an auxiliary short-circuited winding or windings. The shorted winding is displaced in magnetic position from the main winding. They are suitable for operation on AC only.

The shaded-pole motor is not usually reversible. It is used for fans, timers, and electric clocks.

The shading coil looks like a piece of large-diameter copper wire wound around or through the laminations that enclose the coil. The shorted winding makes the magnetic field between the pole piece and rotor slightly out of phase with the main portion of the pole piece in a four-pole, shaded-pole motor. This has the effect of producing a two-phase voltage situation at every pole piece. There is very little torque generated by this phase difference, but it is enough to get the motor started if it is not overloaded. The rotor is similar to the squirrel cage motor in construction.

Shaded-pole motors are of three basic types and are built in a wide variety of ratings with any number of performance characteristics. They may be made in sub-fractional and fractional horsepower only. Horsepower ratings vary from 0.00250 to .1 horsepower. Some serious design characteristics prevent them from being made in larger horsepower sizes. The larger horsepower ratings are available in two-pole and four-pole configurations.

Speeds for the shaded-pole motor are stated with no-load conditions and range from 2700 to 3500 rpm. Full-load speeds range from 1600 to 3000 rpm. The most common 60-Hz models have a full load speed of from 1450 to 1620 rpm. Reduction gears are usually attached; this way the speed may be reduced as much as 450:1. Timer motors usually have an attached reduction gearbox so that water softeners and stove ovens may cycle on and off at a given period of time. General Electric makes four-pole motors in 1000, 1100, and 1500 rpm ratings.

There are three types of shaded-pole motors. They are the salient-pole, skeleton, and distributed-winding types. In the skeleton design a squirrel-cage-type rotor is used. Oil to the rotating shaft is supplied by soaking the bearing in oil or using a wick-type oiler. A spring causes the rotor to be slightly offset from the pole, and it is easily compressed as the coil is energized. The sucking effect of the coil causes the rotor to be pulled inside the hole in the laminated core. This type of motor is used in can openers, small knife sharpeners, clocks, and timers. It is self-starting and will start again if the power fails.

The distributed-winding motor has stator laminations similar to those used for single-phase or polyphase induction motors. The main winding is also similar to the single-phase motor.

If this type of motor is properly lubricated, in most cases the skeleton type where the bearings are sealed, it will last in continuous operation for over 25 years. It can last indefinitely if left plugged in and not overloaded. Efficiency of shaded-pole motors suffers from the presence of winding harmonics. The third harmonic causes some vibration and noise. If used on a fan it can be heard even as the fan blade rotates at rated speed.

The leads on shaded-pole motors are color-coded. Each manufacturer has its own system of colored wiring. One-speed motors have two wires; two-speed motors have three wires; and three-speed motors have four leads. Shaded-pole motors do not need much maintenance. They do, however, need oiling and you need to keep the surface area around the coils free of dust and dirt that may cause the rotor to jam. More uses for this type of motor are being found every day.

Review Questions

1. How does the shaded-pole motor differ from a split-phase motor?
2. What is meant by the term shorted winding on a shaded-pole motor?
3. Why isn't the shaded-pole motor reversible?
4. What are the three basic types of shaded-pole motors?
5. What type of motor does the rotor in a shaded-pole motor resemble?
6. What are the horsepower ratings available in the shaded-pole motor?

7. What are the speeds available in shaded-pole motors?

8. What configuration do the larger horsepower motors take?

9. What is a reduction gear? Why are they needed with this type motor?

10. Who makes the four-pole, shaded-pole motor in three speeds?

11. Describe the skeleton-type, shaded-pole motor.

12. Where are skeleton-type motors used?

13. What type induction motor does the distributed-winding motor resemble?

14. How long can a shaded-pole motor last if not overloaded?

15. Why are color-coded wires needed for shaded-pole motors?

Chapter 7

Three-Phase Motors

Most industrial motors are three-phase. The main reason for this is that the maintenance of a three-phase motor is practically nil. Industrial motors do not have the starting devices that single-phase motors have. The three phases of AC that supply power for the motor produce the phase shift needed to get the motor started and to keep it running once it is started. Figure 7-1 shows the three phases of AC as they appear in a graphic form. Notice how each phase is displayed 120° from the other. All commercial power generated in the United States is generated as three-phase. It is converted to single-phase because the three separate phases can be divided and sent into three different subdivisions or locations. It is cheaper to distribute single-phase AC than three-phase AC. Three-phase power requires at least three, and sometimes four, wires for proper distribution.

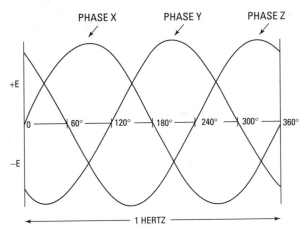

Figure 7-1   Three-phase alternating current.

Uses of Three-Phase Motors

These three-phase motors (3φ) are ideal for machine-tool and general uses where dust and dirt are prevalent. Polyphase motors have operating characteristics that enable them to operate any device that may be powered by equivalently rated single-phase motors. The three-phase motors you see in Figure 7-2 are available in sizes of $\frac{1}{4}$, $\frac{1}{3}$, $\frac{1}{2}$, and $\frac{3}{4}$ horsepower. They may be used for pumps, compressors, fans, blowers, conveyers, farm machinery, saws, and machine tools.

NEMA 48-FRAME
ROLLED STEEL CONSTRUCTION

NEMA 56H-, 143T, 145T-FRAME
ROLLED STEEL CONSTRUCTION

NEMA 143T- to 184T-FRAME
CAST-ALUMINUM CONSTRUCTION

NEMA 213T- 326T-FRAME
CAST-ALUMINUM CONSTRUCTION

Figure 7-2 Fully enclosed three-phase motors.

The motors shown in Figure 7-2 have ball bearings with permanent lubrication. They are cooled by a fan and are fully enclosed, with a service factor of 1.0.

Look at Table 7-1 and see how the size of the motor increases with horsepower. The $\frac{1}{4}$-horsepower motor weighs 19 lb; the $\frac{3}{4}$-horsepower motor weighs 30 lb. Speeds are 850, 1140, 1725, 1425, and 3450 rpm.

Also available are three-phase motors with a nonventilated construction (Figure 7-3). They are totally enclosed to eliminate cleaning and clogging problems, and are well suited for operation in the lint-laden atmospheres of the textile industry and other industrial areas having similar environmental conditions.

Table 7-1   Three-Phase Motor Weight and Horsepower

HP	Speed (rpm)	Volts	NEMA Frame	Bearings	Therm. Prot.	F.L. Amps at 230 V	Est. Shpg. Wt. (Lbs.).
1/4	1725	230/460	48	Ball	None	1.5	19
	1140	230/460	56	Ball	None	1.3	22
	850	230/460	56	Ball	None	2.0	27
1/3	3450	230/460	48	Ball	None	1.7	19
	1725	230/460	56	Ball	None	1.5	22
	1140	230/460	56	Ball	None	1.7	25
1/2	3450	230/460	48	Ball	None	2.0	21
		230/460	56	Ball	None	2.0	21
	1725	230/460	56	Ball	None	2.2	22
	1425	220/380	56	Ball	None	2.3	20
	1140	230/460	56	Ball	None	2.6	27
3/4	3450	230/460	56	Ball	None	2.6	24
	1725	200	56	Ball	None	3.6	24
		230/460	56	Ball	None	3.2	24
	1425	220/380	56	Ball	None	2.6	27
	1140	230/460	56	Ball	None	3.1	30
		230/460	143T	Ball	None	3.1	27

(Courtesy General Electric)

This type of motor is built to handle heavy thrust loads. The grease used in the ball bearings is resistant to oxidation and moisture. These motors should last for 10 years under normal operating conditions. They are rated for continuous duty in temperatures up to 108° F (40° C). These motors are made by General Electric in 1/4-, 1/3-, 1/2-, and 3/4-horsepower sizes. They have a speed of 1725 rpm. They weigh from 15 pounds for the 1/4-horsepower motor to 33 lb for the 3/4-horsepower motor.

Figure 7-3 Totally enclosed, nonventilated, three-phase motor.

Explosion-proof, three-phase motors are used in hazardous locations where a spark could start a fire or cause an explosion. The explosion-proof motor is ideal for use in locations such as dry-cleaning and dyeing plants, paint and varnish factories, alcohol and acetone plants, gasoline refineries, hospitals and laboratories, flour and feed mills, grain elevators, starch, sugar, coke or coal plants, and other locations requiring motors that are UL listed for hazardous locations. Figure 7-4 shows two explosion-proof motors.

NEMA 56-FRAME

NEMA 145T- through 286T- FRAME

Figure 7-4　Explosion-proof, fan-cooled, three-phase motor.

How Three-Phase Motors Work

The stator has windings around it that are placed 120° apart. The rotor is a form-wound type or a cage type. The squirrel cage rotor is standard for motors smaller than 1 horsepower, which we are concerned with here. Figure 7-1 shows how the three-phases are produced by the generator. The rotor will rotate with the rotating field produced by the stator. The stator is nothing more than the primary of a three-phase transformer. The magnetic field produced by the stator revolves and cuts across the rotor conductor. This induces voltage and causes the rotor current to flow. Hence, motor torque is developed by the interaction of the rotor current and the magnetic revolving field (see Figures 7-5 and 7-6). Figure 7-5, parts A and B show how the field rotates. The large motor stator and rotor (Figure 7-5, part C) are shown here to illustrate the details a little more clearly for the large industrial type of three-phase motor.

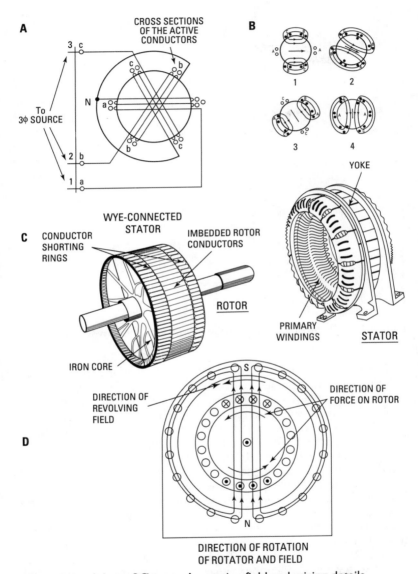

Figure 7-5 A large 3∅ motor's rotating field and wiring details.

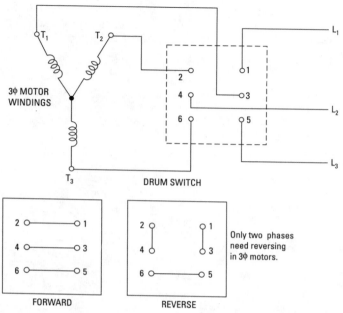

Figure 7-6   Reversing a three-phase motor with a drum switch.

The purpose of the iron rotor is to reduce the air-gap reluctance and to concentrate the magnetic flux through the rotor conductors. Induced current flows in one direction in half of the rotor conductors and in the opposite direction in the remainder. The shorting rings on the ends of the rotor complete the path for rotor current. In Figure 7-5, part D, a two-pole field is assumed to be rotating in a counterclockwise direction at synchronous speed. At the instant pictured, the south pole cuts across the upper rotor conductors from right to left, and the lines of force extend upward. Applying the left-hand rule for generator action to determine the direction of the voltage induced in the rotor conductors, the thumb is pointed in the direction of motion of the conductors with respect to the field. Since the field sweeps across the conductors from right to left, their relative motion with respect to the field is to the right. Hence, the thumb points to the right. The index finger points upward and the second finger points into the page, indicating that the rotationally induced voltage in the upper rotor conductors is away from the observer.

Motor action is analyzed by applying the right-hand rule for motors to the rotor conductors in Figure 7-5, part D, to determine the direction of the force acting on the rotor conductors. For the upper rotor conductors, the index finger points upward, the second finger points into the page, and the thumb points to the left, indicating that the force on the rotor tends to turn the rotor counterclockwise. This direction is the same as that of the rotating field. For the lower rotor conductors, the index finger points upward, the second finger points toward the observer, and the thumb points toward the right, indicating that the force tends to turn the rotor counterclockwise—the same direction as that of the field. Figure 7-7 illustrates the right-hand rule. Take a close look at Figure 7-8, where the rotor is shown in a cutaway view of the rest of the motor.

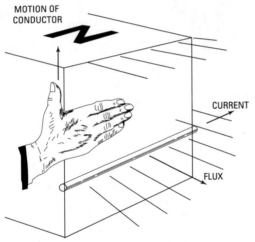

Figure 7-7 The right-hand rule for motors.

The stator of a polyphase (3φ) induction motor consists of a laminated steel ring with slots on the inside circumference. The motor winding is similar to the AC generator stator winding and is generally of the two-layer distributed, preformed type. Stator phase windings are symmetrically placed on the stator and may be either wye or delta connected.

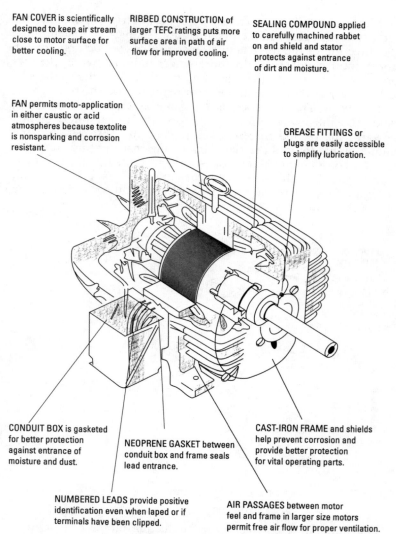

FAN COVER is scientifically designed to keep air stream close to motor surface for better cooling.

RIBBED CONSTRUCTION of larger TEFC ratings puts more surface area in path of air flow for improved cooling.

SEALING COMPOUND applied to carefully machined rabbet on and shield and stator protects against entrance of dirt and moisture.

FAN permits moto-application in either caustic or acid atmospheres because textolite is nonsparking and corrosion resistant.

GREASE FITTINGS or plugs are easily accessible to simplify lubrication.

CONDUIT BOX is gasketed for better protection against entrance of moisture and dust.

NEOPRENE GASKET between conduit box and frame seals lead entrance.

CAST-IRON FRAME and shields help prevent corrosion and provide better protection for vital operating parts.

NUMBERED LEADS provide positive identification even when laped or if terminals have been clipped.

AIR PASSAGES between motor feel and frame in larger size motors permit free air flow for proper ventilation.

Figure 7-8 Cutaway view for the totally enclosed three-phase motor.
(Courtesy General Electric)

Torque

The revolving field produced by the stator windings cuts through the rotor conductors and induces a voltage in the conductors. Rotor currents flow because the rotor end rings provide continuous metallic circuits. The resulting torque tends to turn the rotor in the direction of the rotating field. This torque is proportional to the product of the rotor current, the field strength, and the rotor power factor.

By using the transformer comparison, it is possible to see that the primary is the stator and the secondary is the rotor. At start, the frequency of the rotor current is that of the primary stator winding. The reactance of the rotor is relatively large compared with its resistance, and the power factor is low and lagging by almost 90°. The rotor current therefore lags the rotor voltage by approximately 90°. Because almost half of the conductors under the south pole carry current inward, the net torque on the rotor as a result of the interaction between rotor and the rotating field is small.

As the rotor comes up to speed in the same direction as the revolving field, the rate at which the revolving field cuts the rotor conductors is reduced and the rotor voltage and frequency of rotor currents are correspondingly reduced. Hence, at almost synchronous speed the voltage induced in the rotor is very small. The rotor reactance, X_L, also approaches zero, as may be seen by the relationship

$$X_L = 2\pi f_0 L S$$

where

f_0 is the frequency of the stator current

L is the rotor inductance

S is the ratio of the difference in speed between the stator field and the rotor to the synchronous speed—slip

Slip is expressed mathematically as

$$S = \frac{N_s - N_r}{N_s}$$

where

N_s is the number of revolutions per minute of the stator field

N_r is the number of revolutions per minute of the rotor

$f_0 S$ is the frequency of the induced rotor current

Normal operation occurs between the time the rotor is not turning at all and when it is turning almost at synchronous speed. The motor speed under normal load conditions is rarely more than 10 percent below synchronous speed. At the extreme of 100 percent slip, the rotor reactance is so high that the torque is low because of low power factor. At the other extreme of zero rotor slip, the torque is low because of low rotor current.

Motor Ventilation

Motors are commonly ventilated by placing a fan on the rotor. Fans may be placed at both ends of the motor, or just one fan may be used. Figure 7-9 shows how the air is taken in at one end of the motor and expelled at the other. Another type of construction is shown in Figure 7-10. Here the air is taken in at each end of the motor and is circulated over the end coils and exhausts through the openings in the body. Air does not flow from one end to the other in this type of ventilation.

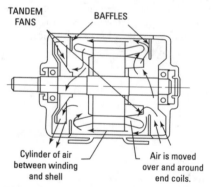

TANDEM FANS   BAFFLES

Cylinder of air between winding and shell

Air is moved over and around end coils.

Figure 7-9   Flow-through ventilation of a motor.

Motor Losses

Some losses are inherent in motor design. Figure 7-11 shows some of the losses due to the construction of the motor, the ventilation method used, and such natural things as eddy current losses, hysteresis losses, and copper losses. Eddy currents are kept to a minimum by laminating the stator pole pieces. Hysteresis losses are kept to a minimum by using silicon steel for the core or pole pieces. Copper losses are kept down by using the proper size of wire for the current needed.

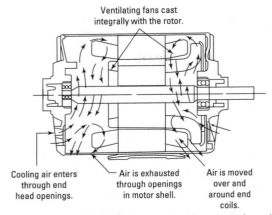

Ventilating fans cast integrally with the rotor.

Cooling air enters through end head openings.

Air is exhausted through openings in motor shell.

Air is moved over and around end coils.

Figure 7-10 Ventilation method by way of cooling air flowing through the intake and out the motor shell.

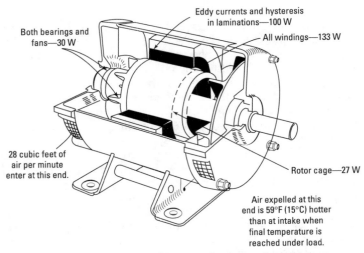

Eddy currents and hysteresis in laminations—100 W

Both bearings and fans—30 W

All windings—133 W

28 cubic feet of air per minute enter at this end.

Rotor cage—27 W

Air expelled at this end is 59°F (15°C) hotter than at intake when final temperature is reached under load.

Figure 7-11 Ventilation and losses of a three-phase motor.

Eddy currents are caused by each piece of metal that will conduct electricity having a small current induced in it from the moving magnetic field. If the piece of metal in the magnetic field is divided into small wafers with insulation between them (in this case a lacquer is usually placed on the lamination sides), then the small pieces of metal offer a larger resistance to eddy currents than a larger piece with low resistance. Eddy currents can cause excess heat in a motor.

Hysteresis losses are caused by the polarity of the material in the pole pieces being magnetized first in one direction and then in the other (in the case of 60 Hz, this means a change of polarity 120 times per second). A material must be used that can easily change its magnetic polarity without undue opposition. After about 30,000 polarity changes per second nothing can keep up, so an air core is used in the coils instead of iron. Of course, this air core is not used with motors in radio frequencies for broadcasting purposes. Silicon steel for years has been the proper material to use in relays, transformers, and motors in order to keep down losses due to hysteresis.

Copper losses are caused simply by the size of the wire. The wire has a definite resistance. If you choose the proper size for the current being handled, it should not cause any undue losses. There is no way to eliminate this effectively since the copper wire is needed to create the current flow and produce a magnetic field.

Reversibility

Three-phase motors can be reversed while running. It is very hard on the bearings and the driven machine, but it can be done by *reversing any two of the three connections*. This is usually done by a switch specifically designed for the purpose.

If a three-phase motor develops an open "leg" on one phase— that means two instead of three wires are coming into the motor terminals with power—it will slow down and hum noticeably. It will, however, continue to run in the same direction. If you try to start it with only two legs (or phases), it will not start but will rotate if started by hand (in fact, it will start in either direction). Once the other phase is connected, it will quickly come up to speed.

The loss of one leg is usually due to a blown fuse in that leg. That is, of course, if there are three individual fuses in the three-phase circuit.

Figure 7-12 shows how the three-phase motor is reversed with a drum switch. The handle on the switch can be moved from FORWARD to REVERSE at any time. The motor will come to a complete stop and then start in the opposite direction when the top two lines are switched. The bottom leg of the power line stays the same in both forward and reverse direction.

Applications of Three-Phase Motors

Three-phase motors are used for machine tools, industrial pumps and fans, air compressors, and air-conditioning equipment. They are

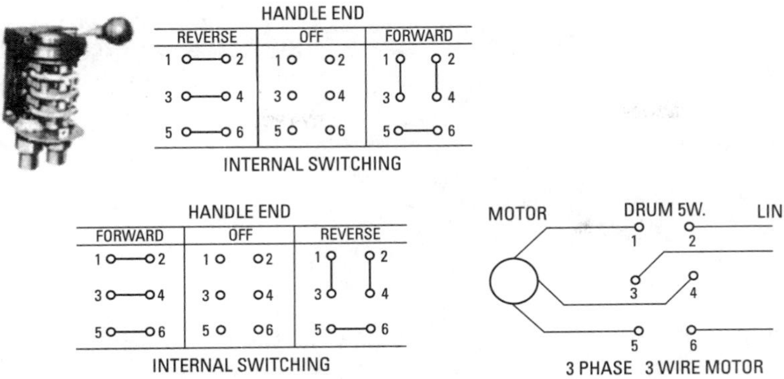

HANDLE END

REVERSE	OFF	FORWARD
1 o—o 2	1 o o2	1 o o 2
3 o—o 4	3 o o4	3 o o 4
5 o—o 6	5 o o6	5 o—o 6

INTERNAL SWITCHING

HANDLE END

FORWARD	OFF	REVERSE
1 o—o2	1 o o2	1 o o 2
3 o—o4	3 o o4	3 o o 4
5 o—o6	5 o o6	5 o—o 6

INTERNAL SWITCHING

MOTOR   DRUM 5W.   LINE

1   2
3   4
5   6

3 PHASE 3 WIRE MOTOR

Figure 7-12 Three-phase drum switch made for reversing the motor rotational direction. *(Courtesy Square D)*

recommended wherever polyphase power supply is available. They provide high starting and breakdown torque with smooth pull-up torque. They are efficient to operate and are designed for 208–230/460-V operation, with horsepower ratings from ¼ to the hundreds. They can be obtained for 50-Hz as well as 60-Hz operation. Usually, the leads are 6 in. and can be connected in a number of methods if six of them are brought out to a terminal or left in a conduit box. Figure 7-13 is a totally enclosed, fan-cooled (TEFC) type of three-phase motor with a rigid mount. Figure 7-14 is a 143T- and 145T-frame, three-phase motor. It is used on hard-to-start

Figure 7-13 Three-phase motor with a rigid mount. *(Courtesy Leeson)*

Figure 7-14 Three-phase motor, totally enclosed, fan-cooled. *(Courtesy Doerr)*

equipment. Note the difference in the construction of the end bells. They can be bought with a resilient mount as well as with the rigid mount shown.

Summary

Most industrial motors are three-phase. The main reason for this is that the maintenance of a three-phase motor is practically nil. Industrial motors do not have the starting devices that single-phase motors require. The three-phases of AC that supply power for the motor produce the phase shift needed to get the motor started and to keep it running once it is going. All commercial power is generated in the United States as three-phase. Three-phase power is converted to single-phase when distributed to homes and offices throughout the country.

These motors are very desirable when it comes to operating in a dusty and dirty atmosphere. Polyphase motors (3φ) have operating characteristics that enable them to operate any device that may be powered by equivalently rated single-phase motors.

These motors are available in a variety of fractional horsepower ratings. They are also available in a number of speeds. One of the most commonly utilized speeds is 1725 rpm. Explosion-proof motors are used in hazardous locations. The squirrel-cage rotor is standard for motors smaller than 1 horsepower. The stator in a three-phase motor is nothing more than the primary of a three-phase transformer. A magnetic field produced by the stator revolves and cuts across the rotor conductor. This induces voltage and causes the rotor current to flow; hence, motor torque is developed by the interaction of the rotor current and the revolving magnetic field.

The purpose of the iron rotor is to reduce the air-gap reluctance and to concentrate the magnetic flux through the rotor conductors. Induced current flows in one direction in half of the rotor conductors and in the opposite direction in the remainder. The left-hand rule for generator action determines the direction of the voltage induced in the rotor conductors. Motor action is analyzed by applying the right-hand rule for motors to the rotor conductors.

The stator of a polyphase (3φ) induction motor consists of a laminated steel ring with slots on the inside circumference. The motor winding is similar to the AC generator stator winding and is generally of the two-layer, distributed, preformed type. The windings on the stator may be either wye or delta connected.

The starting torque in a three-phase motor is proportional to the product of the rotor current, the field strength, and the rotor power factor. These motors are commonly ventilated by placing a fan on the rotor. Fan may be placed at one or both ends of the motor. Some losses are inherent in motor design. Construction of the motor, ventilation method used, and such natural things as eddy current losses, hysteresis losses, and copper losses all have to be considered.

Three-phase motors can be reversed while running. It is easily done by reversing any two of the three connections to the stator windings. It is very hard on the bearings and the driven machine.

When one leg of the three-phase is disconnected, the motor does not run at its rated speed, but will continue to run at a slower speed and will make more noise. However, if one leg of the power supply is missing, caused by a blown fuse in the line, for example, the motor can be manually started in either direction. A three-phase drum switch has been made for reversing the motor rotational direction.

Horsepower ratings for three-phase motors are from 3 horsepower to the hundreds. They can be bought for 50-Hz or 60-Hz operation. They are efficient to operate, relatively quiet, and can be obtained in 208–230/460-V ratings. They are very useful in hard-to-start equipment and have a wide variety of possible uses.

Review Questions

1. Why aren't three-phase motors used in homes and offices?
2. What is the main reason for industry using three-phase motors?
3. How is three-phase power different from single-phase?
4. What is the symbol for phase? Where did it come from?
5. Where are explosion-proof motors used?
6. What is the stator in a three-phase motor comparable to?
7. What is the purpose of the iron rotor in a three-phase motor?

8. What is the left-hand rule used for?

9. Where is the right-hand rule helpful?

10. What is the stator of a polyphase motor made of?

11. What is meant by a wye connection?

12. Where is the delta connection used?

13. Why is ventilation important in the operation of a motor?

14. What happens to the motor when one leg of the three-phase power opens in a three-phase motor circuit?

15. What is the range in which three-phase motors can be obtained? What voltage ratings?

Chapter 8

Squirrel-Cage Motors

The most common form of induction motor is the squirrel-cage type. This motor has derived its name from the fact that the rotor, or secondary, resembles the wheel of a squirrel cage. Its universal use lies in its mechanical simplicity, its ruggedness, and the fact that it can be manufactured with characteristics to suit most industrial requirements.

Construction

A squirrel-cage motor consists essentially of two units, namely

1. Stator
2. Rotor

The stator (or primary) consists of a laminated sheet-steel core with slots in which the insulated coils are placed. The coils are so grouped and connected as to form a definite polar area and to produce a rotating magnetic field when connected to a polyphase alternating-current circuit.

The rotor (or secondary) is also constructed of steel laminations, but the windings consist of conductor bars placed approximately parallel to the shaft and close to the rotor surface. These windings are short-circuited, or connected at each end of the rotor, by a solid ring. The rotors of large motors have bars and rings of copper connected at each end by a conducting end ring made of copper or brass. The joints between the bars and end rings are usually electrically welded into one unit, with blowers mounted on each end of the rotor. In small squirrel-cage rotors, the bars, end rings, and blowers are of aluminum cast in one piece instead of welded together.

The air gap between the rotor and the stator must be very small in order for the best power factor to be obtained. The shaft must, therefore, be very rigid and furnished with the highest grade of bearings, usually of the sleeve or ball-bearing type. A cutaway view of a typical squirrel-cage induction motor is shown in Figure 8-1.

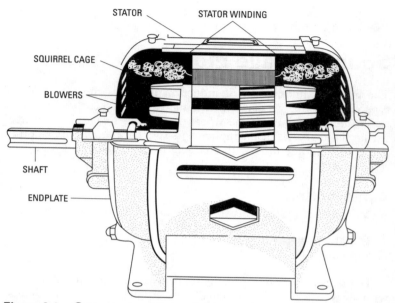

STATOR   STATOR WINDING

SQUIRREL CAGE

BLOWERS

SHAFT

ENDPLATE

Figure 8-1   Cutaway view of a typical squirrel-cage induction motor.

Principles of Operation

In a squirrel-cage motor, the secondary (or squirrel-cage) winding takes the place of the field winding in a synchronous motor. As in a synchronous motor, the currents in the stator set up a rotating magnetic field. This field is produced by the increasing and decreasing currents in the windings. When the current increases in the first phase, only the first winding produces a magnetic field. As the current decreases in this winding and increases in the second, the magnetic field shifts slightly, until it is all produced by the second winding. When the third winding has maximum current flowing in it, the field is shifted a little more. The windings are so distributed that this shifting is uniform and continuous. It is this action that produces a rotating magnetic field.

As this field rotates, it cuts the squirrel-cage conductors, and voltages are set up in these just as though the conductors were cutting the field in a DC generator. These voltages cause currents to flow in the squirrel-cage circuit—through the bars under the adjacent south poles into the other end ring, and back to the original bars under the north poles to complete the circuit.

The current flowing in the squirrel cage, down one group of bars and back in the adjacent group, makes a loop that establishes magnetic fields in the rotor core with north and south poles. This loop consists of one turn, but there are several conductors in parallel and the currents may be large. The poles in the rotor are attracted by the poles of the rotating field set up by the currents in the armature winding and follow them around in a manner similar to the way in which the field poles follow the armature poles in a synchronous motor.

There is, however, one interesting and important difference between the synchronous motor and the induction motor: the rotor of the latter does not rotate as fast as the rotating field in the armature. If the squirrel cage were to go as fast as the rotating field, the conductors in it would be standing still with respect to the rotating field, rather than cutting across the field. Then there could be no voltage induced in the squirrel cage, no currents in it, no magnetic poles set up in the rotor, and no attraction between it and the rotating field in the stator.

Speed

The speed of a squirrel-cage induction motor is nearly constant under normal load and voltage conditions, but is dependent on the number of poles and the frequency of the AC source. This type of motor slows down, however, when loaded with an amount that is just sufficient to produce the increased current needed to meet the required torque.

The difference in speed for any given load between synchronous and load speed is called the *slip* of the motor. Slip is usually expressed as a percentage of the synchronous speed. Since the amount of slip is dependent on the load, the greater the load is, the greater the slip will be, that is, the slower the motor will run. This slowing of the motor, however, is very slight, even at full load, and amounts to from 1 to 4 percent of synchronous speed. Thus, the squirrel-cage type is considered a constant-speed motor.

The slip of an induction motor, as previously defined, is the difference between its synchronous speed and its operating speed and may be expressed in any of the following ways:

1. As a percentage of synchronous speed.
2. As a decimal fraction of synchronous speed.
3. Directly in revolutions per minute.

The most common method used for speed calculation of induction motors is by use of the formula

$$\text{slip } (\%) = \frac{(\text{synchronous speed} - \text{operating speed})100}{\text{synchronous speed}}$$

The synchronous speed of a motor is found by the following:

$$N_s = \frac{\text{frequency} \times 120}{\text{number of poles}}$$

Example A three-phase, squirrel-cage induction motor having four poles is operating on a 60-Hz AC circuit at a speed of 1728 rpm. What is the slip of this motor?

Solution By substituting the values in the previous formulas,

$$N_s = \frac{60 \times 120}{4} = 1800 \text{ rpm}$$

$$\text{Slip} = \frac{(1800 - 1728)100}{1800} = 4\%$$

From the foregoing it follows that this type of motor is not suitable in industrial applications where a great amount of speed regulation is required, because the speed can be controlled only by a change in frequency, number of poles, or slip. Speed is seldom changed by changing the frequency. The number of poles is sometimes changed either by using two or more distinct windings or by reconnecting the same winding for a different number of poles.

Torque

The starting torque of a squirrel-cage induction motor is the turning effort or torque the motor exerts when full voltage is applied to its terminals at the instant of starting. The amount of starting torque that a given motor develops depends, within certain limits, on the resistance of the rotor winding. Starting torque is usually expressed as a percent of full-load torque. An increase of rotor resistance gives an increase in slip and a decrease in efficiency. In fact, all the desirable characteristics are so interrelated that it is impossible to make one of them surpassingly good without adversely affecting the others. Figure 8-2 shows the characteristic curves of one type of squirrel-cage motor.

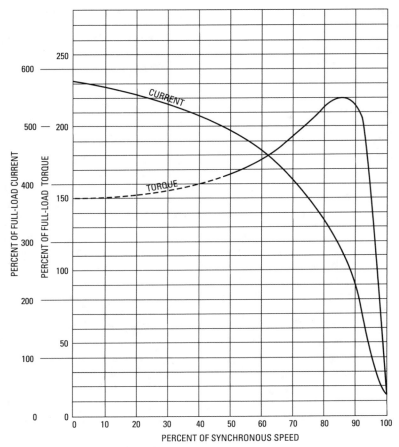

Figure 8-2 Characteristic curves of a normal starting torque, low starting current, squirrel-cage motor.

Motors can be built for high efficiency alone, for high starting torque, or for high power factor, but there is a commercial limit that prohibits a motor from excelling in all characteristics. In order to have a basis for normal starting torques, the National Electrical Manufacturers Association (NEMA) has established the minimum values for motors from 2 poles to 16 poles, inclusive. With full rated voltage applied at the instant of starting, the starting torque will not be less than those listed in Table 8-1.

Table 8-1   Minimum Starting Torques for Squirrel-Cage Motors

No. of Poles	Percent of Full-Load Torque
2	150
4	150
6	135
8	125
10	120
12	115
14	110
16	105

Classification

Squirrel-cage motors are further classified by NEMA according to their electrical characteristics, as follows:

Class A—Normal torque, normal starting-current motors
Class B—Normal torque, low starting-current motors
Class C—High torque, low starting-current motors
Class D—High-slip motors
Class E—Low starting-torque, normal starting-current motors
Class F—Low starting-torque, low starting-current motors

Class A Motors

The Class A motor is the most popular, employing a squirrel-cage winding that has relatively low resistance and reactance (Figure 8-3). It has a normal starting torque and low slip at the rated load and may have sufficiently high starting current to require, in most cases, a compensator or resistance starter for motors above 7½ hp (5.595 kW).

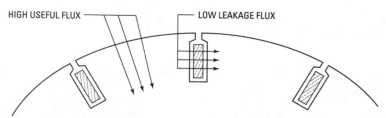

HIGH USEFUL FLUX ———   ⌐— LOW LEAKAGE FLUX

Figure 8-3   Rotor construction of a Class A squirrel-cage motor to obtain normal torque with normal starting current. In this type, the rotor bars are placed close to the surface of the rotor, resulting in relatively low reactance.

Class B Motors

Class B motors are built (Figure 8-4) to develop normal starting torque with relatively low starting current, and can be started at full voltage. The low starting current is obtained by the design of the motor to include inherently high reactance. The combined effect of induction and frequency on the current is termed *inductive reactance*. The slip at rated load is relatively low.

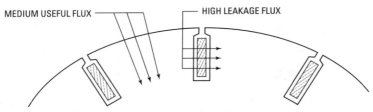

Figure 8-4 To obtain normal torque with a low starting current (Class B motor), the rotor is constructed with deep and narrow rotor bars, resulting in high reactance during starting.

Class C Motors

Class C motors are usually equipped with a double squirrel-cage winding, and combine high starting torque with a low starting current. These motors can be started at full voltage. The low starting current is obtained by design (Figure 8-5) to include inherently high reactance. The slip at rated load is relatively low.

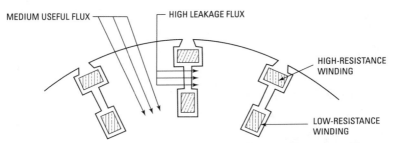

Figure 8-5 Rotor construction of a Class C motor. Two sets of bars are used, with the outer bars having a high resistance to produce a high starting torque with low starting current. At running speed, nearly all of the rotor current flows in the inner windings.

Class D Motors

Class D motors are provided with a high-resistance, squirrel-cage winding (Figure 8-6), giving the motor a high starting torque, low starting current, high slip (15 to 20 percent), and low efficiency.

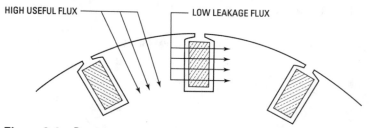

Figure 8-6 Rotor construction of high-slip motor (Class D). To obtain low starting torque with low starting current, thin rotor bars are used to make the leakage flux in the rotor low and the useful flux high.

Class E Motors

Class E motors have low slip at rated load. The starting current may be sufficiently high to require a compensator or resistance started above 7½ hp (5.595 kW) (Figure 8-7).

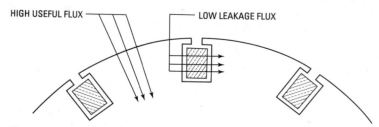

Figure 8-7 Rotors of Class E motors are constructed with a low-resistance winding placed to offer a low reactance.

Class F Motors

Class F motors combine a low starting current with a low starting torque, and may be started on full voltage. The low starting current is obtained by design (Figure 8-8) to include inherently high reactance.

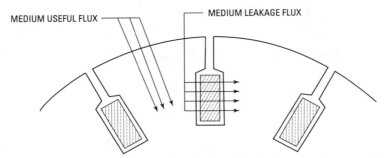

MEDIUM USEFUL FLUX

MEDIUM LEAKAGE FLUX

Figure 8-8 Rotor construction of a low starting torque, low starting current, Class F motor. A low-resistance rotor winding is placed to offer high reactance during the starting period.

Double Squirrel-Cage Motor

Double squirrel-cage motors are so called because they are designed with two separate squirrel-cage windings, one within the other. Thus, a motor of this type is a combination that has both a high- and a low-resistance squirrel-cage winding. The outer cage winding is made of high-resistance material, while the inner cage winding is made of a metal that has low resistance. In starting, the high-resistance winding gives the motor a high torque, while at full-load speed the low-resistance winding carries most of the current.

An example illustrating how the two squirrel cages operate is shown in Figure 8-9. From the shape of the rotor slots, it is apparent that the bars of the inner cage are surrounded entirely by iron except for the constricted portion of the slot between the two cages, which constitutes an air gap. The bars of the outer cage are surrounded by iron at the sides only, and have two air gaps in the magnetic path around them; since this path is much less perfect magnetically than that around the inner conductors, its inductance is lower.

Figure 8-9, part A, shows the condition just as the motor starts. At this instant, the rotating field produced by the stator current sweeps across both sets of rotor conductors at the full line frequency and induces currents in them. Since the outer conductors have a relatively low inductance, considerable current is set up, even at full line frequency. In the inner conductors, however, the current is greatly impeded by the combined action of the higher inductance of this winding and the high frequency of

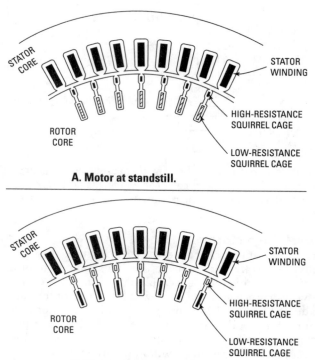

STATOR
CORE

STATOR
WINDING

HIGH-RESISTANCE
SQUIRREL CAGE

LOW-RESISTANCE
SQUIRREL CAGE

ROTOR
CORE

A. Motor at standstill.

STATOR
CORE

STATOR
WINDING

HIGH-RESISTANCE
SQUIRREL CAGE

LOW-RESISTANCE
SQUIRREL CAGE

ROTOR
CORE

B. Motor at normal speed.

Figure 8-9   Current relation in the rotor and stator windings of a double squirrel-cage induction motor. The high-resistance rotor winding carries most of the current induced in the rotor at the instant of start, thus giving high starting torque with low starting current. At full running speed, the low-resistance rotor winding carries most of the current, resulting in high efficiency.

the current. In fact, the choking action of the self-induction at line frequency is so great that very little current can flow through this winding at the start. The relative density of the currents in the two sets of conductors is shown by their amount of shading.

As the rotor gains in speed, the frequency of the currents induced in it decreases, and the relation between the currents in the two cages gradually and automatically changes to that shown for normal speed in Figure 8-9, part B. For this speed, the rotor currents are proportional to the slip, and they alternate at only about 1 hertz. At this low frequency, the higher inductance of the inner-cage winding is of relatively little importance and produces little choking effect.

The resistances of the two cages are now the chief limiting factor in the rotor currents. Consequently, the inner low-resistance cage carries most of the total rotor current, with the advantageous results already mentioned. The starting torque of double squirrel-cage motors is greater than that of ordinary squirrel-cages, but less than that of motors with a single high-resistance squirrel-cage winding.

Multispeed Squirrel-Cage Motors

In certain industrial applications where two, three, or four different constant speeds are required, it is the usual practice to wind the stator with two or more separate independent windings, whose poles may be changed by changing the external connections, giving a limited number of different speeds. The speeds obtained can always be measured in a constant ratio, such as 1800, 1200, 900, 600, and so on, according the speed produced by the individual set of windings.

Multispeed motors can be classified into three groups according to output: constant torque, constant horsepower, and variable torque and horsepower. The relation between torque, horsepower, and speed is

$$\text{horsepower} = \frac{\text{torque} \times \text{speed}}{5252}$$

where torque is measured in pound-feet and speed in revolutions per minute.

The horsepower output of the constant-torque motor varies directly as the speed. With the constant-horsepower type, the torque is inversely proportional to the speed, the horsepower being the same at each speed. In the case of the variable-torque and horsepower type, both the horsepower and the torque decrease with a reduction in the speed. The latter motor is usually employed on loads that vary as the square or cube of the speed, such as fans and blowers, centrifugal pumps, etc. To convert horsepower to kilowatts (kW), multiply by 0.746.

Starting

The condition at the starting of a motor is similar to that of a transformer with a short-circuited secondary, since the rotor, by construction, is being short-circuited by means of heavy metal bars, as previously described.

When the rated voltage is applied to the stator windings of a squirrel-cage motor, a heavy current is drawn from the line. It ranges from four to seven times the full-load current, depending on the type of motor. This high current is momentary only, and falls off rapidly as the motor increases its speed. The magnitude of this current depends on the electrical design of the motor and is independent of the mechanical load. The duration of the starting current, however, depends on the time required for acceleration, which in turn depends on the nature of the driven load.

Squirrel-cage motors in industrial plants are usually started by one of the following methods:

1. Directly across the line
2. By autotransformers (compensators)
3. By resistance in series with the stator winding
4. By means of a step-down transformer

Squirrel-cage motors of 5 hp (3.73 kW) or less are generally started with line switches, connecting them directly across the line. The switches are usually equipped with thermal devices that open the circuit when overloads are carried beyond predetermined limits. Large motors usually require various voltage-reducing methods of the type previously mentioned.

All types of squirrel-cage motors may be thrown directly across the line—provided, of course, that the starting currents do not cause voltage fluctuations that will interfere with power service elsewhere, and also that the starting currents do not exceed those permitted by local power regulations. When so started, a magnetic contactor is usually employed, operated from a start-stop, push-button station, or automatically from a float switch, thermostat, pressure regulator, or other pilot-circuit control device.

Across-the-Line Starting

There are several across-the-line starting methods whose use depends on the duty of the motor and the method of control desired. It is customary to employ a straight single-throw switch or contactor for motors up to 5 hp (3.73 kW), with fuses rated for the starting current.

The methods of starting shown in Figures 8-10 and 8-11 require two sets of fuses and a double-throw switch. The high-current fuses are used for starting only, while the running fuses have a lower current rating. These methods of starting are used for motors up to 250 V.

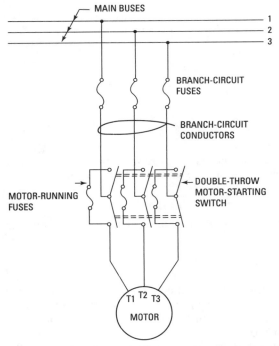

Figure 8-10 Across-the-line starting method employing a double-throw switch. The switch is thrown to the lower position to start the motor, thus cutting out the running fuses. After running speed is reached, the switch is thrown to the upper position.

With large motors having a voltage rating of from 440 to 600 V, oil circuit breakers or magnetic switches are generally used. Both the oil circuit breakers and magnetic switches are provided with overload and undervoltage protection. Various types of

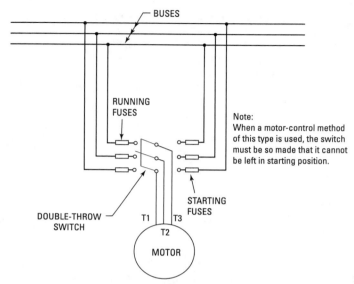

BUSES

RUNNING
FUSES

Note:
When a motor-control method
of this type is used, the switch
must be so made that it cannot
be left in starting position.

STARTING
FUSES

DOUBLE-THROW
SWITCH

T1   T3
T2
MOTOR

Figure 8-11   An across-the-line starting arrangement with two sets of
fuses, one set for starting and one set for normal running operation.

across-the-line starters used with squirrel-cage motors are shown in
Figures 8-12 to 8-15.

Autotransformer Starting

The autotransformer, or compensator, method of starting employs
two or three autotransformers for the reduction of voltage to the
motor terminal. The autotransformer type of starter has the advantage of drawing less current from the line for a given reduction in
voltage to the motor terminals. This type of starter may employ oil
circuit breakers or switches, as illustrated in Figures 8-16 to 8-18.

With reference to Figure 8-17, the compensator consists of an
autotransformer, a manually operated set of contacts, and a temperature overload relay. When the operating handle is pushed away
from the operator, the autotransformer is connected to the power
source, and, at the same time, the motor that is being controlled is
connected to the taps on the autotransformer. These taps apply 50,
65, or 80 percent of the line voltage to the motor. This reduced voltage materially limits the torque on the motor in starting. In fact, the
torque varies as the square of the applied voltage. If, for example,

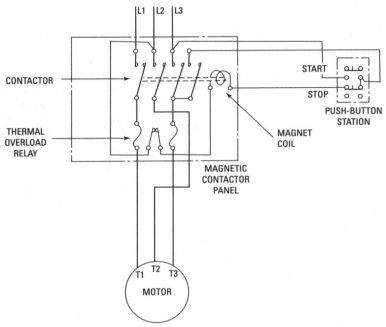

Figure 8-12   An across-the-line starting method with remote control and thermal overload protection.

the 50-percent tap is used, the torque applied to the motor on starting will be only 25 percent of the full-load torque; if the 80-percent tap is used, the torque will be 64 percent of the full-voltage starting torque.

After the motor has accelerated, the operator quickly pulls the handle back to the running position, which connects the motor to full line voltage. The handle is held in the running position by an undervoltage trip coil. If the overload relay trips out, if the line voltage fails momentarily, or if the cover is removed, the under-voltage coil is de-energized, allowing the handle to return to the off position. To restart the motor, the overload relay must be reset by pushing a reset button on the outside of the case. Then the operating handle must be thrown first to the starting and next to the running position. It is not possible to throw the handle from the off position directly to the running position (this would start the motor on full voltage), because the handle is so interlocked that it must first be

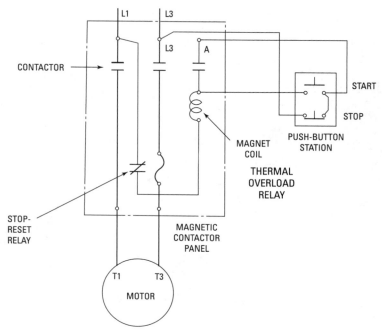

Figure 8-13 Diagram of an across-the-line starting method for single-phase motors.

pushed to the starting position before it can be pulled through the off position to the running position.

Resistance Starting

The resistance-type starter, Figure 8-19, employs a heavy-duty resistance in series with the line conductors. The drop of voltage through this resistance produces a reduced voltage at the motor terminals. The starting current of the motor is reduced in direct proportion to the reduction of the voltage applied to the motor terminals.

It should be noted, however, that in the case of a resistance starter being used to reduce the voltage to half-value, for example, the current drawn from the line will also be half of the full voltage value, whereas if an autotransformer starter is used, the current drawn from the line varies as the square of the voltage ratio of the transformer. Thus, if half-voltage is applied to the motor terminal from the secondary of the autotransformer, the current drawn from the line will only be one-fourth the full-voltage value.

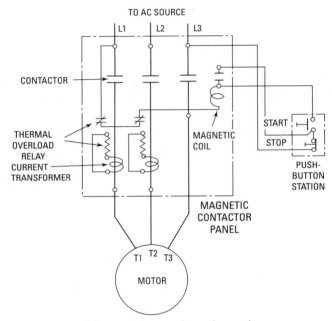

TO AC SOURCE

L1　L2　L3

CONTACTOR

THERMAL
OVERLOAD
RELAY
CURRENT
TRANSFORMER

MAGNETIC
COIL

START
STOP

PUSH-
BUTTON
STATION

MAGNETIC
CONTACTOR
PANEL

T1　T2　T3

MOTOR

Figure 8-14　Magnetic starter for a three-phase motor.

Step-Down Transformer Starting

The voltage step-down transformer method of starting is rarely used because it is more expensive than the autotransformer method. A step-down voltage transformer consists of two complete coils or windings, whereas an autotransformer requires but one tapped coil or winding. The comparison is shown in Figure 8-20.

Control Equipment

Selecting the right squirrel-cage motor for the machine in question is but one step toward meeting the motor-application problem. Selection of the proper control apparatus is just as important as the selection of the motor itself. The intelligent choice of motor control involves a complete knowledge of the types of control apparatus available, as well as their functions.

The functions of the control apparatus for squirrel-cage motors are

1. Starting the motor on (a) full voltage or (b) reduced voltage

2. Stopping the motor

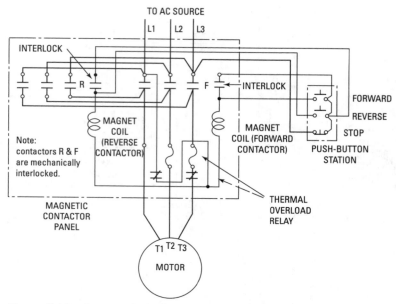

Figure 8-15 An across-the-line reversing type starter.

3. Disconnecting the motor upon failure of voltage
4. Limiting the motor load
5. Changing the direction of rotation of the rotor
6. Starting and stopping the motor (a) at fixed points in a given cycle of operation, (b) at the limit of travel of the load, or (c) when selected temperatures or pressures are reached

Where multispeed motors are involved, the following functions may be added:

7. Changing the speed of rotation (rpm)
8. Starting the motor with a definite speed sequence

Stopping the Motor

Control devices permit motors to be stopped as follows:

1. Under the direction of the operator
2. Under the control of a pilot-circuit device, such as a thermostat, pressure regulator, float switch, limit switch, or cam switch

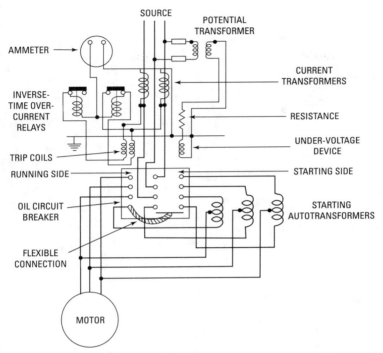

Figure 8-16   Diagram of a double-throw, motor-starting oil circuit breaker with overcurrent protection, used in starting large squirrel-cage motors.

3. Under the control of protective devices that will disconnect the motor under overload conditions detrimental to the motor, or upon failure of voltage or a lost phase

Protecting the Motor
To protect motors against damage during severe momentary or sustained overloads, overcurrent devices are employed to disconnect them from the line. These devices are of two general types:

1. Thermal overload relay
2. Dashpot overload relay

The characteristics of the thermal-type overload relay are illustrated in Figure 8-21, which indicates an inverse relationship

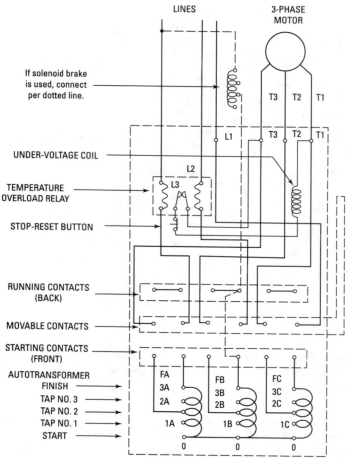

Figure 8-17 Diagram of an autotransformer type, manual reduced-voltage starter for a three-phase, squirrel-cage motor.

between current and time; —that is, the greater the current, the sooner the relay operates to disconnect the motor from the line. When the relay operates, it opens an independent, or pilot, circuit, which causes the main line contacts to drop out, thus disconnecting the motor from the line. Dashpot overload relays have a similar

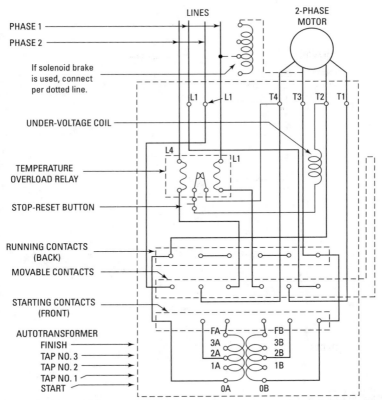

Figure 8-18 Diagram of an autotransformer type, manual reduced-voltage starter for a two-phase, squirrel-cage motor.

inverse time-current relationship. Thermal and dashpot overload relays are of two types:

1. Hand reset

2. Automatic reset

As the names denote, the hand-reset type must be reset by hand after having tripped (usually by pressing a button projecting through the enclosing case), whereas the automatic-reset types reset themselves automatically. Dashpot overload relays are generally of the automatic-reset type, but they can be made of the hand-reset

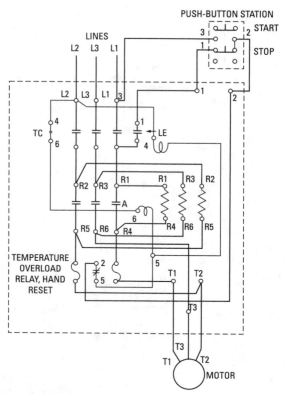

Figure 8-19 Diagram of a resistance-type starter for a three-phase, squirrel-cage motor.

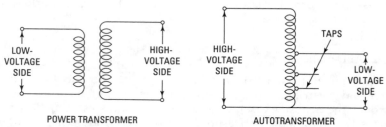

Figure 8-20 Comparison of the copper requirements of a single-phase power transformer and an autotransformer for motor starting.

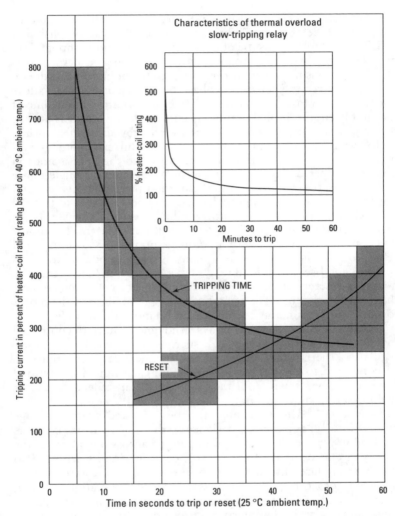

Figure 8-21   Characteristic curve of a typical thermal overload relay.

type by providing a hand-reset attachment that will prevent automatic resetting.

Standard polyphase motor control devices, almost without exception, are equipped with two thermal-overload relays of the hand-reset type. Thermal-overload relays of the automatic-reset type are also used. Their purpose is to prevent burnout of the motor windings; hence, they should be selected carefully to accomplish that end. Thermal relays must be of the proper rating, and dashpot relays must be properly set to disconnect motors from the line before operating conditions tax the motor windings to the point of breakdown.

Overload relays protect polyphase motors against phase failure due to a blown fuse or some other power interruption in one line of the supply circuit. If phase failure occurs while the motor is at rest, the motor will trip the overload relay, thereby disconnecting the motor from the line. If phase failure occurs while the motor is running, the motor will continue to run, provided the load is not too great. The current taken by the motor when running single-phase will be from two to three times the normal three-phase value, and unless the motor is very lightly loaded, will be sufficient to trip the overload relays before the motor windings can be damaged. If the load is very light, the overload relays will not trip, and the motor will continue to operate without damage until shut down, but will refuse to start the next time an attempt is made to operate it.

After hand-reset overload relays have been tripped, they must be manually reset by an operator before the control will again connect the motor to the line. The operator, noting the failure of the motor to start, will presumably investigate and remove the trouble.

Automatic-reset overload relays in magnetic starters controlled by two-wire pilot-circuit devices, such as float switches, pressure regulators, thermostats, snap switches, etc., will cause the motor to be connected to the line again and again; each time a heavy load current will be drawn from the line, which will reheat the motor windings until they burn out.

Operators and motors must also be protected against voltage failure. Control apparatuses designed so that the main-line contacts are held in place by a magnet coil operated across one phase of the supply circuit can have their control or pilot circuits arranged so that, upon failure of the voltage, the motor will not restart until the operator goes through the necessary starting operations. Three-wire, pilot-circuit control is required for such control devices to cause and maintain the interruption of power to the main circuit. Motor applications involving a control apparatus that starts and stops the motor automatically by means of two-wire, pilot-circuit devices, such as thermostats, float switches, and pressure regulators, cannot be arranged for low-voltage protection.

To reverse squirrel-cage motors, it is necessary only to interchange two leads: (a) any two leads to three-phase motors; (b) two leads of either phase to two-phase, four-wire motors, leaving one phase intact; (c) the two outside leads to two-phase, three-wire motors, leaving the common lead intact. Reversal is accomplished by means of either a small drum controller for small motors or a reversing magnetic contactor for larger motors such as the one shown previously in Figure 8-15.

Control for Multispeed Motors

Multispeed motors involve all the devices heretofore mentioned to effect starting and stopping under the six conditions listed. Additional control devices are required, however, for changing speeds, which is accomplished by regrouping the motor windings. Two types of controls used are

1. Manually operated speed-setting drums
2. Combinations of magnetic contactors

Magnetically operated controllers can be arranged so that the multispeed motors will always start up through low speed, subsequently being transferred to a higher operating speed. The auxiliary control device accomplishing this is known as a *compelling relay*. It may be operated either manually or automatically by means of a time-delay device.

Application

As pointed out in the beginning of this chapter, the squirrel-cage motor, because of its simplicity of construction and because it can be built with electrical characteristics to suit almost any industrial requirement, is one of the most widely used machines. As noted from the characteristics and classification, squirrel-cage motors as a rule are not suitable where a high starting torque is required, but are most suitable where the starting-torque requirements are of a medium or low value.

In selecting a motor for a certain application, there are, in addition to torque and starting-current considerations, a large number of factors that affect an economical and efficient operation. The selection is determined in whole or in part by the user on the basis of available data, such as speed range and regulation, mechanical arrangement, available voltage, direction of rotation and reversing, operating schedule, method of control, surrounding atmosphere, and so on.

Since the power factor and efficiency of any induction motor are lower at light loads than at heavy loads, it is obvious that in selecting a motor for a definite load, the size should be such as to permit the operation of the motor as near as possible to full load. Also, low-speed motors have a lower power factor and weigh more per horsepower than high-speed motors. Therefore, in choosing a motor for a certain duty, size and speed should be carefully considered so as to give the most economical and satisfactory service.

For the largest group of applications, such as for fans, pumps, compressors, conveyors, etc., which are started and stopped infrequently and have low-inertia loads so that the motor can accelerate in a few seconds, the conventional general-purpose NEMA Class A motor can be used.

If a motor is to be installed in a location where there is a limitation on the starting current, the modified general-purpose NEMA Class B motor can be used. If the starting current is still in excess of what can be permitted, then reduced-voltage starting is employed.

On those applications where reduced-voltage starting does not give sufficient torque to start the load with either NEMA Class A or B motors, Class C motors with their high inherent torque, reduced starting current, and reduced-voltage starting may be used.

Conveyors and Compressors
For applications such as conveyors and compressors, which sometimes require a starting torque of at least twice full-load torque, NEMA Class C motors may be used with full-load voltage starting.

Large Fans
The large fan type of drive is one that requires special consideration. These drives, once they are accelerated, run continuously at full load; therefore, it is desirable to have the best possible efficiency and power factor. Some of the fans, however, have extremely high moments of inertia (WR^2) and motors with normal starting-torque characteristics may require from 30 seconds to 1 minute to accelerate. Starting current flowing in both the rotor and stator for this long period of time may generate sufficient heat to damage the windings.

To meet this application with a squirrel-cage motor, special NEMA Class B motors are used to reduce the starting current to a minimum. The rotors of these special motors are designed with a large mass of material, especially in the end rings, so that it is possible for the motor rotor to absorb the tremendous losses during the accelerating period without reaching excessive temperatures. Once the motor reaches its full-load speed, the losses return to normal and the rotor rapidly cools down to its normal operating temperatures.

Flywheels, Presses, and Bolt Headers

Drives that have high external inertia, quite often in the form of a flywheel, and pulsating load, instead of a continuous full-load torque, also require special consideration. Typical examples of this type of application are presses and bolt headers. In this type of application, no work is being done most of the time; then a peak load occurs that may require torques of many times the full-load torque of the motor.

Under these conditions, the running efficiency of the motor at full load is not important, because the motor never operates at that point. Therefore, it is deliberately designed with more than normal secondary resistance so that it has a tendency to slow down as the load comes on the drive. This tendency for the motor to slow down permits the flywheel to give up energy to absorb the peak load, with the motor exerting no more than full-load torque. However, the energy taken out of the flywheel during the work stroke must be returned to it by the motor. Thus, the motor must have sufficient torque to accelerate the flywheel before the next work stroke is made.

Since presses work at widely varying rates, say from 2 or 3 to as high as 100 to 150 work strokes per minute, the correct amount of rotor resistance will vary, depending on the number of strokes per minute. There is ample time for the flywheel to slow down 10 to 15 percent during the work stroke and give up a considerable part of its kinetic energy. Therefore, motors used on this type of press should have high slip. One of the standard ratings listed by the motor manufacturers is a motor having 8 to 13 percent slip.

As the number of strokes per minute increases, the length of time of the working stroke decreases, and there is not much time available for the flywheel to slow down. Neither is there much time for the motor to reaccelerate the flywheel between strokes, so the amount of slowdown of the flywheel is usually between 5 and 10 percent. The standard line of press motors with 5 to 8 percent slip has been designed for this application. These presses usually make 10 to 40 strokes per minute.

On smaller presses, making 100 to 150 strokes per minute, there is very little time for the flywheel to accelerate or decelerate. For these, a standard motor that has approximately 3 percent slip is entirely adequate; there is no reason for supplying a high-slip motor. Quite often, someone tries to use NEMA Class C motors on these high-torque applications, and if there is not trouble due to overheating, trouble may develop in the form of mechanical failure of the rotor due to the unequal heating in the double-deck winding, as previously explained (see Table 8-2).

Table 8-2 Squirrel-Cage Motor Troubleshooting Chart

Symptom and Possible Cause	Possible Remedy
Motor Will Not Start	
(a) Overload control tripped	(a) Wait for overload to cool. Try starting again. If motor still does not start, check all the causes as outlined in this table.
(b) Power not connected	(b) Connect power to control and control to motor. Check clip contacts.
(c) Faulty (open) fuses	(c) Test fuses.
(d) Low voltage	(d) Check motor nameplate values with power supply. Also check voltage at motor terminals with motor under load to be sure wire size is adequate.
(e) Wrong control connections	(e) Check connections with control wiring diagram.
(f) Loose terminal lead connection	(f) Tighten connections.
(g) Driven machine locked	(g) Disconnect motor from load. If motor starts satisfactorily, check driven machine.
(h) Open circuit in stator or rotor winding	(h) Check for open circuits.
(i) Short circuit in stator winding	(i) Check for shorted coil.
(j) Winding grounded	(j) Test for grounded winding.
(k) Bearings stiff	(k) Free bearings or replace.
(l) Grease too stiff	(l) Use special lubricant for special conditions.
(m) Faulty control	(m) Check control wiring.
(n) Overload	(n) Reduce load.
Motor Noisy	
(a) Motor running single phase	(a) Stop motor, then try to start. (It will not start on single phase.) Check for "open" in one of the lines or circuits.
(b) Electrical load unbalanced	(b) Check current balance.

Table 8-2 *(continued)*

Symptom and Possible Cause	Possible Remedy
(c) Shaft bumping (sleeve-bearing motors)	(c) Check alignment and condition of belt. On pedestal-mounted bearing, check end play and axial centering of rotor.
(d) Vibration	(d) Driven machine may be unbalanced. Remove motor from load. If motor is still noisy, rebalance rotor.
(e) Air gap not uniform	(e) Center the rotor and, if necessary, replace bearings.
(f) Noisy ball bearings	(f) Check lubrication. Replace bearings if noise is persistent and excessive.
(g) Loose punchings or loose rotor on shaft	(g) Tighten all holding bolts.
(h) Rotor rubbing on stator	(h) Center the rotor and replace bearings if necessary.
(i) Object caught between fan and end shields	(i) Disassemble motor and clean. Any rubbish around motor should be removed.
(j) Motor loose on foundation	(j) Tighten hold-down bolts. Motor may possibly have to be realigned.
(k) Coupling loose	(k) Insert feelers at four places in coupling joint before pulling up bolts to check alignment. Tighten coupling bolts securely.

Motor Running Temperature Too High

(a) Overload	(a) Measure motor loading with ammeter. Reduce load.
(b) Electrical load unbalance (fuse blown, faulty control, etc.)	(b) Check for voltage unbalance or single phasing. Check for "open" in one of the lines or circuits.
(c) Restricted ventilation	(c) Clean air passages and windings.
(d) Incorrect voltage and frequency	(d) Check motor nameplate values with power supply. Also check voltage at motor terminals with motor under full load.

(continued)

Table 8-2 (continued)

Symptom and Possible Cause	Possible Remedy
(e) Motor stalled by driven machine or by tight bearings	(e) Remove power from motor. Check machine for cause of stalling.
(f) Stator winding shorted or grounded	(f) Test winding for short circuit or ground.
(g) Rotor winding with loose connections	(g) Tighten, if possible, or replace with another rotor.
(h) Motor used for rapid reversing service	(h) Replace with motor designed for this service.
(i) Belt too tight	(i) Remove excessive pressure on bearings.
Bearings Hot	
(a) End shields loose or not replaced properly	(a) Make sure end shields fit squarely and are properly tightened.
(b) Excessive belt tension or excessive gear slide thrust	(b) Reduce belt tension or gear pressure and realign shafts. See that thrust is not being transferred to motor bearings.
(c) Bent shaft	(c) Straighten shaft or replace.
Sleeve Bearings Hot	
(a) Insufficient oil	(a) Add oil. If oil supply is very low, drain, flush, and refill.
(b) Foreign material in oil, or poor grade of oil	(b) Drain oil, flush, and relubricate, using industrial lubricant recommended by a reliable oil company.
(c) Oil rings rotating slowly or not rotating at all	(c) Oil too heavy; drain and replace.
(d) Motor tilted too far	(d) Level motor or reduce tilt and realign, if necessary.
(e) Oil rings bent or otherwise damaged in reassembling	(e) Replace oil rings.
(f) Oil ring out of slot	(f) Adjust or replace retaining clip.
(g) Motor tilted, causing end thrust	(g) Level motor, reduce thrust, or use motor designed for thrust.
(h) Defective bearings or rough shaft	(h) Replace bearings. Resurface shaft.

Table 8-2 *(continued)*

Symptom and Possible Cause	Possible Remedy
Ball Bearings Hot	
(a) Too much grease	(a) Remove relief plug and let motor run. If excess grease does not come out, flush and relubricate.
(b) Wrong grade of grease	(b) Add proper grease.
(c) Insufficient grease	(c) Remove relief plug and regrease bearing.
(d) Foreign material in grease	(d) Flush bearings and relubricate; make sure that grease supply is clean. Keep can covered when not in use.
(e) Bearings misaligned	(e) Align motor and check bearings housing assembly. See that the bearings races are exactly 90° with shaft.
(f) Bearings damaged	(f) Replace bearings.

Troubleshooting with the Split-Core Volt-Ammeter

The operating conditions and behavior of electrical equipment can be analyzed only by actual measurement. A comparison of the measured terminal voltage and current will check if equipment is operating within electrical specifications.

The measurement of voltage and current requires the use of two basis instruments: a voltmeter and an ammeter. To measure voltage, the test leads of the voltmeter are in contact with the terminals of the line under test. To measure current, the conventional ammeter must be connected in series with the line so that the current will flow through the ammeter.

The insertion of the ammeter means shutting down the equipment, breaking open the line, connecting the ammeter, starting up the equipment, reading the meter, and then going through as much work to remove the ammeter from the line. Additional time-consuming work may be involved if the connections at the ammeter have to be shifted to a higher or lower range terminal.

Split-Core AC Volt-Ammeter

These disadvantages are practically eliminated by use of the split-core AC volt-ammeter. This instrument combines an AC voltmeter and AC split-core ammeter into a single pocket-size unit with a convenient range switch to select any of the multiple voltage ranges or current ranges. See Figure 8-22. With the split-core ammeter, the line to be tested does not have to be disconnected from the power source.

This type of ammeter uses the transformer principle to connect the instrument into the line. Since any conductor carrying alternating

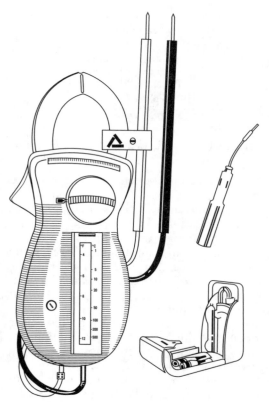

Figure 8-22   Clamp-on volt/ampere/ohmmeter with rotary scale.
(Amprobe)

current will set up a changing magnetic field around itself, that conductor can be used as the primary winding of the transformer. The split-core ammeter carries the remaining parts of the transformer, which are the laminated steel core and the secondary coil. To get transformer action, the line to be tested is encircled with the split-type core by simply pressing the trigger button. See Figure 8-23. Aside from measuring terminal voltages and load currents, the split-core ammeter-voltmeter can be used to track down electrical difficulties in electric motor repair.

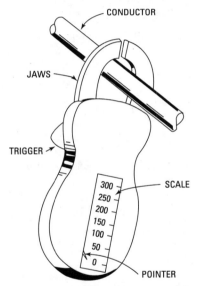

Figure 8-23   The clamp-on volt/ammeter/ohmmeter with parts labeled. *(Amprobe)*

Testing for Grounds

To determine whether a winding is grounded or has a very low value of insulation resistance, connect the unit and test leads as shown in Figure 8-24. Assuming the available line voltage is approximately 120 V, use the unit's lowest voltage range. If the winding is grounded to the frame, the test will indicate full line voltage.

A high resistance ground is simply a case of low insulation resistance. The indicated reading for a high resistance ground will be a little less than line voltage. A winding that is not grounded will be evidenced by a small or negligible reading. This is due mainly to the capacitive effect between the winding and the steel lamination.

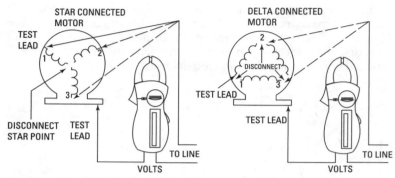

Figure 8-24 Find the location of a grounded phase of a motor.
(Amprobe)

To locate the grounded portion of the winding, disconnect the necessary connection jumpers and test. Grounded sections will be detected by a full line voltage indication.

Testing for Opens

To determine whether a winding is open, connect test leads as shown in Figures 8-25 and 8-26. If the winding is open, there will be no voltage indication. If the circuit is not open, the voltmeter indication will read full line voltage.

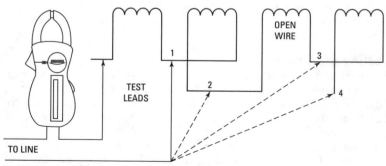

Figure 8-25 Isolating an open phase. *(Amprobe)*

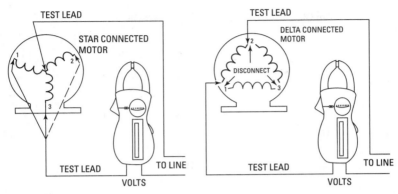

Figure 8-26   Locating an open in a motor. *(Amprobe)*

Checking for Shorts

Shorted turns in the winding of a motor behave like a shorted secondary of a transformer. A motor with a shorted winding will draw excessive current while running at no load. Measurement of the current can be made without disconnecting lines. This means you engage one of the lines with the split-core transformer of the tester. If the ampere reading is much higher than the full load ampere rating on the nameplate, the motor is probably shorted.

In a two- or three-phase motor, a partially shorted winding produces a higher current reading in the shorted phase. This becomes evident when the current in each phase is measured.

Testing Squirrel-Cage Rotors

In some cases, loss in output torque at rated speed in an induction motor may be due to opens in the squirrel-cage rotor. To test the rotor and determine which rotor bars are loose or open, place the rotor in a growler. Engage the split-core ammeter around one of the lines going to the growler winding, as shown in Figure 8-27. Set the switch to the highest current range. Switch on the growler and then set the test unit to the appropriate current range. Rotate the rotor in the growler and take note of the current indication whenever the growler is energized. The bars and end rings in the rotor behave similarly to a shorted secondary of a transformer. The growler windings act as the primary. A good rotor will produce approximately the same current indications for all positions of the rotor. A defective rotor will exhibit a drop in the current reading when the open bars move into the growler field.

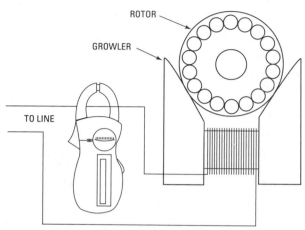

Figure 8-27   Testing a squirrel-cage rotor. *(Amprobe)*

Testing the Centrifugal Switch in a Split-Phase Motor

A faulty centrifugal switch may not disconnect the start winding at the proper time. To determine conclusively that the start-winding remains in the circuit, place the split-core ammeter around one of the start-winding leads. Set the instrument to the highest current range. Turn on the motor switch. Select the appropriate current range. Observe if there is any current in the start-winding circuit. A current indication signifies that the centrifugal switch did not open when the motor came up to speed. See Figure 8-28.

Test for Short Circuit Between Run- and Start-Windings

A short between run- and start-windings may be determined by using the ammeter and line voltage to check for continuity between the two separate circuits. Disconnect the run- and start-winding leads and connect the instrument, as shown in Figure 8-29. Set the meter on voltage. A full-line voltage reading will be obtained if the windings are shorted to one another.

Summary

The squirrel-cage motor is the most common form of induction motor. The motor derives its name from the fact that the rotor, or secondary, resembles the wheel of a squirrel cage.

A squirrel-cage motor consists essentially of a *stator* (or primary) and a *rotor* (or secondary). The stator consists of a laminated sheet-steel

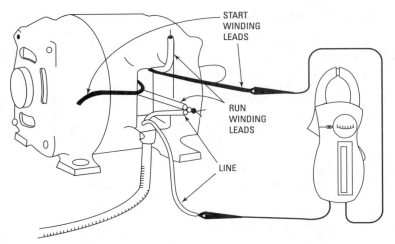

Figure 8-28   Testing a centrifugal switch on a motor. *(Amprobe)*

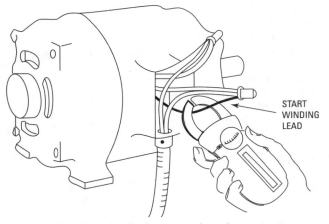

Figure 8-29   Test for finding a winding short circuit. *(Amprobe)*

core with slots containing the insulated coils. The coils are grouped and connected to a polyphase alternating-current circuit.

The rotor is also constructed of steel laminations, but the windings consist of conductor bars placed approximately parallel to the shaft and close to the rotor surface. These windings are short-circuited, or connected together at each end of the rotor, by a solid ring.

The squirrel-cage (or secondary) winding takes the place of the field winding in the synchronous motor. As in a synchronous motor, the currents in the stator set up a rotating magnetic field that is produced by increasing and decreasing currents in the windings. When the current increases in the first phase, only the first winding produces a magnetic field. As the current decreases in the first winding and increases in the second, the magnetic field shifts slightly, until it is all produced by the second winding. When maximum current is flowing in the third winding, the field is shifted a little more. The windings are so distributed that the shifting is uniform and continuous, which produces a rotating magnetic field.

The speed of a squirrel-cage motor is nearly constant under normal load and voltage conditions, but is dependent on the number of poles and frequency of the AC source. This type of motor slows, however, when loaded just enough to produce the increased current for the required torque.

The starting torque of a squirrel-cage induction motor is the turning effort or torque that the motor exerts when full voltage is applied to its terminals at the instant of starting. The amount of starting torque depends, within limits, on the resistance of the rotor winding, and it is usually expressed as a percentage of full-load torque.

The squirrel-cage motor is simple in construction and can be built to suit almost any industrial requirement; therefore, it is one of the most widely used machines. As a rule, squirrel-cage motors are not suitable where high starting torque is required, but they are most suitable for medium- or low-starting-torque requirements.

Review Questions

1. What are the two basic units in a squirrel-cage motor?
2. What is the basic operating principle of the squirrel-cage motor?
3. What is the basic difference between the synchronous motor and the squirrel-cage motor?
4. How is starting torque expressed?
5. Why is the squirrel-cage motor used widely in industry?
6. List five industrial applications for the squirrel-cage motor.
7. What is the most common form of the induction motor?

8. What type of bearing is used in the squirrel-cage motors?

9. What determines the speed of a squirrel-cage motor?

10. How are squirrel-cage motors classified by the NEMA?

11. How is normal torque with a low starting current in the Class B motor obtained?

12. How much slip does a Class E motor have at rated load?

13. What design feature characterizes the double squirrel-cage motor?

14. How do you wind a motor to make it capable of two, three, or four different constant speeds?

15. How are multi-speed motors classified according to output?

Part III

Special-Purpose Motors

Chapter 9

Three-Speed and Instant Reversing Motors

The adjustable-speed motor, the dust-tight motor, the explosion-proof motor, the high-slip motor, the instant-reversing motor, the farm-duty motor, and the air-conditioner fan motor are all examples of special motors. In some cases three-speed and four-speed motors are also special types. No special motor should be operated at values other than those on the nameplate. Improper usage accounts for much of the repair or operational problems in electric motors. Look over the following before proceeding.

Understanding the Nameplate

Every motor has a nameplate (Figure 9-1). This nameplate identifies the motor according to *type*. The motor type is usually coded by the manufacturer. In many cases you almost need the manufacturer's catalog to identify the specifics of the motor. However, by observation and what you have learned so far in this book you should be able to determine some of the characteristics of the motor.

Nameplate Characteristics

The "NO" on the nameplate usually refers to the manufacturer's order or catalog number. "Volts" will usually be 120, 220, 208. Industrial applications will show 440 and 600 V. "PH" stands for phase. It will be either single-phase (1) or three-phase (3).

"AMP" refers to the amperes the motor draws under normal operating conditions. "HZ" is the frequency of the power needed to operate the motor properly; it will be either 50 or 60 (some very old motors will have 25). Aircraft motors use 400 Hz for power generated on board. Some older nameplates will have "cycle" instead of "Hz" or "Hertz."

The rpm ("RPM") may be 1735 or 3450, depending upon the motor and its capabilities. In some cases the speed may be something other than these two. They may also be designated as 1140/1060/990/920, 1060/925/840/775, or 1075/1000/960/935. This means that the motor is wound for five different speeds. The leads must be identified to obtain the proper speed by switching. Most of these multispeed motors are permanent split-capacitor types.

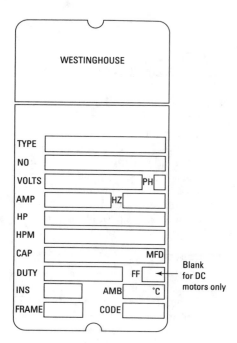

Figure 9-1　Nameplates on motors look much like this one. Some slight variations do appear. *(Courtesy Westinghouse)*

The marking "CAP" refers to the capacitor size. It will be stamped in microfarads for the motor. If this is a split-phase motor, it will not, of course, have a capacitor. "MFD" stands for microfarads.

The word "DUTY" means it is designed for continuous ("CONT") or intermittent operation ("INT").

The letters "FF" on the nameplate stand for "form factor." This is a measure of departure from pure DC. It is defined as the root-means-square (RMS) value of the current divided by the average value of the current. For half-wave rectified current, the form factor is 1.57. For full-wave rectified current, the form factor is 1.11. Pure DC has a form factor of 1.0, or unity. FF applies to DC motors only.

Form factor is an important consideration with motors designed to operate on direct current (DC). When operated from rectified power versus pure DC, the increase in motor heating for a constant output is approximately proportional to the square of the form factor. For example, a motor operating from half-wave rectified DC will have approximately 2½ times the heat rise of the same motor operating on unit form factor DC.

In order to accommodate the increased heating effect of high form factor current, continuous-duty applications generally require a larger (and more costly) motor to drive a given load. Stated another way, a designer may save by using a low-cost-to-high-form-factor control, but will have to give up much of the savings by using a larger motor to keep operating temperatures of the motor within design limits.

The letters "INS" stand for the insulation used for the motor wiring. There are three classes of insulation that may appear in this box: Class A is good for 221° F (105° C); Class B for 266° F (130° C); Class F for 311° F (155° C). This is the maximum temperature at which the motor should be operated. This is normally based on a maximum ambient of 40° C (104° F), but one can expect these windings to get hot since higher insulation ratings are used. If the motor is overloaded, the windings' temperature rise will probably increase and the windings will overheat. Lack of air intake, obstructions to the ventilation flow, and excessive deviations from the nameplate parameters will result in a higher temperature rise. Operating at high temperatures will shorten the life of the motor since it affects the insulation and the lubrication of the bearings.

The letters "AMB" are used to indicate the temperature in which the motor is expected to operate. AMB means ambient.

Frame ("FR") usually refers to the NEMA system of standardization of motor-mounting dimensions. See Appendix A for MENA frame dimension standards. Code letter designation location on nameplates varies with manufacturers, but indicates the locked-rotor current by code letter (see Table 9-1).

Table 9-1   Locked-Rotor Indicating Code Letters

Code Letter	Kilovolt-Amperes per Horsepower with Locked Rotor
A	0–3.14
B	3.15–3.54
C	3.55–3.99
D	4.0–4.49
E	4.5–4.99
F	5.0–5.59
G	5.6–6.29
H	6.3–7.09

(continued)

Table 9-1 *(continued)*

Code Letter	Kilovolt-Amperes per Horsepower with Locked Rotor
J	7.1–7.99
K	8.0–8.99
L	9.0–9.99
M	10.0–11.19
N	11.2–12.49
P	12.5–13.99
R	14.0–15.99
S	16.0–17.99
T	18.0–19.99
U	20.0–22.39
V	22.4–and up

Other Types of Nameplates

General Electric and Westinghouse each have a different style of nameplate. Other manufacturers of motors also have nameplates that are different in shape and other ways, as well. However, the information is basically the same. In Figure 9-2 you will find an example of a nameplate for a GE capacitor-start AC motor. This can be interpreted as follows: "HP ¼" means that horsepower is ¼. "FR 48" refers to the frame number, which is 48. "MOD 5KC35KG 145" means the model number can be found in the GE catalog under this number. Pertinent information will be given in the catalog for this particular model. "RPM 1725" gives the revolutions per minute of the motor: 1725. "PH 1" means that the phase is single. "219500" is the motor serial number. "SF 1.35" is the service factor. "Temp. Rise 40° C" says that the motor can be 40° C above the temperature of the room in which it is operating and still be operating normally. "V 115" is the line voltage for this type of motor. "CODE" is not filled in but it usually means the type of motor designated by the manufacturer. "A 5.2" current is needed to produce the ¼ horsepower. "CY" is frequency of 60–600 Hz (new term is "HZ"). "Time Rating CONT" describes the motor as rated for continuous duty. "SER. No. ZRD-2" is the service number.

Westinghouse has a slightly different nameplate on its latest models (Figure 9-3). The serial number ("S#") is 313P156. Catalog number is AD77 (SER). "HP 2" means that it has two horsepower. The type is FJ. FJ means it is a single-phase motor. RPM, or the speed, is 1725 rpm. Service factor ("SF") is 1.15. V or voltage required for

GENERAL ELECTRIC
A-C MOTOR

HP 1/4   FR 48
MOD 5KC35KG 145
RPM 1725   PH 1

LUBRICATION
AFTER 3 YRS NORMAL OR
1 YR HEAVY DUTY SER-
VICE ADD OIL ANNUALLY.
NO REOILING NORMALLY RE-
QUIRED FOR LIGHT DUTY
WITH TOTAL OPERATING
TIME UNDER 25,000 HRS.
USE ELECTRIC MOTOR OR
AE 10 OIL
DO NOT OVER OIL

TO DISASSEMBLE
MOTOR REMOVE BEARING
END CAP ON TERMINAL
TO THEN REMOVE
CLIP BY
ASING TAB

219500
SF 1.35   TEMP 40°C
V 115   CODE
A 5.2   CY 60
TIME RATING CONT
SER 7RD-2

Figure 9-2   General Electric nameplate.

this motor can be either 115 or 230. The "PH1" means that the phase is single. "A" is used to indicate that at 115 V the current drawn would be 26 amps and at 230 V the current would be 13 amps. As the voltage is doubled, the current is halved. This is simple to understand when you know it takes only so many watts to make the motor operate. The watts are found by multiplying the amperes times the volts. In this case, 115 multiplied by 26 equals 2990 W. If 230 is multiplied by 13, it too will equal 2990 W. Note here that 1 horsepower normally calls for 746 W. If this is a

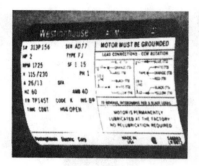

Figure 9-3 Westinghouse nameplate.

2-horsepower motor, it should draw 1492 W from the AC line. Note: the extra wattage difference between the 746 per horsepower and that actually needed to produce the power was consumed by the resistance of the windings and impedance created by the coiled windings and dissipated in the form of heat. "SFA" is left blank. "HZ 60" means that the frequency of the line voltage should be 60 Hz. "AMB 40" means that the temperature of the case or frame of the motor can be 40° C over the surrounding air temperature. Frame number ("FR") is shown as TP145T. Code is K. The NEMA letter designations following the frame number mean:

C—face mount

H—has 2F dimension larger than same frame without H suffix

J—face mount to fit jet pumps

K—has hub for sump-pump mounting

M—flange mount for oil burner, 5½-in. rabbet diameter

N—flange mount for oil burner, 6⅜-in. rabbet diameter

T,TS—integral HP motor dimension standards set by NEMA in 1964

Y—nonstandard mounting: see manufacturer's drawing

Z—nonstandard shaft extension (N, U dim.)

"INS" refers to the insulation related in terms of A, B, F, H. See the index for the temperatures associated with each letter. "TIME" can be "CONT" for continuous or "INT" for intermittent operation. "HSG OPEN" means the housing or frame has an opening for air to flow through. Also note that the nameplate has the wiring diagram for connecting the motor to run either clockwise or counterclockwise. *To reverse, interchange red and black leads.* The nameplate also tells you that the motor is permanently lubricated at the factory so no relubrication is required.

Troubles

Now that you know what the nameplate symbols mean, take a closer look at what may happen in the way of trouble if you do not follow the parameters indicated by the nameplate. In most instances where there is motor trouble, one or more of the nameplate specs has been ignored.

1. Do not operate motors at other than ± 10 percent of the nameplate voltage.
2. Do not operate motors on nominal power source frequencies that are other than that indicated on the nameplate.
3. Do not overload the motor in excess of nameplate output rating.
4. Do not exceed temperature of nameplate insulation class.
5. Do not change the value of capacitance indiscriminately.
6. Do not subject the motor to duty cycles for which it was not designed.

This last paragraph applies mainly to permanent split-capacitor motors. Motor-start capacitors, used with split-phase motors (see Chapter 4), are usually specified to achieve maximum starting torque and/or minimum locked current, and deviations are not usually made by the user. Deviating to a higher value of capacitance will provide increases in starting torque and, in some cases, speed, but at the risk of certain hazards such as higher winding temperatures, shorter motor life, nuisance operation of the overload protectors, and increases in the level of noise and vibration.

Table 9-2 indicates the problems caused by variations in the nameplate parameters of an electric motor. Keep in mind that a motor is designed for certain conditions. If these conditions are exceeded, you can expect trouble. This may have been the case when you are called to look at a motor that has been abused. If you repair it and then reinstall it in the same conditions, the same thing may happen again. Investigate the operational conditions for the motor before replacing it.

Table 9-2   Performance Parameters Adversely Affected by Nameplate Deviations

Nameplate Parameters	Torque	Speed	Tempera-ture	Noise	Vibra-tion	Thermal O/L Protectors	Current Sensitive	Centrifugal Cut-Outs	Capa-citor Life	Motor Life
Voltage	×	×	×	×	×	×	×		×	×
Frequency	×	×	×	×	×	×	×	×	×	×
Horsepower (Torque)		×	×	×	×	×	×			×
Temperature	×	×			×		×			×
Capacitor	×	×	×	×	×	×			×	×
Duty			×			×	×	×	×	×

(Courtesy Bodine Electric Co).

Bearings

No matter which type of special-application motor you may choose, it will have bearings. The type of bearings it uses has much to do with maintenance and noise level. Ball bearings, for instance, are rugged, precision-made, and noisier than sleeve bearings. The ball bearing is called for when heavy axial and radial thrust loads are encountered. The rugged, single-shielded ball bearings used in motors are usually oversized and can withstand the thrusts and shocks encountered in heavy-duty industrial applications. Bearings are preloaded with spring washers to take end play out of the bearing and contribute to quieter operation. The open ball bearings shown in Figure 9-4 are repackable; that is, they can be repacked with grease if necessary. The covered, or enclosed, type also shown in the illustration is not easily repacked.

Large babbitt-lined sleeve bearings are precision-machined to extremely close tolerances for accurate alignment, quiet operation,

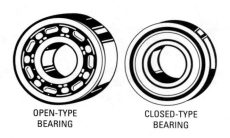

OPEN-TYPE
BEARING

CLOSED-TYPE
BEARING

Figure 9-4   Motor ball bearings.

Figure 9-5   Motor sleeve bearings.

and long, dependable operation (Figure 9-5). Accuracy of alignment helps ensure that the shaft is supported by a maximum bearing surface, thus reducing bearing wear to a minimum.

Wide-angle bearing oil grooves distribute the lubricant evenly over the entire bearing surface. This eliminates the possibility of "dry spots" and metal contact. This even, continuous distribution of freshly filtered oil cleans as it lubricates. That means it is capable of getting rid of any abrasive foreign particles.

Three-Speed Motors

The permanent-split capacitor motor can be easily changed in speed by simply flicking a switch from *Lo* to *Med* to *Hi* speed. Figure 9-6 shows how the three windings are connected to allow for operation on three different speeds. Note that the three windings are all in the circuit when the *Lo* speed is selected. More windings mean slower motor speed.

These split capacitor types are usually found on fans and other devices that need very little horsepower. In fact, take a look at

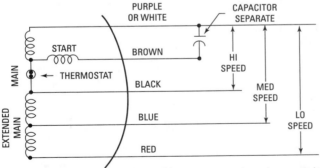

Note: Some motors will have blue as hi, orange as med., and green as lo.

Figure 9-6   Color-coded wires show the relationship to speed of the multispeed motor.

Table 9-3 to see how the millihorsepower compares with horse-power in fractional form. *Milli* means $\frac{1}{1000}$ of a horsepower. Therefore, we can see that a $\frac{1}{20}$-horsepower motor is about 50 millihorsepowers in size. Today there is a demand for conversion to metric. If you care to convert to the metric method of indicating power rating of a motor, just use the term *watts*. A 1-horsepower motor will draw 746 watts from the power source. That means that a 1-horsepower motor will be rated at 746 watts metric. A $\frac{1}{1000}$-horsepower motor will be rated as 0.746 watt in metric.

Table 9-3   Motor Power Output Comparison

Watts Output	MHP*	HP**
.746	1	$\frac{1}{1000}$
1.492	2	$\frac{1}{500}$
2.94	4	$\frac{1}{250}$
4.48	6	$\frac{1}{170}$
5.97	8	$\frac{1}{125}$
7.46	10	$\frac{1}{100}$
9.33	12.5	$\frac{1}{80}$
10.68	14.3	$\frac{1}{70}$
11.19	15	$\frac{1}{65}$
11.94	16	$\frac{1}{60}$
14.92	20	$\frac{1}{50}$
18.65	25	$\frac{1}{40}$
22.38	30	$\frac{1}{35}$
24.90	33	$\frac{1}{30}$
26.11	35	
20.80	40	$\frac{1}{25}$
37.30	50	$\frac{1}{20}$
49.70		$\frac{1}{15}$
60.17		$\frac{1}{12}$
74.60		$\frac{1}{10}$

*Millihorsepower

**Fractional horsepower

NOTE: Watts output is the driving force of the motor as calculated by the formula:

$$\frac{TN}{112.7}WO$$

where T = torque (oz. ft.), N = speed (rpm).

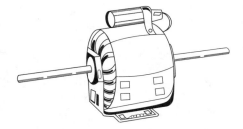

Figure 9-7   Permanent-split capacitor fan motor.

Figure 9-7 shows a permanent-split capacitor motor used for an air-conditioner blower. It is made in ⅙ to ½ horsepower for use on 115 or 230 V of 60 Hz. It has an automatically operated thermal protector built in. Note the location of the capacitor on top of the motor. It is not necessary to locate it on the motor. It is possible to place the capacitor somewhere in the air-conditioner cabinet and connect it to the capacitor by longer leads. It, too, is multispeed in design.

In Figure 9-8 you see a ¼-horsepower, two-speed, yoke-mounted fan motor. It, too, is a permanent-split capacitor motor. This one is totally enclosed and has a 1100- and 800-rpm speed capability. It was especially made for column-mounted air circulators. It relies on air passing over the motor for cooling. Since it is designed for use on fans, it is only proper to expect it to operate normally under those conditions. If it is used elsewhere, the limitations should be noted and taken into account when it comes to proper ventilation. Note the pull-chain-operated switch to turn it on or change the speed.

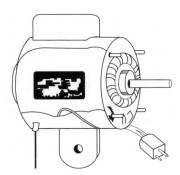

Figure 9-8   A ¼-horsepower, two-speed, yoke-mounted, enclosed motor.

Another split capacitor blower motor is shown in Figure 9-9. It has the capability for four speeds. Take a look at the speeds and volts as well as the horsepower shown in the illustration. It is designed for direct-drive furnace blowers and other air-over fans. Speed is selected by making the proper connection when the motor is hooked up to the furnace. Instructions that come with the motor tell which color wire produces which speed.

Figure 9-9   Direct-drive furnace blower motor with leads. This is a three-speed motor that pulls up to 6.2 amps on 115 V, 60 Hz. Thermally protected, the motor has a Frame 42 design. *(Courtesy Westinghouse)*

Instant-Reversing Motors

Reversing a motor can be rather hard on the motor and component parts. It takes a great deal of effort to stop the motor from its top speed and then cause it to run in the opposite direction. The bearings take quite a beating, as do the start switch and start capacitor. Some motors have a built-in relay for reversing, and others have to rely on a reversing switch (Figure 9-10).

Drum switches are manually operated, three-position, three-pole switches that carry a horsepower rating and are used for manually reversing single- or three-phase motors. They are available in several sizes and can be spring-return-to-off (momentary contact) or maintained contact. Separate overload protection, by manual or magnetic starters, must usually be provided since drum switches do not include this feature.

Figure 9-10   Drum switch. *(Courtesy Square D)*

Reversing Starter

Reversing the direction of motor shaft rotation is often required. Three-phase, squirrel-cage motors can be reversed by reconnecting any two of the three line connections to the motor. By interwiring two contactors, an electromagnetic method of making the reconnection can be obtained.

Figure 9-11 shows that the contacts *(F)* of the forward contactor, when closed, connect lines 1, 2, and 3 to motor terminals *T1*, *T2*, and *T3*, respectively. As long as the forward contacts are closed, mechanical and electrical interlocks prevent the reverse contactor from being energized.

When the forward contactor is deenergized, the second contactor can be picked up, closing its contacts *(R)*, which reconnect the lines to the motor. Note that by running through the reverse contacts, line 1 is connected to motor terminal *T3*, and line 3 is connected to motor terminal *T1*. The motor will now run in the opposite direction.

Whether operating through the forward or reverse connector, the power connections are run through an overload relay assembly, which provides motor overload protection. A magnetic reversing starter, therefore, consists of a starter and contactor, suitably interwired, with electrical and mechanical interlocking to prevent the coil of both units from being energized at the same time.

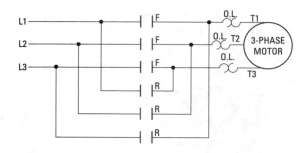

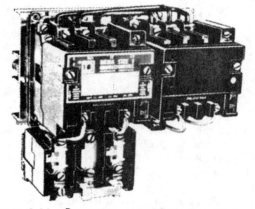

Figure 9-11 Reversing start schematic and starter. *(Courtesy Square D)*

Applications

One of the many uses for an instantly reversing motor is shown in Figure 9-12. The ⅓-horsepower motor shown here is used for parking lot gates. The motor has three leads; one is common and the other two are used for reversing and forward. Simple toggle or individual single-pole relays are used to control direction of rotation.

Severe-Duty Motors

Severe-duty motors (Figure 9-13) are used in abnormal conditions. Typical applications include dairy and food-processing equipment where hosing down of motors is routinely done, installations associated with plating equipment, mining equipment (not requiring explosion-proof construction), tropical applications, pumps or

Figure 9-12   An instant-reversing motor is needed for such jobs as parking gate operation. *(Courtesy Doerr)*

other devices located in unusually humid, wet, salty, acidic, alkaline, or dirty locations. This motor is not suitable for explosive and combustible-dust atmospheres. Use an explosion-proof type of motor for these applications.

The housing is rolled steel. A special sealing compound is used between the end shields and the motor stator. The nameplate is

Figure 9-13   Severe-duty motor. This one is enclosed so that hosing down will not affect the interior.

usually stainless steel; the bearings are usually ball and are lubricated for many years of service.

In some cases, such as in the process industries, the conditions for motor operation are so demanding that an aluminum motor has been designed (Figure 9-14). These motors are totally enclosed, with tight rabbet fits between the frame and end shields. The windings, bearings, air gap, and other internal parts are protected from corrosion.

Figure 9-14   Special aluminum housing for a severe-duty motor.

Foreign material is kept out by a durable neoprene slinger on the motor shaft that throws off water or any contaminant before it reaches the bearing. A gasket locks dust and moisture out of the conduit box. A lead seal gasket is used between the box and the frame to isolate completely the interior of the motor. Epoxy paint is used to coat the motor. This type of motor may be capacitor-start or three-phase. It comes in horsepower ratings from $\frac{1}{4}$ to $\frac{1}{2}$ in capacitor-start, single-phase motors, and from $\frac{1}{3}$ to 50 horsepower in 3ϕ. All single-phase motors are 1725 rpm, but the 3ϕ can be both 1725 and 3450 rpm.

Farm-Duty Motors

Heavy-duty motors designed specially for severe farm duty are capacitor-start, induction-run types that furnish high starting torque with normal current. They are gasketed throughout for environmental protection and have double-sealed ball bearings with a water flinger on the shaft end to protect the motor and bearings from contaminants. An oversize conduit box makes wiring easy. Grounding provision is also included. Low-temperature, thermal-overload protectors can be used, but a manual reset overload button (with rubber weather boot) is standard for maximum operator safety.

In Table 9-4 are shown some of the variations of the capacitor-start and three-phase motor available for farm use by only one of the many motor manufacturers.

The farm-duty motor is shown in Figure 9-15. It is used for pumps, conveyors, poultry equipment, and other farm machinery. The motor shown in Figure 9-16, however, is designed for use in direct- and belt-driven fans where the motor is located in the air

Table 9-4   Single-Phase, Capacitor-Start, Induction-Run, Totally-Enclosed, Fan-Cooled 104° F (40° C) Ambient

HP	RPM	Frame	Voltage	Hz	Overload Protector
Rigid Mount					
1/3	1725	E56	115/208–230	60	Manual
1/2	1725	D56	115/208–230	60	Manual
3/4	1725	E56	115/208–230	60	Manual
1	1725	F56	115/208–230	60	Manual
1½	1725	H56H	115/208–230	60	Manual
2	1725	J56H	230	60	Manual
C-Face Round Body without Base					
1/3	1725	E56C	115/208–230	60	Manual
1/2	1725	D56C	115/208–230	60	Manual
3/4	1725	E56C	115/208–230	60	Manual
1	1725	F56C	115/208–230	60	Manual
1½	1725	H56C	115/208–230	60	Manual
2	1725	J56C	230	60	Manual
Three-Phase, Totally-Enclosed, Fan-Cooled, 104°F (40° C) Ambient					
Rigid Mount					
1/3	1725/1425	C56	208–230/460	60/50	Manual
1/2	1725/1425	D56	208–230/460	60/50	Manual
3/4	1725/1425	D56	208–230/460	60/50	Manual
1	1725/1425	E56	208–230/460	60/50	Manual
1½	1725	F56	208–230/460	60	Manual
2	1725	F56	208–230/460	60	Manual
C-Face Round Body without Base					
1/3	1725	C56C	208–230/460	60/50	Manual
1/2	1725/1425	D56C	208–230/460	60/50	Manual
3/4	1725/1425	D56C	208–230/460	60/50	Manual
1	1725/1425	E56C	208–230/460	60/50	Manual
1½	1725	F56C	208–230/460	60	Manual
2	1725	G56C	208–230/460	60	Manual

(Courtesy Leeson)

Figure 9-15 Farm-duty motor with box mounted on the side and capacitor on top. *(Courtesy Leeson)*

Figure 9-16 Belted fan and blower motor. *(Courtesy Leeson)*

stream, such as on the poultry-house and barn exhaust fans. They operate without trouble in lint, dust, and dirt. All of these have ball bearings since the noise level is not important. Two speeds and two voltages are available. Keep in mind that most farms have only single-phase AC and must rely on this type of motor to serve their needs.

Table 9-5 shows the variety of motors, speeds, and voltages available in the single-phase design for farm use.

Table 9-5   Farm-Duty Motors (Single-Phase)

HP	Speed (rpm)	Volts	NEMA Frame	Bearings	Therm. Prot.	Full-Load Amps
¼	1725	115	48	Ball	None	4.5
		115	48	Ball	Auto	4.5
	1140	115	56	Ball	None	5.3
		115	56	. Ball	Auto	5.3
⅓	1725	115	48	Ball	None	5.3
		115	48	Ball	Auto	5.3
		230	48	Ball	None	2.7
		230	48	Ball	Auto	2.7
	1140	115	56	Ball	None	7.0
½	1725	115	56	Ball	None	8.0

Rotation—Shaft rotation easily reversed by electrical reconnection. (Courtesy General Electric)

High-Slip Motors

High-slip motors have high starting torque. They are designed for use where frequent or protracted starting under heavy loads is required. These could be used in hoists, cranes, punch presses, elevators, oilwell loads, shears, and bending brakes.

Characteristics

The high—8 to 13 percent—running slip and varying-speed characteristic qualifies these motors for drives with high inertia. They are made in sizes from 3 to 7½ horsepower. This is well above our preset limit of 1 horsepower for the book. However, they are mentioned here to indicate that such a motor does exist and can be used for very exact purposes where other motors may be overloaded. They are made in 3ϕ only since they are intended for industrial applications. The voltages are 230–460 at 60 Hz. Most have Class B insulation (Figure 9-17).

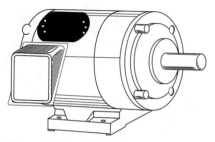

Figure 9-17   High-slip motor.

Motor Slip

When the induction motor is started without a load, the magnetic field of the stator revolves past the rotor at synchronous speed. This produces a maximum current in the rotor conductors and causes the rotor to revolve in the direction of the rotating magnetic field. As the rotor speed increases, the relative speed of the moving magnetic field decreases. Since the rotor tends to "catch up" with the rotating field, there is no longer any relative motion between the rotor and the magnetic field. This means that there is no cutting of lines of force or induced current in the rotor. That, in turn, means that there is no torque or turning effort in the rotor. As you can see from this, the speed of the rotor is always less than that of the rotating field. That means a motor with 1800 rpm designed as the rotating field will have about 1725 rpm for the rotor speed, and a 3600-rpm motor really has a speed of 3450 rpm.

The percent of slip in an induction motor is defined as the difference between the synchronous speed and the actual speed. The 1725 rpm of the 1800-rpm motor calls for a 5-percent slip ratio. The same percentage is found for the 3450- or 3600-rpm motor. Therefore, you can see that the 8- to 13-percent slip motor is really a high-slip or high-torque motor.

Explosion-Proof Motors

Fractional-horsepower explosion-proof motors are available in ¼ to 1 horsepower. They are made in split-phase and capacitor-start designs. They have sealed ball bearings and are totally enclosed, non-ventilated types in the ¼-horsepower size. The ⅓- to 1-horsepower motors are totally enclosed and fan-cooled (Figure 9-18).

Explosion-proof motors are made to meet Underwriters Laboratories Standards for use in hazardous (explosive) locations, and this is indicated on the nameplate. Certain locations are hazardous because the atmosphere does or may contain gas, vapor, or dust in explosive quantities. The NEC (National Electrical Code)

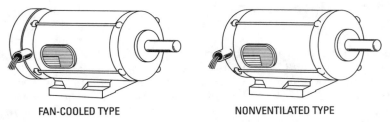

FAN-COOLED TYPE   NONVENTILATED TYPE

Figure 9-18   Explosion-proof motors, vented and nonventilated types.

divides these locations into classes and groups according to the type of explosive agent that may be present. Some of these classes and groups are listed below. For a complete list, refer to *Article 500* of the latest edition of the National Electrical Code.

Class I (Gases, Vapors)

Group A—Acetylene. Motors are not available for this group.

Group B—Butadiene, ethylene oxide, hydrogen, propylene oxide. Motors are not available for this group.

Group C—Acetaldehyde, cyclopropane, diethyl ether, ethylene, isoprene.

Group D—Acetone, acrylonitrile, ammonia, benzene, butane, ethylene dichloride, gasoline, hexane, methane, methanol, naphtha, propane, propylene, styrene, toluene, vinyl acetate, vinyl chloride.

Class II (Combustible Dusts)

Group E—Aluminum, magnesium, and other metal dusts with similar characteristics.

Group F—Carbon black, coke, or coal dust.

Group G—Flour, starch, or grain dust.

Other Motors

Open, drip-proof motors are designed for use in areas that are reasonably dry, clean, well-ventilated, and usually indoors. If installed outdoors, it is recommended that the motor be protected with a cover that does not restrict the flow of air to the motor.

Totally enclosed motors are suitable for use where exposed to dirt and moisture, and in most outdoor locations, but not for very moist or hazardous (explosive vapor, dust-filled atmosphere) locations. Severe-duty enclosed motors are suited for use in corrosive or excessively moist locations.

Metric Motor Ratings

For some time the United States has been considering the idea of switching to the metric system for measurements. However, the electrical field has always been, in most instances, metric. The volt, ampere, ohm, and other units of measurement are already used in metrics, too. Only a few terms in the electrical field are still American in their measurements. One of these is the horsepower. The metric unit for measuring power is the watt. It takes 746 W to equal 1 horsepower. Therefore, the newer (SI) International Standard for measurement is the watt or kilowatt. It takes 1000 W to equal 1 kW. That means 746 W equal 0.746 kW. There are some standard motor sizes that can be quickly converted to the metric rating of kilowatt instead of horsepower (see Table 9-6).

Table 9-6   Relationship of Horsepower to Kilowatt for Motor Ratings

Horsepower	Kilowatt (kW)*
$\frac{1}{20}$	0.025
	0.035
	0.05
	0.071
$\frac{1}{8}$	0.1
$\frac{1}{6}$	0.14
$\frac{1}{4}$	0.2
$\frac{1}{3}$	0.28
$\frac{1}{2}$	0.4
1	0.8
$1\frac{1}{2}$	1.1
2	1.6
3	2.5
5	4.0
7.5	5.6
10	8.0

*James W. Polk, A Preview of Metric Motors, Westinghouse Electric Corporation.

Summary

There are some specially designed motors that serve definite purposes. They may be adjustable-speed, two-speed, four-speed, dust-tight, explosion-proof and instant-reversing. The nameplate specifies how they are to be operated in regards to voltage, load, and operating environment.

Every motor has a nameplate. This nameplate identifies the motor according to type. The motor type is usually coded by the manufacturer. For instance, the word duty means the motor is designed for continuous (CONT) or intermittent operation (INT).

General Electric and Westinghouse have different styles in nameplates and other manufacturers have nameplates of various sizes and shapes. The information on the nameplates, however, is essentially the same.

Some troubles with motors can be avoided if specific warnings from the manufacturer are heeded. For instance, do not operate motors at more than 10 percent or less than 10 percent of the nameplate stated voltage. Do not subject the motors to duty cycles for which they were not designed.

Motor bearings are chosen for the job the motor is intended to perform. The type of bearings it uses has much to do with the maintenance and noise level. Ball bearings, for instance, are noisier than sleeve bearings, but they are called for in some instances where rugged motor life is expected.

The permanent-split capacitor motor can be easily changed in speed by simply flicking a switch from *Lo* to *Med* to *Hi* speed.

Milli-horsepower motors are small. A 1/20-horsepower motor is equal to about 50 milli-horsepower. Today there is a demand for conversion to metric. If you care to convert to the metric method of indicating power rating of a motor, just use the term *watts*. A 1-horsepower motor draws 746 W from the power source. That means a 1-horsepower motor will be rated at 746 W in metric. A 1/1000-horsepower motor will be rated as 0.746 watt in metric.

Drum switches are manually operated, three-position, three-pole switches that carry a horsepower rating and are used for manually reversing single- or three-phase motors. They are available in several sizes and can be spring-return-to-off (momentary contact) or maintained contact.

A magnetic reversing starter consists of a starter and contactor, suitably interwired, with electrical and mechanical interlocking to prevent the coil of both units from being energized at the same time.

Severe-duty motors are operated in abnormal conditions. They may be used in dairy and food-processing equipment where hosing down of motors is routinely done and in installations associated with plating equipment, mining equipment, tropical applications, pumps, or other devices located in unusually humid, wet, salty, acidic, alkaline, or dirty locations. This motor is not, however, suitable for explosive and combustible-dust atmospheres.

Heavy-duty motors designed specially for severe farm duty are capacitor-start, induction-run types that furnish high starting torque with normal current. Farm motors have only single-phase AC available in most instances, with REA supplying power. Two speeds and two voltages are available for poultry-house and barn exhaust fans. They can operate without trouble in lint, dust, and dirt.

High-slip motors have high starting torque. They can be used in hoists, cranes, punch presses, elevators, oil well loads, shears, and bending brakes. They are made in 3φ only and in 3 to 7½ horsepower sizes. Explosion-proof motors are made to meet UL's standards for use in hazardous or explosive locations. The locations for which they are suitable are listed in the National Electrical Code. These locations are divided into classes and groups according to the materials (gas, vapor, dust, or other explosive-prone atmospheres). They are available in vented and non-vented designs.

Open, drip-proof motors are designed for use in areas that are reasonably dry, clean, well ventilated, and usually indoors. If installed outdoors, it is recommended that the motor be protected with a cover that does not restrict the flow of air to the motor.

Review Questions

1. List three specially designed motors and specify their usual applications.
2. What is the purpose of a motor nameplate?
3. What does CONT on a nameplate mean?
4. How much fluctuation in voltage can most motors stand and still operate properly?
5. What is a motor's duty cycle?
6. How are bearings chosen for motors?
7. What is a milli-horsepower?
8. Why is the watt now used in motor measurements?
9. How many watts in a horsepower?
10. What is a drum switch used for in a motor circuit?
11. How are motors reversed?
12. What is a severe condition for a motor?
13. Why are high-slip motors preferred for some industrial jobs?
14. Where is an explosion-proof motor used?
15. Where do you use drip-proof motors?

Chapter 10

Steppers and Synchronous Motors

Computers and programmable controllers use electric motors that can be controlled by electronics. These motors have taken various forms, but still rely upon the basic principles of motor operation, with some additional characteristics developed for electronic signal control purposes.

Robotics has demanded more accuracy in motor and motion control. Programmable controllers have also provided the necessary signals for making small precision motors do what is intended as an end product or motion.

Some of these motors are very expensive and are found in precision measuring devices and strip recorders. They are usually made to require little or no maintenance; however, in some instances it may become necessary for you to open the motor and check for various conditions. In order to know what you are doing, it is also best to know the circuitry, the intended purpose of the motor, and how it is supposed to operate to determine if it is malfunctioning.

The intent of this chapter is to introduce some of the characteristics and applications of specialty motors that use electronics to control their actions.

Shaded-Pole Unidirectional Synchronous Motors

The shaded-pole, one-direction-of-rotation (unidirectional) synchronous motor is used for timing purposes (Figure 10-1), such as driving clocks, elapsed time indicators, repeat cycle timers, stop clocks, and potentiometers. This type of motor is also found as a drive device for encoder disc chart recorders and medical instrumentation, and as an agitator drive for blood analyzers. It is an inexpensive type of motor that can be utilized in visual displays as a driver for cycle changes. It is also used for fiber optic displays, an oil pump drive on copying machines, and as a code wheel drive on remote fire alarm systems.

The instant start-stop characteristic of this motor is useful for integrating time intervals. It provides an ideal drive for elapsed time indicators used for billing on a time-usage basis, for stop clocks measuring precise time intervals, and for chart drives. In many applications, the fast start-stop ability eliminates the need for a

Figure 10-1 Shaded-pole, unidirectional synchronous motor.
(Courtesy of Hayden Switch & Instrument, Inc.)

clutch. When voltage is applied, the motor accelerates from dead stop to full speed in less than 0.01 of a second with no load. For this reason, the inertia of any member rigidly coupled to the output shaft should be kept to a minimum or a resilient coupling used.

Direction of rotation may be left (counter-clockwise) or right (clockwise), and it may be obtained with a 115-V or 20-V coil. Note the two leads for this particular type of motor. The torque available for various voltages and speeds is shown in Table 10-1.

Table 10-1 Speed-Torque, Shaded-Pole, Unidirectional Synchronous Motor

Volts	Output Speed (rpm)				Rotation
	360	300	60	3.6	
115	0.08 Oz-In	0.1 Oz-In	0.5 Oz-In	5.0 Oz-In	Left or Right (ccw) (cw) Depends on Motor Selected
20	0.08 Oz-In	0.1 Oz-In	0.5 Oz-In	5.0 Oz-In	Left or Right Depends on Motor Selected

Oz-In = Ounce-Inch (unit for torque measurement)
Gear train for this type of motor is designed for no more than 5.0 oz-in. Do Not Exceed.
Ambient Temperature 0° to +50°C continuous duty.

Capacitor Start-and-Run, Bidirectional Synchronous Motor

This motor looks like the previous one except that it has three wires coming from the case. This motor is easily reversed by switching a capacitor from one coil to the other. The small size of the motor enables one to design equipment to fit the motor and capacitor (Figure 10-2). The 115-V, 60-Hz motor requires a 0.35-MF capacitor whereas the 20-V, 60-Hz type uses a 9.0-MF capacitor.

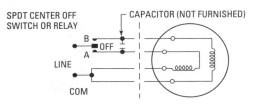

Figure 10-2 Capacitor start-and-run, bidirectional synchronous motor wiring diagram. *(Courtesy Hayden Switch & Instrument, Inc.)*

Fast start-stop and reversing characteristics eliminate the need for clutching, braking, or pre-starting in many applications. This motor is ideally suited for control functions where shafts, potentiometers, indicators, and the like must be precisely positioned. The motor can "inch" in either direction in small increments and stop without coasting.

This type of motor can be found in chart drives with portable pen drives for plotting curves. Spectrographs use them to drive potentiometers for speed control analyzers. Remote temperature controls on master panels for heat-treating also use this type of motor. Industrial processing equipment will use it where remote speed control is called for. Chart recorders use it to drive the code wheel. These motors are also used for ink flow control on printing presses and lens system drives for focus control on 35-mm slide and 35-mm motion picture projectors. Industrial packing equipment uses the motor for pot control.

The typical drive circuit is shown in Figure 10-3. Note the triacs and the two inputs, A and B. In Table 10-2, the A and B represent these two inputs. Power input for these motors is about two watts. Acceleration from dead stop to full speed is 0.01 second with no load.

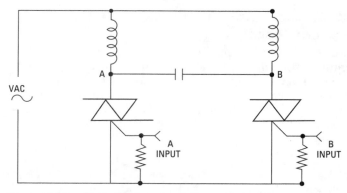

Figure 10-3 Typical drive circuit for capacitor start-and-run, bidirectional synchronous motor. *(Courtesy Hayden Switch & Instrument, Inc.)*

Table 10-2   Step-Angle, Torque, and Pulse Rate

Step Angle	Torque (Oz-In)	
	1–20 Hz	40 Hz
36°	0.3	0.2
30°	0.3	0.2
6°	1.7	1.0
3.6°	3.0	2.0

Motors are available on special order to operate on 12 volts.

Single-Phase, Stepper Motor (Unipolar Drive)

This motor has only two leads (Figure 10-4). It operates in only one direction depending on the motor selected, either left rotation or right rotation, but it is not reversible as made (Figure 10-5). A motor of this type comes in handy for a number of uses as it is small and converts electrical pulses into discrete angular steps of the output shaft without need for control logic (Figure 10-6, part A). For each impulse, the rotor turns 360°, 180° when power is applied (external pulse). By means of magnetic detenting, the rotor turns an additional 180° when power is then removed (internal pulse). No power is consumed between pulses, making the motor ideal for use in battery-operated systems.

This motor has a special capability for operation in applications where the power supply is limited and command pulses are widely spaced, since a low duty cycle will reduce the average power

Figure 10-4   Single-phase, stepper motor, unipolar drive.
(Courtesy Hayden Switch & Instrument, Inc.)

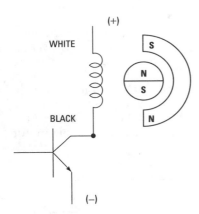

Figure 10-5   Typical drive circuit for single-phase stepper. *(Courtesy Hayden Switch & Instrument, Inc.)*

drain to a few milliwatts. For example, the motor could drive a battery-operated digital clock for automotive or aircraft applications where average power consumption must be low. A counter reading 23 hours 59 minutes and seconds in increments of 5 seconds could be driven by the two-wire stepper motor geared to step the seconds drum in 12°, or 5 second increments, from an accurate pulse source such as a crystal-controlled oscillator circuit.

The peak power requirement would be approximately 2 W for a pulse duration of 0.010 seconds or an average power consumption of only 4 mW.

$$\frac{2 \times 0.010}{5.0} = 4\text{mW}$$

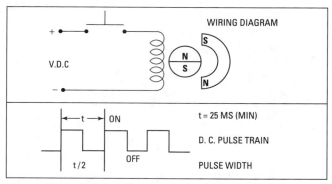

Figure 10-6 Wiring diagram and pulse needed for operation.
(Courtesy Hayden Switch & Instrument, Inc.)

In cases where the power supply is not capable of providing 2 watts peak power but is capable of handling the low 4mW average wattage requirement, a system developed by Tri-tech, Inc. might be of interest. In the circuit shown in Figure 10-7, a capacitor is charged at a low rate between pulses and discharged into the stepper motor by a short negative-going command pulse from a time base of 5-second intervals. In the interval between pulses, the capacitor is recharged, storing up energy to drive the stepper motor when the next signal pulse is received.

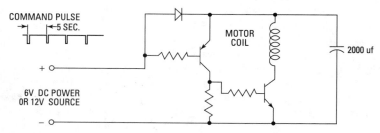

Figure 10-7 Placement of capacitor in circuit. *(Courtesy Hayden Switch & Instrument, Inc.)*

The two-wire stepper motor is capable of many modes of operation. The significant thing to remember is that energy is consumed only while the stepper motor coil is energized and only for a short period of time, as short as 10 milliseconds. The average power consumed depends on the pulse interval. If this is long with respect to the 10 ms interval, then the average power will be quite low. This type of motor can find use in disc or strip chart recorders, fire alarm reporting systems and other devices operated from a battery power source.

The step rate is from 0 to 80 total pulses per second (pps), 40 external and 40 internal, with a 12.5-millisecond minimum pulse width for each. For maximum motor efficiency, a symmetrical input (equal ON and OFF times) is always desirable (Figure 10-8). Internal pulse is produced by the collapsing magnetic field generated by the first (external) pulse.

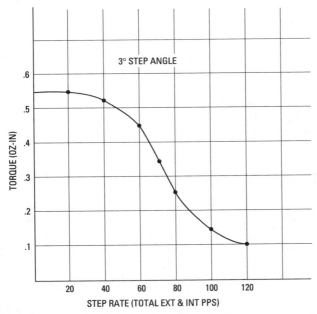

Figure 10-8 Step rate versus torque graph. (*Courtesy Hayden Switch & Instrument, Inc.*)

Uses for this type of motor include electronic speedometer/ odometer drive, bidirectional tape footage counter using two motors with mechanical differential, driver for watt-hour meters, counter drive for flow meters, rotary stepping-switch drives, marine odometers and trip odometers, as well as recorder tape drives. They operate well on 24 V DC or 12 V DC depending on the coil used in the motor.

Single-Phase, Stepper Motor (Bipolar Drive)

The unidirectional-series, single-phase, bipolar-drive, stepper motor exhibits all the qualities of the synchronous motor and even looks like it. Its low wattage, high frequency response, compact size, and low cost are significant when considering it for various purposes. Pulses may be derived from simple DPDT circuitry (Figure 10-9) arranged to reverse the polarity of the voltage applied to the two motor leads (Figure 10-10). This motor is available in both 12- and 24-V sizes. They will rotate either left or right depending upon the design. Figure 10-11 shows the 3° step angle on the torque-step rate chart.

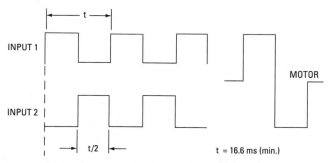

Figure 10-9 Typical pulse for single-phase, stepper motor, bipolar drive.
(Courtesy Hayden Switch & Instrument, Inc.)

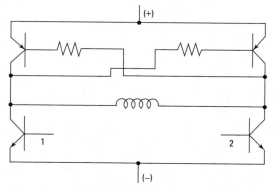

Figure 10-10 Typical drive circuit for single-phase, stepper motor, **bipolar drive.** *(Courtesy Hayden Switch & Instrument, Inc.)*

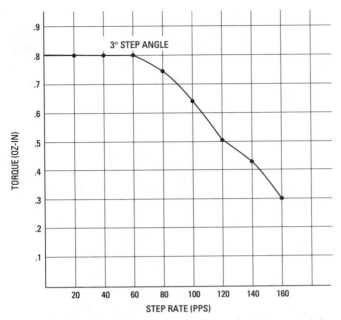

Figure 10-11   Graph for torque vs. step rate for single-phase, stepper motor, bipolar drive. *(Courtesy Hayden Switch & Instrument, Inc.)*

There are the 18°, 15°, 3°, and 0.18° step-angles per pulse with a maximum torque of 5 oz-in. The maximum step rate is 120 pulses per second. Input power is about 2 W.

This type of motor is used for paper tape drive for thermal printers, chart drives, and DC counter drives.

Two-Phase, Stepper Motor (Unipolar Drive)

This motor resembles the previous one except that it has three leads (white-black-black) coming from the enclosure, and it provides an economical solution to instrumentation problems requiring a small compact drive system. For each pulse, the rotor revolves 180° when coil B is energized, and another 180° when coil A is energized in sequence. The step rate is from 0 to 120 pps.

Simple control circuitry for this type of motor is shown in Figure 10-12. Note where the black and white leads are connected to the circuitry, with the black leads each connected to a collector of a different transistor.

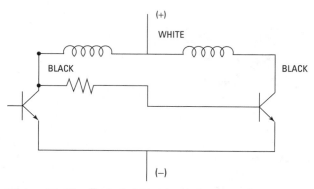

Figure 10-12 Typical drive circuit for two-phase, stepper motor, unipolar drive. *(Courtesy Hayden Switch & Instrument, Inc.)*

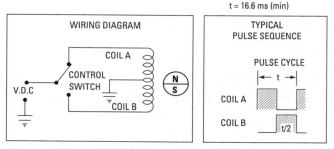

Figure 10-13 Wiring diagram and typical pulse sequence for two-phase stepper. *(Courtesy Hayden Switch & Instrument, Inc.)*

The wiring diagram indicates the coil and control switch arrangement in reference to the DC input source. The typical pulse is shown in Figure 10-13.

This type of motor is made for 12 or 24 V depending upon the selection of coils. Maximum torque is 5 oz-in. Step angle per pulse is 18°, 15°, 3°, or 0.18°, depending on the motor selected. Left or right rotation is available.

Figure 10-14 shows the torque for a 3° step angle motor with variations in torque as the step rate is changed from 20 to 160 pps.

Some of the practical applications for the motor are as counter drivers, indexing control requirements, drivers for rotary stepping switches, marine odometers, seismograph recorders, and as a driver for watt-hour meters.

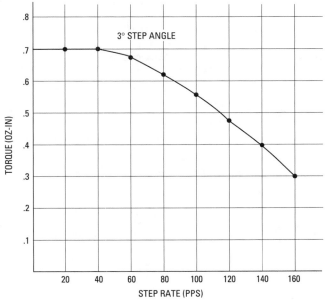

Figure 10-14 Graph for torque vs. step rate for two-phase, stepper motor. *(Courtesy Hayden Switch & Instrument, Inc.)*

Four-Phase, Stepper Motor (Unipolar Drive)

This stepper motor has two center tandem coils and two permanent magnet rotors in a tandem magnetic structure. Each half-coil is considered a separate phase (Figure 10-15).

The motor resembles all the previous ones in appearance, but has more wires coming from the enclosure. The wires are white for the common, green, black, blue, and red (Figure 10-16). Energizing two phases at a time in the sequence shown in Figure 10-16 will cause the rotor to turn in 90° steps. This may be done conveniently with two flip-flop circuits arranged to provide the desired sequence of coil excitation. With the appropriate logic circuitry, the reversible stepper can be used for step servo positioning, for instrument drives in self-balancing potentiometer circuits, for constant or variable speed drives for tapes, charts, cams, actuators, or drums, and for similar applications. Figure 10-17 shows the pulse sequence for clockwise and counter-clockwise rotation. Power input is about 1.5 W per phase.

(+)
WHITE
WHITE
BLACK
RED
BLUE
GREEN
1
3
2
4
(–)

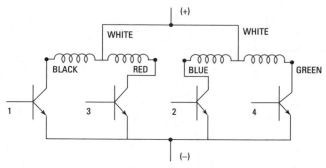

Figure 10-15 Typical drive circuit for four-phase, stepper motor, unipolar drive. *(Courtesy Hayden Switch & Instrument, Inc.)*

1
BLACK
WHITE
3
RED
2
BLUE
WHITE
GREEN
4
COM
N
S

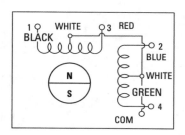

Figure 10-16 Wiring code for four-phase, stepper motor, unipolar drive. *(Courtesy Hayden Switch & Instrument, Inc.)*

PULSE SEQUENCE
STEP
CW
CCW
GREEN
BLACK
BLUE
RED
1
4
2
3
3
2
4
1
WHITE COMMON

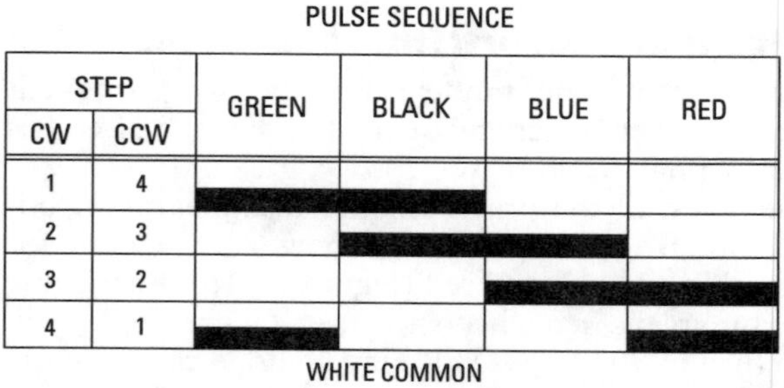

Figure 10-17 Pulse sequence for four-phase, stepper motor, unipolar drive. *(Courtesy Hayden Switch & Instrument, Inc.)*

The motor is available in either 12 or 24 volts. The step angle per pulse is 9°, 7.5°, 1.5° or 0.09°, depending on the motor selected. Note in Figure 10-18 that the maximum torque is about 1.4 oz-in for the 1.5° series of motors. Maximum step rate for this motor is 240 pps.

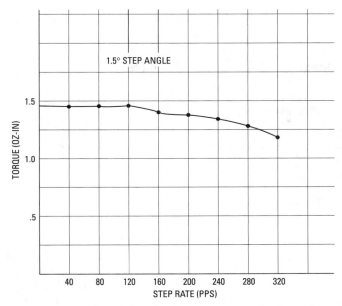

Figure 10-18 Graph for torque vs. step rate for four-phase, stepper motor, unipolar drive. *(Courtesy Hayden Switch & Instrument, Inc.)*

You can understand the popularity of these motors when you consider that they weigh 2 to 2.5 oz with a diameter of only 1 in. and a maximum width of about 1.25 in.

Open-Frame, Synchronous and Stepper Motor

When more torque is needed, the open-frame motor is used, as it is larger and capable of driving heavier loads (Figures 10-19 and

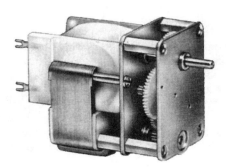

Figure 10-19 Open-frame, unidirectional, shaded-pole, stepper motor. *(Courtesy Hayden Switch & Instrument, Inc.)*

Figure 10-20 Open-frame, reversible, shaded-pole, step servo motor. *(Courtesy Hayden Switch & Instrument, Inc.)*

10-20). All types of these motors have a barium ferrite rotor and a laminated silicon steel stator structure forming two split poles. The motor shaft and the output shaft are journaled in sintered bronze bearings. The gears are molded nylon, turning on fixed studs driven into the gear train plates. The motor pinion and the final gear have brass inserts and are driven on the motor shaft and the output shaft, respectively. This provides a strong gear train capable of carrying 10 oz-in continuous duty or 30 oz-in intermittent duty.

Basically, the open-frame, two-wire, stepper motor (Figure 10-19) is the same as the series two-wire motor previously mentioned (Figure 10-4) and enclosed in a 1-inch diameter can. The 360° rotation per pulse cycle is the same, as is the two-wire control circuitry (Figure 10-21). The applications are also the same, except that this motor is used where the added weight and space needed for mounting is not a problem and where more torque is needed to operate the driven device.

Keep in mind that the rotor and output shaft bearings are sintered bronze and vacuum impregnated so that no further lubrication is needed. The integral gear reducer, with nylon gears and pinions, provides extremely long life and requires no lubrication. The lubricant supplied is usually for operation over an ambient temperature range of 0–71° C. Special lubrication can be obtained for operation over a wider ambient temperature range.

Stepper motors, as well as relays, solenoids, and other devices with coils, are inductive loads. That means that suppression should be provided for contact or transistor protection and to minimize the generation of transients. Simple diode suppression across the coil should not be used as this adversely affects performance. Instead, a combination of a blocking diode and a zener diode in series across the coil is recommended. A typical circuit for the stepper and voltage and current waveforms for one pulse cycle are shown in Figure 10-22.

The average weight for these motors is 6 oz, and the power consumption averages 3 W on 24 V DC. Figure 10-23 shows the dimensional data for this type of open-frame motor. Table 10-2 illustrates the torque in oz-in for the various step angle motors. Note how the 1- to 20-Hz and 40-Hz pulsing makes a difference in the torque provided. These motors are available in right rotation or left rotation, but are not reversible.

Reversible Step-Servo Motor

The series step-servo motor has two permanent magnet rotors in a tandem laminated magnetic structure with two center tapped coils (Figure 10-20). For purposes of identification, each half-coil is considered a separate coil (Figure 10-24). With connections shown in Figure 10-25, energizing the coils individually in sequence (1-2-3-4) will cause the rotor to turn right in 90° steps for one revolution (360°).

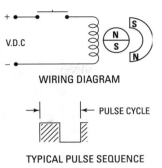

WIRING DIAGRAM

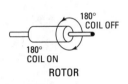

TYPICAL PULSE SEQUENCE

ROTOR

Figure 10-21 Wiring diagram and typical pulse sequence for a two-wire stepper motor. *(Courtesy Hayden Smith & Instrument, Inc.)*

Normally, this type of operation is not practical since it is simpler to drive the step servo with two flip-flop circuits that, at any instant, energize two coils. This mode of operation is shown in Figure 10-26. The sequence of operation then becomes 1-4, 1-2, 2-3, 3-4 for right rotation and 1-4, 3-4, 2-3, 1-2 for left rotation.

The step servo has no magnetic detent when deenergized and consequently the rotor position does not shift when power is removed from the system. The unit will operate with a rated load up to 50 complete cycles per second or 200 changes of state per second, each resulting in a 90° rotor step. Maximum slewing rate is approximately twice this figure. The motors are available in 9°, 7.5° and 1.5° step angles. Typical torque output with the 9° step-angle motor varies from 0.1 oz-in at 200 steps per second to 0.2 oz-in at 1 step per second (Table 10-3). Power required is 8 W nominal.

With appropriate solid-state logic circuitry, the series step-servo motor provides a reliable variable-speed drive for strip chart recorders and other devices requiring a range of speed on the order of 200:1 or higher. If an accurate frequency standard is available, solid-state divider circuits will enable stepper operation at predetermined sub-multiples of the referenced frequency.

254 Chapter 10

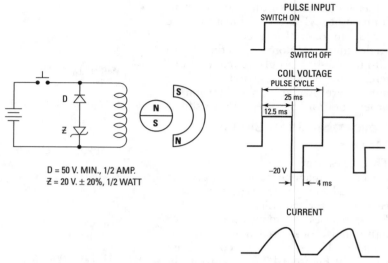

Figure 10-22 Arc suppression and pulses for a stepper motor.

(Courtesy Hayden Switch & Instrument, Inc.)

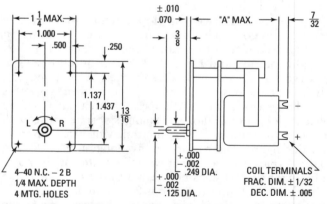

Figure 10-23 Dimensional data for a two-wire, stepper motor.

(Courtesy Hayden Switch & Instrument, Inc.)

Again, with the appropriate solid-state logic circuitry, the series stepper can be operated as a reversible step-servo motor for instrument drives in self-balancing potentiometer circuits and similar applications.

Figure 10-27 shows the coil connections and the dimensions of the motor. Left rotation, of course, refers to counter-clockwise (ccw) rotation and the right rotation refers to clockwise rotation (cw) when the motor shaft is viewed from the end as shown in Figure 10-27.

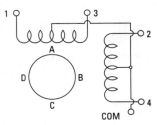

Figure 10-24   Wiring diagram for reversible step-servo motor. *(Courtesy Hayden Switch & Instrument, Inc.)*

High-Torque, Reversible AC Motor

This is an instantly reversible motor. Figure 10-20 shows what it looks like, and Figure 10-28 shows the wiring diagram. It is reversed by switching the capacitor from one coil to the other. This

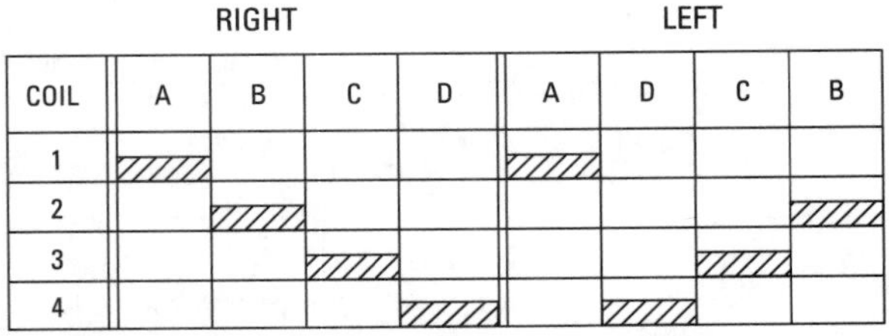

	RIGHT				LEFT			
COIL	A	B	C	D	A	D	C	B
1	/////				/////			
2		/////						/////
3			/////				/////	
4				/////	/////			

Figure 10-25   Mode of operation, sequence of operation.
(Courtesy Hayden Switch & Instrument, Inc.)

	RIGHT				LEFT			
COIL	A	B	C	D	A	D	C	B
1	/////////				/////			/////
2		/////////					/////////	
3			/////////			/////////		
4	/////			/////	/////////			

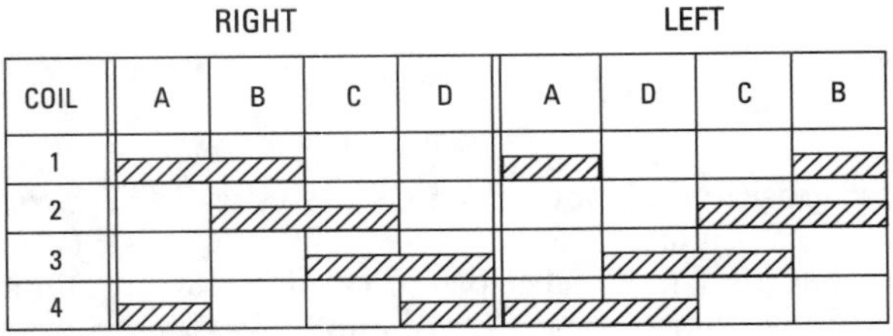

Figure 10-26   Mode of operation, sequence of operation.
(Courtesy Hayden Switch & Instrument, Inc.)

Table 10-3   Step Angle/Dimension B

Voltage 24 Volts DC ± 19%

Step Angle per change of state	9°	7.5°	1.5°
Dimension B (Figure 10-27)	0.672 in.	0.672 in.	1.137 in.

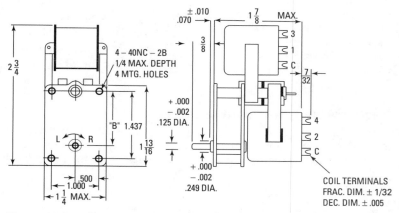

Figure 10-27   Dimensional data for reversible step-servo motor.
(Courtesy Hayden Switch & Instrument, Inc.)

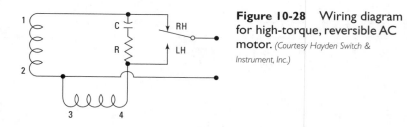

Figure 10-28   Wiring diagram for high-torque, reversible AC **motor.** *(Courtesy Hayden Switch & Instrument, Inc.)*

makes it ideal for positioning potentiometers, shafts, indicators, and similar services since it starts and stops within one cycle of AC power. The series motor can "inch" in either direction in small increments and stop without coast. With split coils, this motor becomes a step-servo motor capable of bidirectional operation up to 200 pps.

This motor operates on 115 V ± 10 percent and consumes 2.5 W. Torque at 3600 rpm is 0.07 oz-in. The rotor and output shaft bearings are porous bronze and are lubricated for life at the factory for operation over an ambient temperature range of 0° to 71° C. The capacitor is 0.27 MF and the resistor used in the circuit is 4.5k ohms at 5 W. When checking the coil terminal numbers, start at the top with 1 and 2. Coils 3 and 4 are on the bottom. Table 10-4 shows the speed and torque for four different speed motors. Gear train is rated at 10 oz-in continuous and 30 oz-in intermittent duty.

Table 10-4 Speed/Torque Characteristic for 115 V, 60-Hz Motors

Speed	Torque
60 rpm	4 oz-in
30 rpm	8 oz-in
20 rpm	12 oz-in
10 rpm	24 oz-in intermittent

Gear trains are rated at 10 oz-in continuous and 30 oz-in intermittent duty.

Three-Wire, Stepper Motor (Open Frame)

The three-wire stepper is slightly different from the two-wire. Take a look at the wiring diagram in Figure 10-29. Magnetic detenting occurs at 0° and 180° rotor positions whether or not the windings are energized.

The three-wire stepper produces 360° rotation with each pulse. The typical pulse sequence is also shown in Figure 10-29. This comes in handy when used in pulse counters, stepping switches, pulse storage devices, and remote positioning. Note how arc suppression is accomplished in Figure 10-30. Simple diode suppression across the coils should not be used as this adversely affects performance. Instead, a combination of blocking diodes and a zener diode is recommended. Note the typical voltage and current waveforms for one pulse cycle.

Power consumed is 4 W at 60 Hz and double that or 8 W at 1 Hz. The motor weighs 6 oz and can be obtained for 12-V operation when specially ordered.

Table 10-5 shows the four step angles available in these motors and the torque generated at two different frequencies.

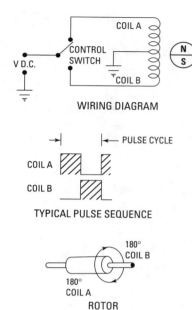

Figure 10-29 Wiring diagram and pulse sequence for a three-wire, stepper motor. *(Courtesy Hayden Switch & Instrument, Inc.)*

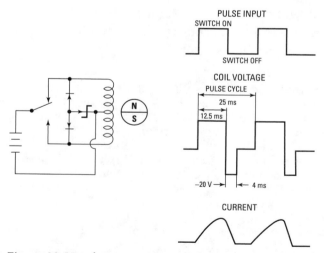

Figure 10-30 Arc suppression and pulses for operation of three-wire, stepper motor. *(Courtesy Hayden Switch & Instrument, Inc.)*

Table 10-5   Step Angle/Torque

Step Angle	Torque (Oz-In)	
	1–20 Hz	60 Hz
36°	0.6	0.3
30°	0.6	0.3
6°	3.5	1.5
3.6°	6.0	3.0

Figure 10-31 shows that the center tap is the bottom coil terminal and not the middle one. Dimension A is either 1.4375 or 1.750, depending on the step angle. The larger step angles of 36° and 30° have the smaller size while the 6° and 3.6° step-angle motors are a little larger at 1.75 in. Dimension B is 0.672 in. on the larger step-angle motors and 1.137 in. on the 6° and 3.6° motors.

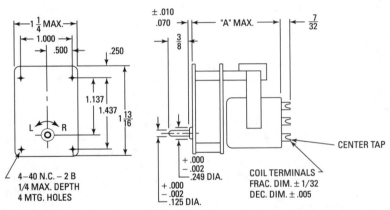

Figure 10-31   Dimensional data for three-wire stepper motor. Note the center tap location. *(Courtesy Hayden Switch & Instrument, Inc.)*

Summary

Motors that are controlled by electronic means are many and varied. Robots and computers have put a different demand on electric motors. Precision motors do what is intended as an end product or motion. Some of these motors are very expensive and are found in precision measuring devices and strip recorders. They are made to require little or no maintenance. It is best to know the circuitry involved in the control of the motor in order to determine if proper operation is being achieved or the motor is malfunctioning...and needs to be replaced.

Shaded-pole, one-direction-of-rotation motors are referred to as being unidirectional, synchronous, shaded-pole. They are used for repeat cycle timers, elapsed time indicators, stop clocks, and to drive devices for encoding discs, chart recorders, and medical instrumentation. It is an inexpensive type motor that can be used in visual displays as a driver for cycle changes. It is also used for fiber optic displays and many other devices.

Capacitor start-and-run, bidirectional, synchronous motors are easily reversed by switching a capacitor from one coil to the other. These motors can inch in either direction in small increments and stop without coasting. They can be used for a number of purposes, such as ink flow control on printing presses and lens focus control on motion picture projectors.

Single-phase, stepper motors (unipolar drive) are not reversible. They convert electrical pulses into discrete angular steps of the output shaft without need for logic control. These motors operate on a minimum of power. They are ideal for operation on batteries for they draw very little current. The two-wire, stepper motors are capable of many modes of operation. They operate well on 24 V DC or 12 V DC, depending on the coil used in the motor.

Single-phase, stepper motors (bipolar drive) operate on 12 and 24 V. They rotate either left or right depending on the design. The maximum step rate is 120 pulses per second. Input power is about 2 W. They are used for chart drives and DC counter drives.

The two-phase, stepper motors (unipolar drive) resemble the previous one except that they have three leads (white-black-black) coming from the enclosure, and it provides an economical solution to instrumentation problems requiring a small compact drive system. They are made for 12 or 24 V DC operation, and are capable of left (ccw) or right (cw) rotation. These motors are used for counter drivers, indexing controls, drivers for rotary stepping switches, marine odometers, and as drivers for watt-hour meters.

Four-phase, stepper motors (unipolar drive) have two center-tapped coils and two permanent-magnet rotors in a tandem magnetic structure. Each half-coil is considered a separate phase. The motors resemble all the previous ones in appearance, but have more wires coming from the enclosure. The wires are white for the common, green, black, blue, and red. They can be obtained in 12-V or 24-V DC configurations. Power input is about 1.5 W per phase. They are capable of clockwise and counterclockwise operation.

Open-frame, synchronous, and stepper motors are available when more torque is needed. They are capable of handling heavier loads. All types of these motors have a barium ferrite rotor and a laminated silicon steel stator structure forming two split poles. The motor shaft and the output shaft are journaled in sintered bronze bearings.

Stepper motors, as well as relays, solenoids, and other devices with coils, are inductive loads. That means arc suppression should be provided for contacts or transistor protection and to minimize the generation of transients. Simple diode suppression across the coil should not be used as this adversely affects performance. Instead, a combination of a blocking diode and zener diode in series across the coil is recommended. The average weight of these motors is 6 oz and power consumption averages 3 W on 24 V DC. They are available in right or left rotation, but are not reversible.

The reversible-step, servo-motor has no magnetic detent when deenergized. Consequently, the rotor position does not shift when power is removed from the system. This unit will operate with a rated load up to 50 complete cycles per second or 200 changes of state per second each resulting in a 90° rotor step. Maximum slewing rate is approximately twice this figure. Power requirements are about 8 W. Remember left rotation means counterclockwise (ccw) and right rotation means (cw) clockwise rotation.

High-torque, reversible, AC motors are instantly reversible. They are reversed by switching the capacitor from one coil to the other. They can start and stop with one cycle of AC power. The series motors can inch in either direction in small increments and stop without coasting. They operate on 115 V AC and consume 2.5 W. Torque is 0.07 oz.-in. at 3600 rpm.

Three-wire, stepper motors (open-frame) are slightly different from the two-wire. Three-wire motors produce 180° rotor positions whether or not the windings are energized. They produce 36° rotation with each pulse. A combination of blocking diodes and a zener diode is recommended for arc suppression. Power consumed is 4 W at 50 Hz and double that at 1 Hz. These motors weigh 6 ounces and can be obtained for 12-V operation when specially ordered.

Review Questions

1. Where are motors controlled by electronics used today?

2. What are shaped-pole, unidirectional motors used for?

3. How are capacitor start-and-run, bidirectional, synchronous motors reversed?

4. Where are the capacitor start-and-run motors used?

5. What motor is ideal for use with a battery power source?

6. What is meant by left rotation of a motor?

7. What is meant by right rotation of a motor?

8. What are the colors used for two-phase, stepper motor power leads?

9. What colors do the four-phase, stepper motors use for power leads?

10. Why are open-frame, synchronous and stepper motors used?

11. What type of loads to a power source do stepper motors, relays, and solenoids present?

12. Why shouldn't a simple diode across the coil be used for arc suppression in most motors?

13. What is a zener diode?

14. How are high-torque, reversible, AC motors reversed?

15. How do two-wire, stepper (open-frame) and three-wire, stepper motors differ?

Chapter 11

Wound-Rotor Motors

As far as the stator is concerned, the squirrel cage motor and the wound-rotor motor are identical. The main difference between the two motors lies in the rotor winding.

In the squirrel cage motors previously described (Chapter 8), the rotor winding is practically self-contained; it is not connected either mechanically or electrically to the outside power-supply or control circuit. It consists of a number of straight bars uniformly distributed around the periphery of the rotor and short-circuited at both ends by end rings to which the bars are integrally joined. Since the rotor bars and end rings have fixed resistances, such characteristics as starting and pull-out torques, rate of acceleration, and full-load operating speed cannot be changed for a given motor installation.

In wound-rotor motors, however, the rotor winding consists of insulated coils of wire that are not permanently short-circuited, but are connected in regular succession to form a definite polar area having the same number of poles as the stator. The ends of these rotor windings are brought out to collector rings, or slip rings, as they are commonly termed.

The currents induced in the rotor are carried by means of slip rings (and a number of carbon brushes riding on the slip rings) to an externally mounted resistance, which can be regulated by a special control, as indicated in Figures 11-1 and 11-2. By varying the amount of resistance in the rotor circuit, a corresponding variation in the motor characteristics can be obtained. Thus, by inserting a high external resistance in the rotor circuit at starting, a high starting torque can be developed with a low starting current. As the motor accelerates to full speed, the resistance is gradually cut out until, at full speed, the resistance is entirely cut out and the rotor windings are short-circuited.

Characteristics

Varying the values of resistance in the rotor circuit can affect the characteristics of the motor as follows:

1. Variation in the starting torque and current
2. Smooth acceleration
3. Variation in operating speed

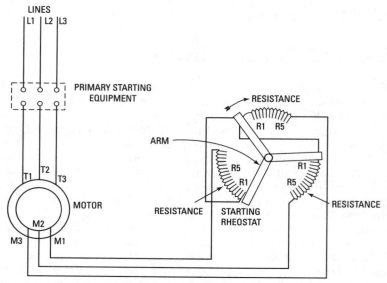

Figure 11-1 Diagram of a starter or controller for a wound-rotor induction motor. The resistances in each phase of the rotor are gradually cut out to increase the motor speed.

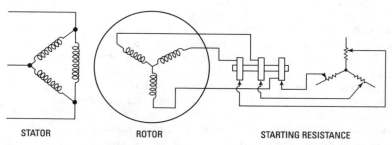

Figure 11-2 Schematic representation for starting resistance in each phase of a wound-rotor 3ϕ motor.

The operating speed, of course, depends entirely on the amount of resistance incorporated in the control equipment.

Figure 11-3 shows the speed-torque and corresponding speed-current curves obtainable for a typical wound-rotor induction motor using different values of external rotor resistance. The numbers given

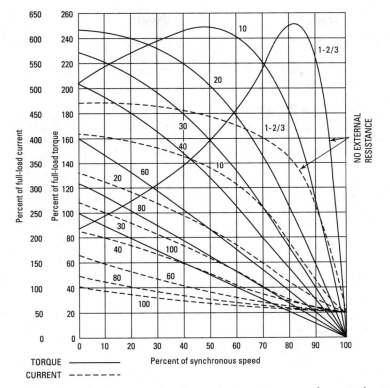

Figure 11-3 Speed-torque and speed-current curves of a typical wound-rotor induction motor.

on the curves indicate the rotor-circuit resistance in the percent value required to give full-load torque at standstill.

The curves in Figure 11-3 illustrate that external resistance in the rotor circuit reduces the speed at which the rotor will operate with a given load torque. If the rotor resistor is designed for continuous duty, a portion of it may be allowed to remain in the circuit, thus obtaining reduced-speed operation. Therefore, the motor has a varying speed characteristic; that is, any change in load results in a considerable change in speed. It should be borne in mind, however, that the efficiency of a wound-rotor motor, including the I^2R losses in the rotor resistance, is reduced in direct proportion to the speed reduction obtained.

Thus, the wound-rotor motor has found its use in cranes, hoists, and elevators. These are operated intermittently and for short periods, where exact speed regulation and loss in efficiency are of little consequence. If, however, lower speed is required over longer periods, poor speed regulation and loss in efficiency may become prohibitive.

Control Equipment

The previous section noted that wound-rotor motors have certain inherent speed-torque characteristics, and that these characteristics can be considerably altered by the secondary control equipment, which introduces varying values of resistance into the rotor circuit. Since the control equipment has this ability, a further study of the most common types of controllers and associated equipment is appropriate.

The functions of controllers used with wound-rotor motors are

1. To start the motor without damage or undue disturbance to the motor, driven machine, or power supply
2. To stop the motor in a satisfactory manner
3. To reverse the motor
4. To run the motor at one or more predetermined speeds below synchronous speed
5. To handle an overhauling load satisfactorily
6. To protect the motor

The various types of controllers utilized may be divided into the following groups, depending on the size and function of the motor:

1. Faceplate starters
2. Faceplate speed regulators
3. Multiswitch starters
4. Drum controllers
5. Motor-driven drum controllers
6. Magnetic starters

Faceplate Starters

A typical wiring diagram of a combined faceplate starter and speed regulator is shown in Figure 11-4. This particular type requires a separate switch for control of the primary circuit. Faceplate starters

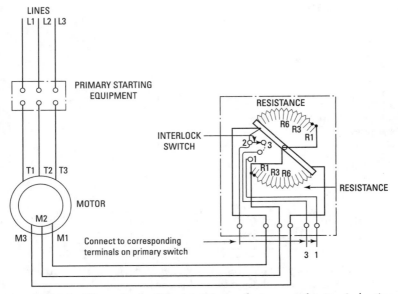

Figure 11-4 Diagram of a faceplate starter for a wound-rotor induction motor.

are usually adaptable for motors up to 50 hp (37.3 kW) where the full-load current in the rotor circuit is less than 150 amperes.

As shown in the diagram, the resistance is made up of three delta-connected sections, but is connected in two phases only, the third being a fixed step. The starting lever is equipped with a spring similar to that used in DC motor starters. The spring returns the lever to the starting (all-resistance) position when it is released. As an additional safety feature, the contact arm or lever cannot be left in the full-in position unless the primary (stator) contactor is closed; nor can it be left in an intermediate position at any time.

Faceplate Speed Regulators

A typical wiring diagram of a combined faceplate starter and speed regulator for use in the rotor circuit of wound-rotor induction motors is shown in Figure 11-5. Starters of this type usually provide for a 50-percent speed reduction when the motor operates under full load at normal speed. They are usually built for motors in sizes up to 40 hp (29.84 kW) and for rotor currents up to 100 amperes. Since the starter is not connected with the primary circuit of the

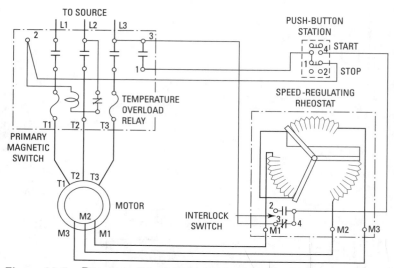

Figure 11-5 Diagram of a secondary, speed-regulating rheostat and a primary magnetic switch for a wound-rotor motor.

motor, a magnetic switch, an oil circuit breaker, or similar device must be installed to control the primary circuit.

Multiswitch Starters

Multiswitch starters are used in the secondary circuits of large, wound-rotor induction motors up to 2000 hp (1492 kW) with rotor currents up to 1000 amperes. A typical wiring diagram of a multiswitch starter is shown in Figure 11-6.

The type of switch to be used in the primary (stator) circuit depends on the voltage of the supply source. The resistor units in the secondary (rotor) circuit are balanced on all steps of the controller. The contact levers are of the double-pole type and are mechanically arranged in such a manner that they must be closed in a predetermined sequence, and only one at a time. Since the switches are designed for hand-over-hand operation, a desirable time element is introduced that prevents too rapid acceleration of the motor. When the final switch has been closed, it is held in place by a magnetic coil, and because of the mechanical interlocking feature, all other switches remain closed.

A failure of the motor supply source will cause the magnet coil to release the switches, returning the starter to the starting position

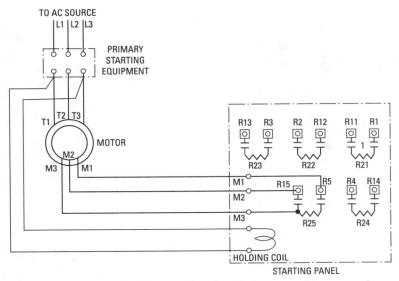

Figure 11-6 Diagram of a typical multiswitch starter for a wound-rotor motor.

with all the resistance in the secondary circuit. This type of starter is designed for starting duty only and is not suitable where speed regulation is required.

Drum Controllers

Drum controllers (Figure 11-7) are frequently employed for starting and for speed control of wound-rotor induction motors. Because of their construction features, they offer a compact, sturdy, and dependable control means, and also add to the safety of the operator. Drum controllers are built to handle both stator and rotor circuits, the cylinder that mounts the contact segments being built in two insulated sections. They can also be built to handle only the rotor circuit, in which case the stator circuit is controlled by a circuit breaker or line starter. Other types are built for operation by means of an independent motor. In addition to starting and regulating, speed-regulating drum collectors are commonly used for speed-reversing duty as well.

On small- and medium-sized motors, the faceplate starters described earlier are commonly used, but with large motors, drum controllers are generally preferred because their contacts are heavier and better able to handle the heavy currents required.

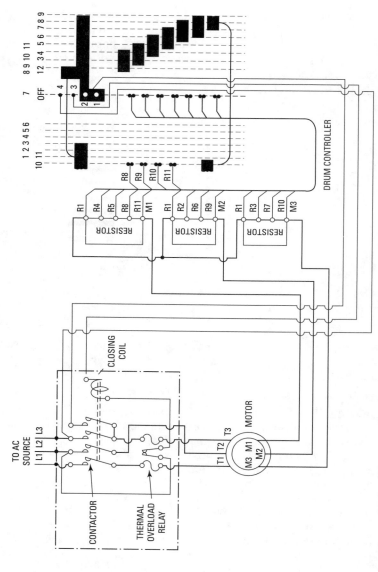

Figure 11-7 Diagram of a nonreversing drum controller for a wound-rotor motor that has a three-phase secondary.

Figure 11-8 shows the connections of a typical drum controller for reversing and speed-regulation duty with a wound-rotor motor. The resistor material is mounted separately from the drum, and connections from the resistor are brought to the drum fingers.

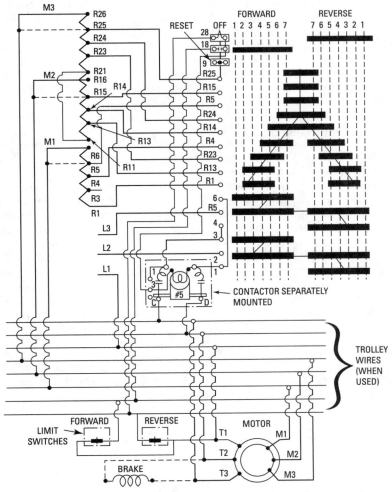

Figure 11-8 Diagram of a typical reversing and speed-regulating drum controller for a wound-rotor, induction motor.

The resistor connections shown in Figure 11-8 are for starting the motor with one phase of its secondary open on the first point of the controller. When higher starting torque is desired, or when connecting a motor that is rated above 80 hp (59.68 kW), connect R_1 to R_{11} at the resistor and the finger marked R_1 on the controller to terminal R_3 on the resistors. Resistor steps R_5 to R_6, R_{15} to R_{16}, and R_{25} to R_{26} are for resistances that remain permanently in the circuit. When these are not furnished, connect M_1, M_2, and M_3 to R_5, R_{15}, and R_{25}, as indicated by the dotted lines.

Motor-driven drum controllers are used in certain drives requiring close automatic speed regulation, such as in large air-conditioning plants, blowers, stokers, etc. In construction, these controllers differ from those previously described in that cam-operated switches are used in place of the segments and fingers.

These drums are built with either 13 or 20 balanced speed-regulating points, and the cam-switch construction saves space when so many balanced points are needed. Each switch carries two contacts, both of which are connected to the middle leg of the resistor. They make contact simultaneously with stationary contacts, one connected to each of the outside legs of the resistor. In this way, the closing of any cam-operated switch short-circuits the resistor at that point. The drum is driven by a small motor connected to the drum shaft through suitable gearing. The pilot motor is energized by some automatic device, such as a thermostat or a regulator.

A typical, motor-driven drum may have 20 balanced speed points and be rated at 600 amperes and 1000 volts. It may have a positioning device that ensures that the pilot motor stops only at positions in which the switches are fully closed. It may also have self-contained limit switches that open to stop the motor at each end of the drum travel.

Magnetic Starters

When used for starting and control of wound-rotor motors, magnetic starters are built in three different forms, depending on the duty of the motor.

- Plain starting
- Speed regulating
- Speed setting

Magnetic starters consist of a magnetic contactor for connecting the stator circuit to the line, and one or more accelerating contactors to commutate the resistance in the rotor circuit. The number of

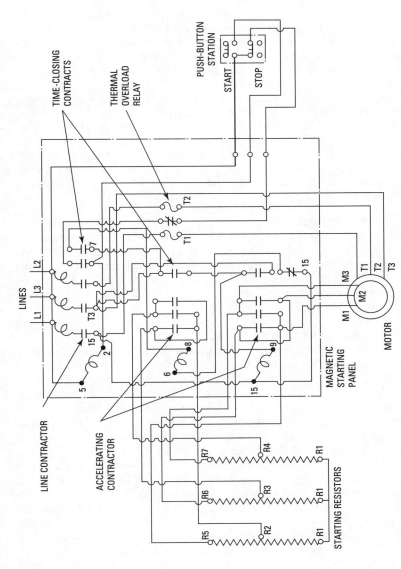

Figure 11-9 Diagram of a magnetic contactor for use with a wound-rotor induction motor.

LINE CONTRACTOR

ACCELERATING
CONTRACTOR

TIME-CLOSING
CONTRACTS

THERMAL
OVERLOAD
RELAY

PUSH-BUTTON
STATION

START

STOP

LINES

L2

L3

L1

T2

T1

T7

T3

15

2

5

8

6

9

15

MAGNETIC
STARTING
PANEL

M3

M2

M1

T1

T2

T3

MOTOR

R7

R4

R6

R3

R5

R2

R1

R1

R1

STARTING RESISTORS

273

secondary accelerating contactors varies with the rating, a sufficient number being employed to assure smooth acceleration and to keep the inrush current within practical limits. The operation of the accelerating contactors is controlled by a timing device, which provides definite time acceleration. For high-voltage service, the primary contactor is usually of the oil-immersed type. The diagram of a typical magnetic starter for use with a wound-rotor induction motor is shown in Figure 11-9.

Magnetic-starting speed regulators are very similar to magnetic starters except that the secondary circuit is controlled by a series of magnetic contactors operated from a dial switch in the control station or by a faceplate type of rheostat mounted in the controller itself. Where magnetic contactors are used, they also act as accelerating contactors on starting. If a rheostat is used, accelerating contactors on the panel accelerate the motor to the speed for which the rheostat is set.

Controller Resistors

All standard wound-rotor phase circuits, whether for two- or three-phase circuits, have secondaries wound for three-phase. The controller resistors used for each phase are identical with the exception of the terminal marking. The resistor for the first phase has its terminals marked consecutively (R_1, R_2, R_3, etc.), the second phase R_{11}, R_{12}, R_{13}, etc., the third phase R_{21}, R_{22}, R_{23}, etc., as shown in Figure 11-10. The actual resistor will consist of one, two, three, or multiples of these frames of tubes or grids.

Secondary resistors for wound-rotor induction motors are, as a rule, designed for star connection. Resistors for most manual controllers may be connected with all three secondary phases closed or with one secondary phase open on the first point of the controller. Resistors for magnetic controllers are connected with all three phases closed in the secondary on the first point. The torque obtained with a resistor of a given class number varies with the connection used on the first point of the controller.

The NEMA resistor classifications for wound-rotor induction motors are given in Table 11-1. For example, Class 114 is intended for starting duty only and for a motor that will be started and brought up to speed in approximately five seconds with a minimum of 75 seconds in between successive starts.

The capacity of resistors depends largely on the ventilating space. As a rule, the resistor frames should never be stacked more than four in height, and when space is available, each frame should be separated from the next by approximately the width of the end

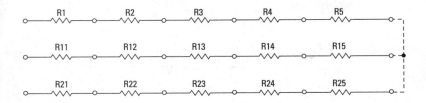

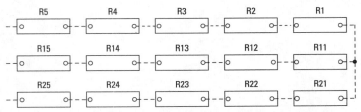

Figure 11-10 Typical connections of secondary resistor units.

frame. Frames may be mounted on the floor, platform, or wall, but in such a way as to obtain free ventilation.

It must also be emphasized that a close periodic inspection of resistors and associated connections should be made. This inspection should include the tightening of loose lock nuts, connections, etc. The collection of dirt and dust should be blown out from between the resistor units.

Application

Wound-rotor motors have the ability to start extremely heavy loads. Hence they are suitable for

- Driving various types of machinery that require development of considerable starting torque to overcome friction
- Accelerating extremely heavy loads that have a flywheel effect or inertia
- Overcoming back pressures set up by fluids and gases in the case of reciprocating pumps and compressors

Double squirrel-cage motors are also applicable on many of the heavier machines involving the problems mentioned above. However, if a considerable length of time is required to accelerate the load

Table 11-1 NEMA Resistor Classification for Wound-Rotor Motors

Percent Full-Load Current on First Point	Starting Torque % of Full Load					Resistor Class Number						
	Series Motors	Compound Motors	Shunt Motors	Wound-Rotor Induction Motors 1 Ph. Stg.	3 Ph. Stg.	5 Sec. on Out of 80 Sec.	10 Sec. on Out of 80 Sec.	15 Sec. on Out of 90 Sec.	15 Sec. on Out of 60 Sec.	15 Sec. on Out of 45 Sec.	15 Sec. on Out of 30 Sec.	Cont.
25	8	12	25	15	25	111	131	141	151	161	171	91
30	30	40	50	30	50	112	132	142	152	162	172	92
70	50	60	70	40	70	113	133	143	153	163	173	93
100	100	100	100	55	100	114	134	144	154	164	174	94
150	170	160	150	85	150	115	135	145	155	165	175	95
200	250	230	200		200	116	136	146	156	166	176	96

to full speed, double squirrel cage rotors may burn out before full speed is attained. For that reason, wound-rotor motors should be used instead. Where high starting torque alone is involved, double squirrel cage motors will qualify, but the fact that all the heat developed in the secondary circuit is confined to the rotor prevents their use if the starting period is too long. Frequent starting has the same effect of overheating double squirrel cage motors, for which reason wound-rotor motors should be used on machinery started frequently.

Operating Speed Variation

Variations in operating speed are essential on many applications. It is often desirable to vary the operating speed of conveyors, compressors, pulverizers, stokers, and so on, in order to meet varying production requirements. Wound-rotor motors, because of their adjustable speeds, are ideally suited for such applications. However, if the torque required does not remain constant, the speed of the wound-rotor motor will vary over wide limits, a characteristic constituting one of the serious objections to the use of wound-rotor motors for obtaining reduced speeds. Another factor that must be taken into consideration when wound-rotor motors are selected for reduced-speed operations is that of lowered motor efficiency.

This type of motor is suitable where the speed range required is small, where the speeds desired do not coincide with the synchronous speed of the line frequency, and where the speed must be gradually or frequently changed from one value to another.

The wound-rotor motor also gives high starting torque with a low current demand from the line, but it is not efficient when used a large proportion of the time at reduced speed, since power corresponding to the percent of drop in speed is consumed in the external resistance without doing any special work.

The wound-rotor motor can operate at any speed from its maximum full-load speed down to almost standstill. Speed reduction below one-half is not recommended because of poor speed regulation (the no-load speed is always synchronous speed) and because of increased heating of the motor due to the decreased ventilation.

When wound-rotor motors are used for cranes, hoists, and elevators—machinery operated intermittently and for short periods—poor speed regulation and loss in efficiency are of little consequence. However, if lowered speed is required over longer periods, poor speed regulation and loss in efficiency may become prohibitive.

Smooth Starting

By the use of external resistors in the slip-ring rotor windings, a wide variation in rotor resistance can be obtained with a resultant variation in acceleration characteristics. Thus, a heavy load can be started as slowly as desired, without jerk, and can be accelerated smoothly and uniformly to full speed. It is merely a matter of supplying the necessary auxiliary control equipment to insert sufficient high resistance at the start, and to gradually reduce this resistance as the motor picks up speed.

Low Starting Current

Many power companies have established limitations on the amount of current that motors may draw at starting. The purpose is to reduce voltage fluctuations and prevent flickering of lights. Because of such limitations, the question of starting current is often the deciding factor in the choice of wound-rotor motors instead of squirrel-cage motors.

Wound-rotor motors, with proper starting equipment, develop a starting torque equal to 150 percent of full-load torque with a starting current of approximately 150 percent of full-load current. This compares very favorably with squirrel-cage motors, one type of which requires a starting current of as much as 600 percent of full-load current to develop the same starting torque of 150 percent.

Troubleshooting

Table 11-2 shows symptoms and possible causes, as well as possible remedies, for common wound-rotor motor problems.

Table 11-2   Wound-Rotor Motor Troubleshooting Chart

Symptom and Possible Cause	Possible Remedy
Motor Runs at Low Speed with External Resistance Cut Out	
(a) Wires to the control unit too small	(a) Use larger cable to the control unit.
(b) Control unit too far from motor	(b) Bring control unit nearer motor.
(c) Open circuit in rotor circuit (including cable to the control unit)	(c) Test by ringing out circuit and repair.
(d) Dirt between brush and ring	(d) Clean rings and insulation assembly.

(continued)

Table 11-2 *(continued)*

Symptom and Possible Cause	Possible Remedy
Motor Runs at Low Speed with External Resistance Cut Out	
(e) Brushes stuck in holders	(e) Use right size brush.
(f) Incorrect brush tension	(f) Check brush tension and correct.
(g) Rough collector rings	(g) File, sand, and polish.
(h) Eccentric rings	(h) Turn down on lathe, or use portable tool to true-up rings without disassembling motor.
(i) Excessive vibration	(i) Balance motor.
(j) Current density of brushes too high (overload)	(j) Reduce load. If brushes have been replaced, make sure they are of the same grade as originally furnished.

Summary

The stator in the wound-rotor motor is identical to the stator in the squirrel-cage motor. The basic difference in the two motors lies in the rotor winding.

In the squirrel-cage motor, the rotor winding is nearly always self-contained; it is not connected either mechanically or electrically to the outside power-supply or control circuit. However, in wound-rotor motors, the rotor winding consists of insulated coils of wire that are not permanently short-circuited, but are connected in regular succession to form a definite polar area that has the same number of poles as the stator. The ends of these rotor windings are brought out to collector rings, or slip rings.

External resistance in the rotor circuit reduces the speed at which the rotor will operate with a given load torque. If the rotor resistor is designed for continuous duty, a portion may be permitted to remain in the circuit to obtain reduced-speed operation. Therefore, the motor has a varying speed characteristic: Any change in load results in a considerable change in speed.

The wound-rotor motor is often used in cranes, hoists, and elevators. These devices are operated intermittently and for short periods of time, where exact speed regulation and loss in efficiency are of little consequence.

Wound-rotor motors can be used to start extremely heavy loads. Hence, they are suitable for: (1) driving various types of machinery that require development of considerable starting torque to overcome friction; (2) accelerating extremely heavy loads that have a flywheel effect or inertia; and (3) overcoming back pressures set up by fluids and gases, as in reciprocating pumps and compressors.

Review Questions

1. What is the main difference between the squirrel-cage motor and the wound-rotor motor?

2. How is the wound-rotor motor provided with a varying speed characteristic?

3. List five functions of controllers used with wound-rotor motors.

4. List five types of controllers.

5. List three types of equipment that are suitable for wound-rotor motors.

6. Which, the squirrel-cage or wound-rotor, has a short-circuited rotor?

7. What is external resistance used for in a wound-rotor motor?

8. Where is the wound-rotor motor used?

9. How does a change in load affect the speed of a wound-rotor motor?

10. What is meant by flywheel effect?

11. What kind of starting torque does the wound-rotor motor have?

12. Why would this type of motor be useful in reciprocating pumps?

13. What is meant by the term wound rotor?

14. How are the current carrying devices connected to a rotor in a wound-rotor motor?

15. What does I^2R loss mean?

Chapter 12

Fractional-Horsepower Motors

Fractional-horsepower motors are manufactured in a large number of types to suit various applications. Because of its use in a great variety of household appliances, the fractional-horsepower motor is perhaps better known than any other type. It is almost always designed to operate on single-phase, AC standard frequencies, and is reliable, easy to repair, and comparatively low in cost.

Single-phase motors were one of the first types developed for use on alternating current. They have been perfected through the years from the original repulsion type into many improved types, such as

- Split-phase
- Capacitor-start
- Permanent-capacitor
- Repulsion
- Shaded-pole
- Universal
- Reluctance-synchronous
- Hysteresis-synchronous

Split-Phase Induction Motors

The *split-phase induction motor* is one of the most popular of the fractional-horsepower types. It is most commonly used in sizes ranging from $\frac{1}{30}$ hp (24.9 W) to $\frac{1}{2}$ hp (373 W) for applications such as fans, business machines, automatic musical instruments, buffing machines, etc. As shown in Figures 12-1 and 12-2, the motor consists essentially of a squirrel-cage rotor and two stator windings (a main winding and a starting winding). The main winding is connected across the supply line in the usual manner and has a low resistance and a high inductance. The starting or auxiliary winding, which is physically displaced in the stator from the main winding, has high resistance and a low inductance. This physical displacement, in addition to the electrical phase displacement produced by the relative electrical resistance values in the two windings, produces a weak rotating field that is sufficient to provide a low starting torque. The characteristics of a resistance-type, split-phase motor are shown in Figure 12-3.

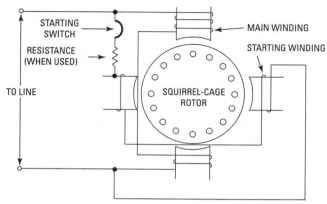

Figure 12-1 Schematic diagram of a split-phase induction motor.

After the motor has accelerated to 75 or 80 percent of its synchronous speed, a starting switch (usually centrifugally operated) opens its contacts to disconnect the starting winding. The function of the starting switch (after the motor has started) is to prevent the motor from drawing excessive current from the line and to protect the starting winding from damage due to heating. The motor may

Figure 12-2 Split-phase induction motor. *(Courtesy Leeson)*

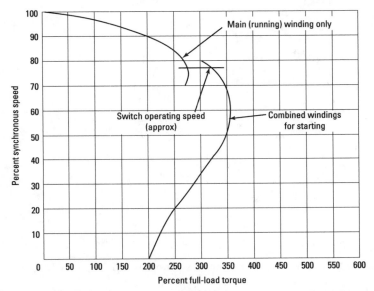

Figure 12-3 Speed-torque characteristics of a split-phase induction motor.

be started in either direction by reversing the connections to *either* the main or the auxiliary winding, *but not to both*.

Starting Switches for AC Motors
Only the AC single-phase motors require centrifugally operated starting switches. Centrifugal force from the motor causes a cone or a slider on the shaft of the motor to be thrown outward from the windings. This movement is harnessed to cause a switch to open. Once the motor is turned off, the switch will close again when the cone or sliding device comes to rest near the windings or closer to the rotor.

Figure 12-4 shows one way of representing the centrifugally operated switch. Note the four connections to the windings of the motor. Trace the connections from 2 to 3 to see how the switch is placed in the circuit with the starting coils.

Figure 12-5 shows an older type of rotor with a centrifugal device mounted on the shaft. It throws out the arms when it comes up to speed. These arms operate the pressure-sensitive switch mounted on the frame of the motor. Figure 12-6 shows the newer type rotor with sliding cone operating switch. Figure 12-7 shows a

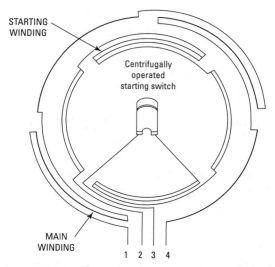

STARTING
WINDING

Centrifugally
operated
starting switch

MAIN
WINDING

1 2 3 4

Figure 12-4 Centrifugally operated starting switch.

different arrangement where springs are used to control the amount of movement in the switch mechanism.

Figure 12-8 is a switch operated by the start mechanism shown previously. Note how the spring action of the switch mechanism can cause the start cone on the rotor to snap back in place once the motor has been turned off and has to come to a resting position (Figure 12-9).

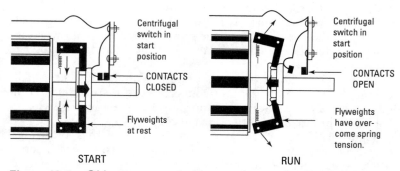

Centrifugal
switch in
start
position

CONTACTS
CLOSED

Flyweights
at rest

START

Centrifugal
switch in
start
position

CONTACTS
OPEN

Flyweights
have over-
come spring
tension.

RUN

Figure 12-5 Older-type rotor with centrifugally operated switch mechanism on shaft.

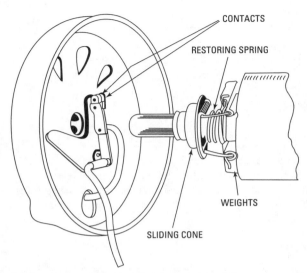

CONTACTS

RESTORING SPRING

WEIGHTS

SLIDING CONE

Figure 12-6   Newer-type rotor with sliding cone operating switch.

Figure 12-7   Location of the centrifugal switch mechanism on a more recent motor model.

285

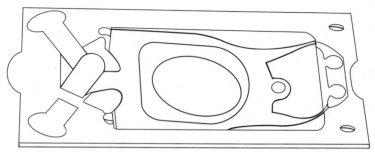

Figure 12-8 Another centrifugally operated switch. Contacts are located on the left. The rest of the metal part serves as a spring to force the start mechanism back toward the rotor when the motor stops.

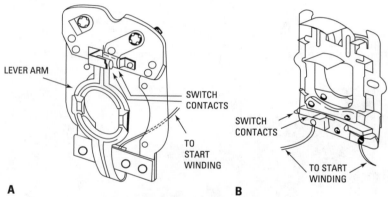

LEVER ARM

SWITCH CONTACTS

TO START WINDING

SWITCH CONTACTS

TO START WINDING

A   **B**

Figure 12-9 (A) Centrifugal switch unit. (B) Another type of centrifugal switch unit.

Switch Designs
These switches take a variety of shapes. They are designed by the motor manufacturer to fit a particular motor, so every manufacturer has a different design. Figure 12-10 shows just five of the many shapes used as replacement switches.

Newer Switches
Modern motors used for clothes washers and dryers are designed with the start switch outside of the motor, so that they can be replaced when they short or opened with little or no effort on the part of the service technician. Figure 12-11 shows a split-phase motor with the start switch mounted on the outside of the motor

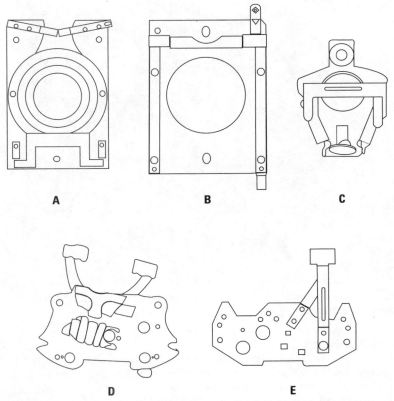

Figure 12-10   Replacement switches for various makes of general-purpose, split-phase, and capacitor-start motors.

frame. The centrifugal force causes the cone to move outward on the shaft. As it moves outward, it puts pressure on a spring-like arm that operates the switch located inside the enclosure sitting on top of the motor.

Resistance-Start Motors

A *resistance-start motor* is a form of split-phase motor that has a resistance connected in series with the auxiliary winding. The auxiliary circuit is opened by a starting switch when the motor has attained a predetermined speed.

Figure 12-11 Centrifugally operated switch located on the outside of the motor frame.

Reactor-Start

A *reactor-start motor* is a form of split-phase motor designed for starting with a reactor in series with the main winding. The reactor is short-circuited, or otherwise made ineffective, and the auxiliary (starting) circuit is opened when the motor has attained a predetermined speed. A circuit arrangement for this type of motor is shown in Figure 12-12. The function of the reactor is to reduce the starting current and to increase the angle of lag of the main-winding current behind the voltage. This motor will develop approximately the same torque as the split-phase motors discussed previously. The centrifugally operated starting switch must be of the single-pole, double-throw type for proper functioning.

Capacitor-Start Motors

The *capacitor-start* motor is another form of split-phase motor that has a capacitor connected in series with the auxiliary winding. The auxiliary circuit is opened when the motor has attained a predetermined speed. The circuit in Figure 12-13 shows the winding arrangement.

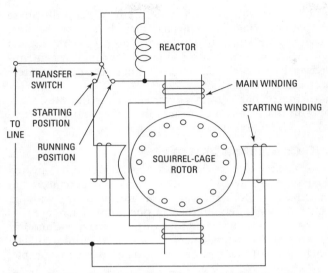

Figure 12-12   Schematic diagram of a reactor-start motor.

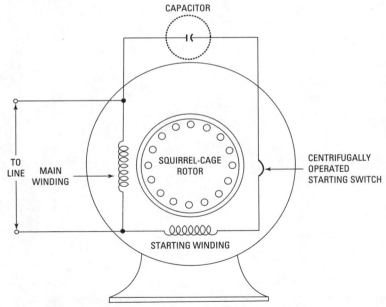

Figure 12-13   Schematic diagram of a capacitor-start motor.

The rotor is of the squirrel-cage type, as in other split-phase motors. The main winding is connected directly across the line, while the auxiliary or starting winding is connected through a capacitor, which is connected into the circuit through a centrifugally operated starting switch. The two windings are approximately 90° apart electrically.

This type of motor has certain advantages over the previously described types in that it has a considerably higher starting torque accompanied by a high power factor.

Permanent-Capacitor Motors

A *permanent-capacitor motor* has its main winding connected directly to the power supply and the auxiliary winding connected in series with a capacitor. Both the capacitor and auxiliary winding remain in the circuit while the motor is in operation. There are several types of permanent-capacitor motors, differing from one another mainly in the number and arrangement of capacitors employed. The running characteristics of this type of motor are extremely favorable,

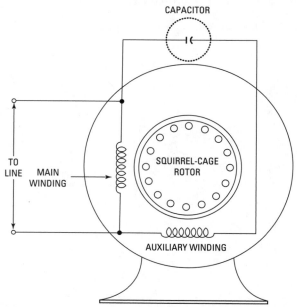

CAPACITOR

TO LINE   MAIN WINDING

SQUIRREL-CAGE ROTOR

AUXILIARY WINDING

Figure 12-14   Schematic diagram of a permanent-capacitor, split-phase motor.

and the torque is fixed by the amount of additional capacitance, if any, added to the auxiliary winding during starting.

The simplest of this type of motor is the low-torque, permanent-capacitor motor shown in Figure 12-14, in which a capacitor is permanently connected in series with the auxiliary winding. This type of motor can be arranged for an adjustable speed by the use of a tapped winding or an autotransformer regulator.

High-torque motors are usually provided with one running and one starting capacitor connected as shown in Figure 12-15, or with an autotransformer connected to increase the voltage across the capacitor during the starting period, as indicated in Figure 12-16.

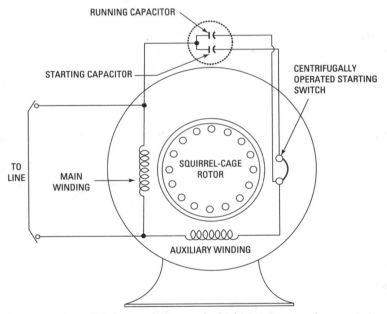

Figure 12-15   Schematic diagram of a high-torque capacitor motor.

Repulsion Motors

Repulsion motors have a stator winding arranged for connection to the source of power and a rotor winding connected to a commutator. As shown in Figure 12-17, brushes on the commutator are short-circuited and are placed so that the magnetic axis of the rotor winding is inclined to the magnetic axis of the stator winding. This type of motor has a varying speed characteristic.

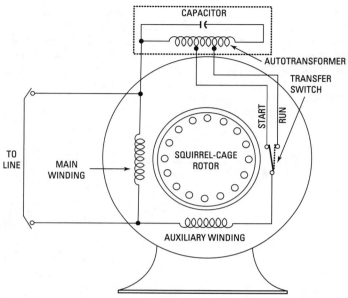

Figure 12-16   Schematic diagram of a high-torque capacitor motor using an autotransformer.

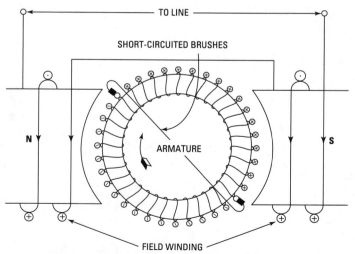

Figure 12-17   Simplified schematic of a series repulsion motor.

Principally, it has a stator like that of a single-phase motor, but has a rotor like the armature of a DC motor, with the opposite brushes on the armature short-circuited, that is, connected by a connector with a negligible resistance.

The brushes are placed so that a line connecting them makes a small angle with the neutral axis of the magnetic field of the stator. The stator induces a current in the armature; the current produces an armature field with poles in the neighborhood of the brushes. These armature fields have the same polarity as the adjacent field poles, and are repelled by them so that this repulsion causes the armature to revolve. The motor derives its name from this action.

A repulsion motor of this original type has characteristics (Figure 12-18) similar to those of the series motor. It has a high starting torque and moderate starting current. It has a low power factor, except at high speeds. For this reason, it is often modified into the *compensated repulsion motor*, which has another set of brushes placed midway between the short-circuited set and connected in series with the stator winding.

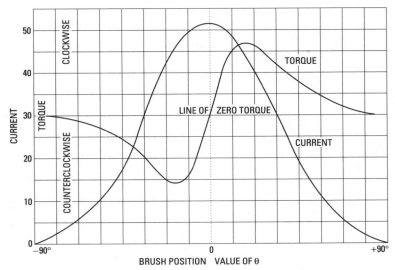

Figure 12-18 Characteristic curves of a repulsion motor showing the effect of shifting brushes on current and torque.

Repulsion-Start Induction Motors

A *repulsion-start induction motor* is a single-phase motor that has the same windings (Figure 12-19) as a repulsion motor. At a predetermined speed, however, the rotor winding is short-circuited, or otherwise connected, to give the equivalent of a squirrel-cage winding. This type of motor starts as a repulsion motor, but runs as an induction motor with constant-speed characteristics.

The repulsion induction motor has a single-phase distributed-field winding, with the axis of the brushes displaced from the axis

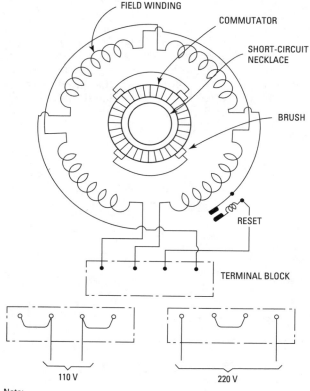

Note:
Top terminal shows connection to motor windings only. For proper connection to AC circuit check voltage source and connect as indicated on lower terminals.

Figure 12-19   Coil relationship in a single-phase, repulstion-start induction motor.

of the field winding. The armature has an insulated winding. The current induced in the armature, or rotor, is carried by the brushes and commutator, resulting in a high starting torque. When nearly synchronous speed is attained, the commutator is short-circuited, so that the armature is then similar in function to a squirrel-cage armature.

The examples in Figures 12-20 and 12-21 show the working principles of the mechanism for simultaneously lifting the brushes and short-circuiting the commutator to change the operation from repulsion to induction. The object of lifting the brushes is to eliminate wear on the commutator during the running periods, since it makes no difference electrically whether the brushes are in contact or not after the motor comes up to speed.

This type of motor has gone through many stages of improvement since its first appearance on the market, although its general principle has remained the same. The general reliability of this type of motor is largely governed by the reliability of the short-circuiting mechanism. For this reason, it has been the constant aim of engineers to improve on the principle and construction of the short-circuiting switch.

Since the motor starts on the repulsion principle, it has the same starting characteristics as the repulsion motor described pre-

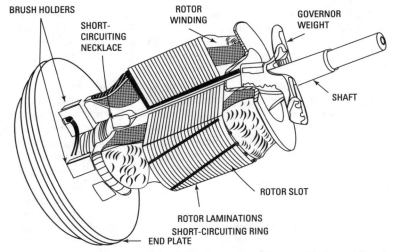

Figure 12-20 The position of the short-circuiting necklace and brush mechanism while the motor is starting.

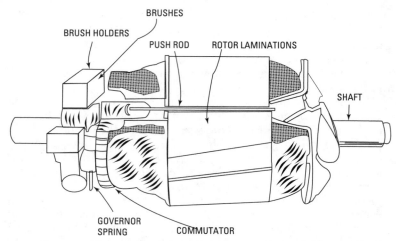

BRUSHES

BRUSH HOLDERS

PUSH ROD   ROTOR LAMINATIONS

SHAFT

GOVERNOR
SPRING   COMMUTATOR

Figure 12-21   The position of the short-circuiting necklace and brush mechanism after the governor weights have operated.

viously, namely, high starting torque and low starting current. As the motor speeds up, the torque falls off rapidly. At some point on the speed-torque curve, usually at about 80 percent of synchronization, the commutator is automatically short-circuited, producing the effect of a cage winding in the armature, and the motor comes up to speed as an induction motor. After the commutator has been short-circuited, the brushes do not carry current and therefore may be lifted from the commutator, but lifting is not absolutely necessary.

The curve in Figure 12-22 shows the speed-torque characteristics of a typical repulsion-start, induction motor. The short-circuiting mechanism operates at point A, at which time the induction-motor torque is greater than the repulsion-motor torque. This means that if the repulsion winding has sufficient torque to bring the load up to this speed, there will be sufficient torque as an induction motor to bring the load up to full speed. The higher the speed at which the short-circuiting mechanism operates, the lower will be the induction-motor current at that point, and, consequently, the less disturbance to the line. After the commutator has been short-circuited, the motor has the same characteristics as the single-phase induction motor previously described.

If the short-circuiting mechanism operates before the repulsion curve crosses the induction-motor curve, and if the torque of the

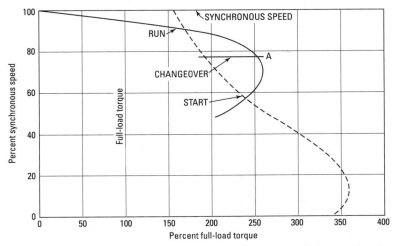

Figure 12-22 Speed-torque characteristics of a typical repulsion-start induction motor.

induction motor is less than that required to accelerate the load, the motor may slow down until the short circuit is removed from the commutator, in which case the motor will again operate on the repulsion principle. The armature will then speed up until the commutator is again short-circuited, after which the armature will slow down until it again becomes repulsion. This cycle will be repeated over and over again until some changes take place.

The efficiency and the maximum running torque of the repulsion-start induction motor are usually less than those of a cage-wound induction motor of comparative size. In other words, a repulsion-start induction motor must be larger than a cage-wound motor of the same rating to give the same performance.

Repulsion-Induction Motors

A *repulsion-induction motor* is another form of repulsion motor that has a squirrel-cage winding in the rotor in addition to the repulsion winding. A motor of this type may have either a constant-speed or a varying-speed characteristic. Specifically, this motor is a combination of the repulsion and induction types and operates on the combined principles of both repulsion and induction. It is sometimes termed a squirrel-cage repulsion motor. In this motor, the desirable starting characteristics of the repulsion motor and the constant-speed characteristic of the induction motor are obtained. It is, of course, impossible to combine the two types of motors and obtain only the desirable characteristics of each.

As shown in Figures 12-23 and 12-24, the field has the same type of windings used in the repulsion-start induction motor, and the armature has two separate and independent windings. They are

- Squirrel-cage winding
- Commutated winding

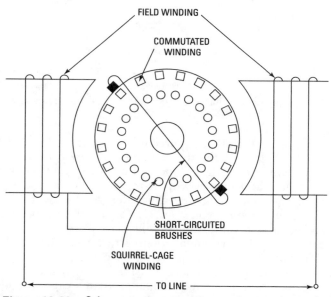

Figure 12-23 Schematic diagram of a repulsion-induction motor.

Both of these armature windings function during the entire period of operation of the motor. There are no automatic devices, such as the starting switch of the split-phase motor or the short-circuited device of the repulsion-start induction motor.

The cage winding is located in slots below those that contain the commutated winding. The slots that contain the two windings may or may not be connected by a narrow slot. Usually, there are the same number of slots for the two windings. It is not, however, absolutely essential that they be the same. Because of its construction,

Figure 12-24 (Top) Note location of shorting rings. (Bottom) Three-phase wound rotor. Note the three rings.

the squirrel-cage winding inherently has a high inductance. Its reactance, with the armature at rest, is therefore high.

The commutated winding has a low reactance, with the result that most of the current will flow in this winding. The ideal condition at starting would be for all of the flux to pass beneath the commutated winding and none of it beneath the cage winding. If this condition could be obtained, this motor would have the same starting characteristics as the repulsion-start induction motor. However, at full-load speed, which is slightly below synchronization, the reactance of the cage winding is low, and most of the mutual flux passes beneath the case winding. Both windings produce torque, and the output of the motor is the combined output of the cage winding and the commutated winding.

The commutation of the motor is good at all speeds. The no-load speed is above synchronization and is limited by the combined effect of the field winding on the commutated winding and the cage winding and by the action of the two armature windings on each other. At synchronous speed, a squirrel-cage motor has no torque. At synchronous speed, and for a short distance above synchronous speed, the torque of a repulsion-induction motor is greater than that of the commutated winding alone. This shows that because of the interaction between the two armature windings, the squirrel cage supplies torque instead of acting as a brake.

At full-load speed, and up to about the maximum running torque point, the torque of this motor is greater than the sum of the torques of the cage winding and the commutated winding. The inherent locked-torque curve of a repulsion-induction motor, shown in Figure 12-25, is similar to that of the repulsion motor. At

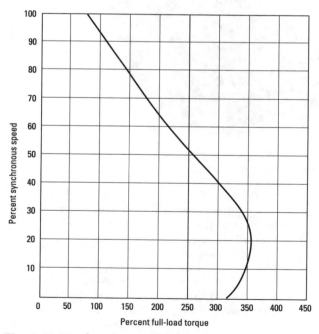

Figure 12-25 Speed-torque characteristic of a typical repulsion-induction motor.

soft neutral, the primary winding carries the squirrel-cage current in addition to the exciting current of the motor.

Since its starting current is low, the repulsion-induction motor may be operated from lighting circuits when driving frequently started devices. The repulsion-induction motor is especially suitable for such applications as household refrigerators, water systems, garage air pumps, gasoline pumps, compressors, and so on.

Shaded-Pole Motors

A *shaded-pole motor* is a single-phase induction motor equipped with an auxiliary winding displaced magnetically from, and connected in parallel with, the main winding. This type of motor is manufactured in fractional-horsepower sizes and is used in a variety of household appliances, such as fans, blowers, hair driers, and other applications that require a low starting torque. It is operated only on alternating current, usually nonreversible, low in cost, and extremely rugged and reliable. The diagram of a typical shaded-pole motor is shown in Figure 12-26.

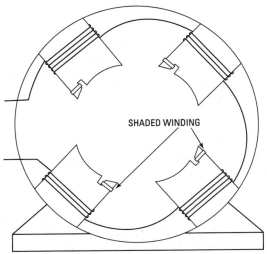

Figure 12-26   A single-phase motor with shading coils for starting.

The shading coil (from which the motor has derived its name) consists of low-resistance copper links, embedded in one side of each stator pole, that are used to provide the necessary starting torque. When the current increases in the main coils, a current is induced in the shading coil. This current opposes the magnetic field building up in the part of the pole pieces they surround. This produces the condition shown in Figure 12-27, where the flux is crowded away from that portion of the pole piece surrounded by the shading coil.

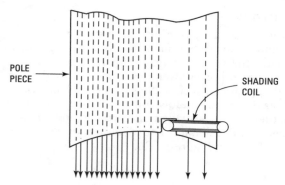

Figure 12-27 Action of a shading coil in an AC motor.

When the main-coil current decreases, the current in the shading coils also decreases until the pole pieces are uniformly magnetized. As the main-coil current and the magnetic flux of the pole piece continue to decrease, the current in the shading coils reverses and tends to maintain the flux in part of the pole pieces. When the main-coil current drops to zero, current still flows in the shading coils to give the magnetic effect that causes the coils to produce a rotating or magnetic field that makes the motor self-starting.

Universal Motors

A *universal motor* is a series-wound or compensated, series-wound motor that may be operated either on direct current or on single-phase alternating current at approximately the same speed and output. These conditions must be met when the DC and AC voltages are approximately the same and the AC frequency is not greater than 60 hertz.

Universal motors are commonly manufactured in fractional-horsepower sizes, and are preferred because of their use on either AC or DC, particularly in areas where power companies supply both types of current.

As previously noted, all universal motors are series wound; therefore, their performance characteristics are very much like those of the usual DC series motor. The no-load speed is quite high, but seldom high enough to damage the motor, as is the case with larger DC series motors. When a load is placed on the motor, the speed decreases and continues to decrease as the load increases. Although universal motors of several types of construction are manufactured, they all have the varying-speed characteristics just mentioned.

Because of the difficulty in obtaining similar performance on AC and DC at low speeds, most universal motors are designed for operation at speeds of 3500 rpm and higher. Motors operating at a load speed of 8000 to 10,000 rpm are common. Small stationary vacuum cleaners and the larger sizes of portable tools have motors operating at 3500 to 8000 rpm.

The speed of a universal motor can be adjusted by connecting a resistance of proper value in series with the motor. The advantage of this characteristic is obvious in an application such as a motor-driven sewing machine, where it is necessary to operate the motor over a wide range of speeds. In such applications, adjustable resistances by which the speed is varied at will are used.

When universal motors are to be used for driving any apparatus, the following characteristics of the motor must be considered:

- Change in speed with change in load
- Change in speed with change in frequency of power supply
- Change in speed due to change in applied voltage

Since most small motors are connected to lighting circuits, where the voltage conditions are not always the best, this last item is of the utmost importance. This condition should also be kept in mind when determining the proper motor to use for any application, regardless of type. In general, the speed of the universal motor varies with the voltage. The starting torque of universal motors is usually much more than required in most applications and not to be considered.

Universal motors are manufactured in two types. They are

- Concentrated-pole, noncompensated
- Distributed-field, compensated

Most motors of low-horsepower rating are of the concentrated pole, noncompensated type, while those of higher ratings are of the distributed-field, compensated type. The dividing line is approximately ¼ hp (186.5 W), but the type of motor to be used is determined by the severity of the service and the performance required. All of the motors have wound armatures that are similar in construction to an ordinary DC motor.

The concentrated-pole, noncompensated motor is exactly the same in construction as a DC motor except that the magnetic path is made up of laminations. The laminated stator is made necessary because the magnetic field is alternating when the motor is operated on alternating current. The stator laminations are punched, with the poles and the yoke in one piece.

The compensated type of motor has stator laminations of the same shape as those in an induction motor. These motors have stator windings in one of two different types.

The noncompensated motor is simpler and less expensive than the compensated motor and would be used over the entire range of ratings if its performance were as good as that of the compensated motor. The noncompensated type is used for the higher speeds and lower horsepower ratings only. Figures 12-28 and 12-29 show the

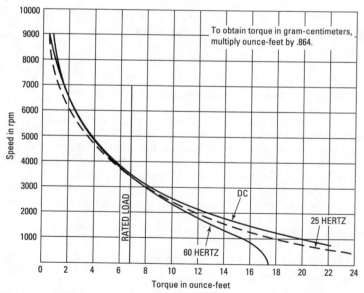

Figure 12-28 Speed-torque characteristics of a typical ¼ hp (186.5 W), 3400 rpm, compensated universal motor.

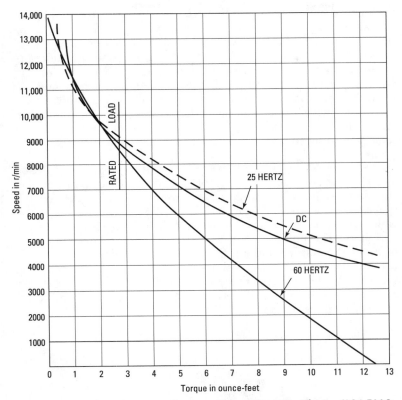

Figure 12-29   Speed-torque characteristics of a typical ¼ hp (186.5 W), 8700 rpm, noncompensated universal motor.

speed-torque curves for a compensated and noncompensated motor, respectively. It can be noted in Figure 12-28 that, although the rated speed is relatively low for a universal motor, the speed torque curves for the various frequencies lie very close together, up to 50 percent above the rated torque load.

In Figure 12-29, the performance of a much-higher-speed, noncompensated motor is shown. For most universal-motor applications, the variation in speed at rated loads, as shown on this curve, is satisfactory. However, the speed curves separate rapidly above full load. If this motor had been designed for a lower speed, the tendency of the speed-torque curves to separate would have been more pronounced. The chief cause of the difficulty in keeping the speeds the same is the reactance voltage that exists when the motors are operated on AC. Most of this reactance voltage is produced in the field windings by the main working field. However, in the noncompensated motor, some of it is produced in the armature winding by the field produced by the armature ampere-turns. The true working voltage is obtained by subtracting the reactance voltage vectorially from the line voltage. If the reactance is high, the performance at a given load will be the same as if there were no reactance voltage and as if the applied voltage had been reduced, with consequent reduction in speed.

Reluctance-Synchronous Motors

The *reluctance-synchronous motor* is really a variation on the classic squirrel-cage rotor. The rotor is modified to provide equally spaced areas of high reluctance (Figure 12-30). Note how the rotor is designed (Figure 12-31). Notches or flats are placed in the rotor periphery. The number of notches corresponds to the number of poles in the stator winding. Salient poles are the sections of the rotor periphery between the high-reluctance areas. These poles create a relatively low reluctance path for the stator flux. They are attracted to the poles of the stator field. The stator field rotates at the synchronous speed.

Figure 12-30   Reluctance-synchronous motor rotor. *(Courtesy Bodine)*

The reluctance-synchronous rotor starts and accelerates like a regular squirrel cage rotor. But as this rotor approaches the rotational speed of the field, a critical point is reached. This is where an increased acceleration takes place and the rotor "snaps" into synchronization with the stator field.

Figure 12-31   Cutaway view of a reluctance-synchronous motor.
(Courtesy Bodine)

A load that is too great will prevent the rotor from pulling in and synchronizing with the rotating field in the stator. Too great a load will cause the motor to operate roughly and produce a non-uniform operation.

A load on the reluctance synchronous motor produces an effect on the magnetic field. The magnetic field of lines coupling the rotor to the stator field is stretched by the application of a load. This increases the coupling angle and continues until eventually the coupling between the rotor and stator breaks. This causes the rotor to pull out of synchronization.

Reluctance-synchronous motors may be designed for operation on three-phase and two-phase power sources. They may also be designed for single-phase operation. In single-phase operation they may take the split-phase, capacitor-start, or permanent-split capacitor configurations. They have the same characteristics as these other nonsynchronous motors.

Note in Figure 12-30 the skew of a rotor that aids in providing improved smoothness of operation when compared to capacitor-start and split-phase motors.

Hysteresis-Synchronous Motors

The *hysteresis-synchronous motor* has a rotor that is made of a heat-treated, cast, permanent-magnet alloy cylinder. It has a non-magnetic support securely mounted to the shaft (Figure 12-32).

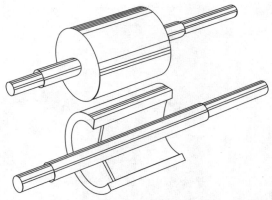

Figure 12-32   Hysteresis-synchronous motor rotor views. Solid and cutaway views showing the supported cylinder. *(Courtesy Bodine)*

The stator is wound much like that of the conventional squirrel-cage motor. The rotor design gives the hysteresis-synchronous motor its characteristics.

This type of motor starts on the hysteresis principle and accelerates at a fairly constant rate until it reaches the synchronous speed of the rotating field. Instead of the permanently fixed poles found in the rotor of the reluctance-synchronous design, the hysteresis-rotor poles are *induced* by the rotating magnetic field. During the acceleration period, the stator field will rotate at a speed faster than the rotor. The poles that it induces in the rotor will shift around its periphery. When the rotor speed reaches that of the magnetic field, the rotor poles will take a fixed position.

The coupling angle in this motor is not rigid. As the load increases beyond the capacity of the motor, the poles on the periphery of the rotor core will shift as the rotor speed slips below that of the field. This means they take up new positions. If the load is then reduced to the *pull-in* capacity of the motor, the poles will take up fixed positions until the motor is overloaded once again or stopped and restarted.

The hysteresis motor has the ability to *lock-in* at an infinite number of positions with respect to the stator field. The reluctance rotor locks-in only at the same number of positions as it had poles.

One of the advantages of the hysteresis motor is its ability to pull into synchronization any load that is within its capacity to start and accelerate. It will do this with a more gradual characteristic than other motors.

Advantages

Synchronous motors are known for their constant speed characteristic. They will operate at a fixed speed that is determined by the number of stator poles and the frequency of the power supply. The hysteresis-synchronous motor has a uniform acceleration characteristic. It can pull into synchronization any load that is within its ability to start and accelerate.

Disadvantages

The reluctance motor requires increased acceleration of the rotor at the critical point when it approaches the rotational speed of the field. That means it is possible that while the reluctance motor may start a high-inertia load, it may not be able to accelerate the load enough to pull it into synchronization with the rotating field.

If the reluctance motor does not synchronize, it will continue to operate, but will operate as an ordinary induction motor. This means it will operate at low efficiency and at a very irregular angular

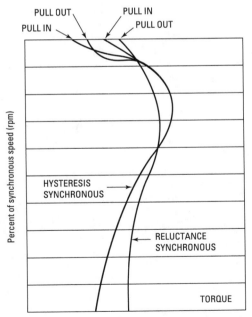

Figure 12-33 Comparison of typical speed curves for hysteresis- and reluctance-synchronous motors of identical frame size. *(Courtesy Bodine)*

velocity. This irregular angular velocity is easily detected by the pounding noise the motor produces. As you can see, it is important that the motor is not overloaded so it can come up to speed and lock-in on the rotating field.

Synchronous motors are usually larger and more costly than nonsynchronous motors. Therefore, they should be chosen for jobs where the load needs to be driven at the exact speed of the rotating field and design speed of the motor.

Figure 12-33 shows the comparison of typical speed curves for hysteresis- and reluctance-synchronous motors of identical frame size.

Speed Control

The control method used in single-phase motors depends on the type of motor selected for a certain application. Speed control is usually accomplished by means of a centrifugal switch on motors that have speeds above 900 rpm. A comparison of typical speed-torque characteristics of fractional-horsepower, single-phase motors is given in Figure 12-34.

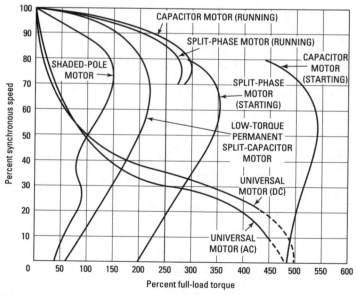

Figure 12-34 Speed-torque characteristics of common types of fractional-horsepower motors.

Where speeds of 900 rpm and below are involved, capacitor motors utilizing voltage relays for changeover are preferable to motors using centrifugal switches because of the sluggishness of a centrifugal switch of normal design at the lower speeds. A voltage relay for changeover is not subject to the same limitations.

Speed control, because of its more general use, has brought about the standardization of fan speeds, permitting the use of high-slip motors. A squirrel-cage motor with 8- to 10-percent slip and with low maximum torque operates satisfactorily, not only for constant-speed drives, but also for adjustable-speed drives through the use of voltage control on the primary winding. Within the working range of a high-slip, squirrel-cage motor that has adjustable voltage applied to the primary winding, the speed-torque characteristics resemble the characteristics of a wound-rotor motor that has different amounts of external resistance in the secondary circuit.

A tapped autotransformer connected to the motor through a multiposition snap switch offers the simplest form of control. The transformer may have a number of taps, only two or three of which are brought out to the switch. This provides two or three speeds, any or all of which may be changed to suit the needs of the application by the proper selection of taps to connect to the switch. Additional contacts on the switch make it a complete starter and speed regulator. Two such transformers connected in open-delta for three-phase, or in each phase of a two-phase circuit, may be used to control the speed of polyphase motors. For polyphase use, the snap switch must have duplicate contacts for the transformers.

In applying speed control by means of line-voltage adjustment, it must be remembered that the speed of the drive will vary with the load. Therefore, this method of control is not suitable for centrifugal fans, especially those with damper control, which affects the fan load. Two-speed, pole-changing motors offer a solution for centrifugal-fan drives, inasmuch as the speed on either pole combination is affected only slightly by the change in load. Motor speeds in the ratio of 2 to 1 are obtainable as 3600/1800 and 1800/900 rpm; 3 to 2 as 1800/1200 rpm; and 4 to 3 as 1200/900 rpm. Other pole combinations are possible.

There is still another means of obtaining speed reduction on induction motors. The equivalent of building the transformer into the motor is obtained by a suitable tap on the primary winding, as shown in Figure 12-35. By this means, normal speed and a single reduced speed are provided. A simple, pole-changing controller is

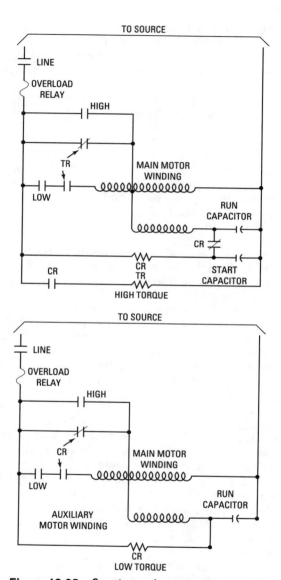

Figure 12-35 Starting and running circuits of low-torque and high-torque, adjustable-speed, capacitor motors with tapped main windings.

all that is required to complete the installation. Such motors are not generally available, but are provided through propeller-fan manufacturers. Each motor is designed for the characteristics of the fan it is to drive, and the tap on the winding is located to give the desired low speed.

When speed controllers are used with low-torque capacitor motors, it must be remembered that the available 50-percent starting torque, with full-voltage applied, is reduced in proportion to the square of the voltage involved. The result is low breakaway and low accelerating torque on the low-speed setting. In manually operated control, one remedy for this limitation is a progressive starting switch that provides full voltage on the first (high-speed) position and reduced voltage on the intermediate and low-speed positions (Figure 12-36).

Where temperature control is used to start and stop the motor, a preset, full-voltage starting controller is usually provided. In this case, a relay is energized by the rising voltage of the motor auxiliary winding as it accelerates; the relay disconnects it to the transformer tap for which the controller is set. In high-torque motors, this relay serves to disconnect the starting capacitor and to change over from full to reduced voltage on the motor.

Under the general subject of adjustable-speed fan motors, the amount of speed reduction is often debated, and although it is generally agreed that 30- to 35-percent speed reduction will meet almost any air-conditioning requirement, as much as 50-percent speed reduction is sometimes specified. In most cases, a lower percentage of speed reduction is found to be acceptable. It is fairly easy to obtain 50-percent reduction on a fan if the characteristic curve follows the law of horsepower load, increasing as the cube of the speed, and if the fan fully loads the motor at full speed. It is quite difficult to obtain 50-percent speed reduction with stability in cases where the drive is 20 to 25 percent over-motored.

In considering the wide range of air-conditioning equipment, from unit heaters and coolers to barn ventilators and incubator ventilating fans, with the many types of equipment between these extremes, the foremost impression is one of economy coupled with simplicity and safety. Simplicity promotes economy. Any additional expense imposed in the interest of safety is abundantly justified from the standpoint of good operating practice. It is not advisable to use a motor of the "long annual service" classification where the operation is infrequent, if a motor for short annual service will meet the requirements at a lower first-cost with no

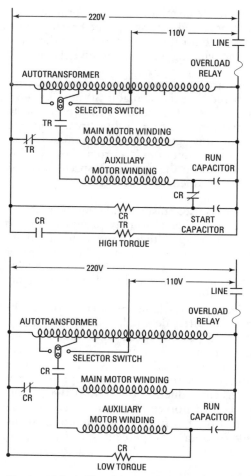

Figure 12-36 Starting and running circuits of low-torque and high-torque, adjustable-speed capacitor motors with manually operated transformer speed regulator.

appreciable increase in operating cost. Similarly, a fractional-horsepower, split-phase motor is less expensive than a capacitor motor.

It is essential to avoid misapplications such as low-torque motors on belted drives or tapped-winding and transformer speed control in place of multispeed motors for centrifugal drives. The drive should always be closely motored with the correct size and type of motor to give the best results.

Application

Although single-phase motors are most commonly used in fractional-horsepower sizes, certain applications and power-supply conditions make use of single-phase motors in integral horsepower sizes. For extremely small capacities up to $\frac{1}{40}$ hp (18.7 W), the shaded-pole type of motor is most frequently used. It provides sufficient torque for fans, blowers, and other similar equipment, and the starting current is not objectionable on lighting lines.

The split-phase motor is the most commonly used in sizes ranging from $\frac{1}{30}$ hp (24.9 W) to $\frac{1}{2}$ hp (373 W), particularly for fans and similar drives where the starting torque is low. In this type of motor, a built-in centrifugal switch is usually provided to disconnect the starting winding as the motor comes up to speed.

In sizes above $\frac{1}{4}$ hp (186.5 W), a capacitor motor with 300 percent starting torque may be used to advantage, especially for pump and compressor drives. It may also be used in the overlapping ratings for fan drives in the lower capacities at higher speeds. However, it does not offer significant advantages over the split-phase motor to warrant its higher cost.

At low speeds (below 900 rpm, where centrifugal switches are less successful), and in ratings from $\frac{3}{4}$ hp (0.5595 kW) to 3 hp (2.238 kW) at all speeds, the capacitor motor finds its widest field of application in fan drives. The running characteristics of this motor are extremely good, and the starting torque is fixed by the amount of additional capacitance added to the auxiliary motor winding during the starting period.

A low-torque capacitor motor, in which no capacitance is added to the auxiliary winding during the starting period, provides approximately 50-percent starting torque. This is considered sufficient for directly connected fans, if the unit is one of constant speed or is always started on the high-speed position of the starting switch. Where the fan is coupled or belted to the motor, the high-torque type of motor with the additional starting capacitance is preferable. The changeover from start to run of the high-torque type may be accomplished either through the use of a centrifugal switch or by electrical means responsive to current decay or voltage rise as the motor approaches normal running speed.

The repulsion-induction and the repulsion-start motors have ratings that parallel those of the capacitor types. These motors have commutators to provide the starting torque. Under normal running conditions, the commutator of the repulsion-start, induction-run type is short-circuited, and the motor operates as an induction motor. This differs from the repulsion-induction motor in which the commutator is not short-circuited. A squirrel-cage winding deep in the rotor is inactive at starting, but takes up the load as the rotor accelerates to full speed, where the normal load is almost equally divided between the repulsion and the squirrel-cage winding. The starting efficiency of both types is high, and the 300 percent or more starting torque that is available makes them suitable for compressor drives.

The universal-type motor, because of its ability to run on DC as well as on AC, is preferred on appliances that must operate on either an AC or a DC circuit. Popular applications for the universal-type motor are in vacuum cleaners, sewing machines, portable drills, saws, routers, home motion-picture projectors, and business machines of all kinds.

Whenever it becomes necessary to substitute a motor on any appliance, a good deal of grief and disappointment may be avoided if a motor of the identical size and type is reinstalled. Particular attention should be observed with respect to the nameplate data that gives all the necessary information. A sound rule to follow is to copy the entire nameplate reading. This observation holds true whether it is a complete motor or only a spare part.

Table 12-1 provides information needed to wire the motor properly so that overheating due to low voltage does not occur.

Table 12-1   Individual Branch Circuit Wiring for Single-Phase Induction Motors

| Motor Data | | Copper Wire Size (Minimum AWG No.) | | | | |
| | | Branch Circuit Length | | | | |
Hp	Volts	0–25 ft	50 ft	100 ft	150 ft	200 ft
⅙	115	14	14	14	12	10
	230	14	14	14	14	14
¼	115	14	14	12	10	8
	230	14	14	14	14	14
⅓	115	14	12	10	8	6
	230	14	14	14	14	12

Table 12-1 (continued)

Motor Data		Copper Wire Size (Minimum AWG No.)				
		Branch Circuit Length				
Hp	Volts	0–25 ft	50 ft	100 ft	150 ft	200 ft
½	115	14	12	10	8	6
	230	14	14	14	14	12
¾	115	12	10	8	6	4
	230	14	14	14	12	10
1	115	12	10	8	6	4
	230	14	14	14	12	10
1½	115	10	10	6	4	4
	230	14	14	12	10	8
2	115	10	8	6	4	
	230	14	12	12	10	8
3	115	6	6	4		
	230	10	10	10	8	8
5	230	8	8	8	6	4

Table 12-2 shows symptoms, possible causes, and remedies for common problems with fractional horsepower motors.

Table 12-2   Fractional-Horsepower Motor Troubleshooting Chart

Split-Phase Induction Motors

Symptom and Possible Cause	Possible Remedy
Failure to Start	
(a) No voltage	(a) Check for voltage at motor terminals with test lamp or voltmeter. Check for blown fuses on meter. Check for blown fuses or open overload device in starter. If motor is equipped with a slow-blow fuse, see that the fuse plug is not open and that it is screwed down tight.
(b) Low voltage	(b) Measure the voltage at the motor terminals with the switch closed. Voltage should read within 10% of the voltage stamped on the motor nameplate. Overload transformers or circuits

(continued)

Table 12-2 (continued)

Split-Phase Induction Motors

Symptom and Possible Cause	Possible Remedy
	may cause low voltage. If the former, check with the power company. Overloaded circuits in the building can be found by comparing the voltage at the meter with the voltage at the motor terminals with the switch closed.
(c) Faulty cutout switch operation	(c) Cutout switch operation may be observed by removing the inspection plate in the front end bracket. The mechanism consists of a cutout switch mounted on the front end bracket, and a rotating part called the governor weight assembly, which consists of a Bakelite® disc that is supported in such a way that it is moved back and forth along the shaft by the operation of the governor weights. At stand-still, the disc holds the cutout switch closed. If the disc does not hold the switch closed, the motor cannot start. This may call for adjustment of the end-play washers. Dirty contact points may also keep the motor from starting. See that the contacts are clean. After the motor has accelerated to a predetermined speed, the disc is withdrawn from the switch, allowing it to open. With the load disconnected from the motor, close the starting switch. If the motor does not start, start it by hand and observe the operation of the governor as the motor speeds up, and also when the switch has been opened and the motor slows down. If the governor fails to operate, the governor weights may be clogged. If it operates too soon or too late, the spring is too weak or too strong. Remove motor to service shop for adjustment. Governor weights are set to operate at about 75% of synchronous speed. Place rotor in balancing machine and, with a tachometer, determine if the governor operates at the correct speed.
(d) Open overload device	(d) If the motor is equipped with a built-in micro switch, or similar overload device, remove the cover plate in the end bracket on which the switch is mounted and see if the switch contacts are closed. Do not attempt to adjust

Table 12-2   (continued)

Split-Phase Induction Motors

Symptom and Possible Cause	Possible Remedy
	this switch or to test its operation with a match. Doing so may destroy it. If the switch is permanently open remove the motor to the service shop for repairs.
(e) Grounded field	(e) If the motor overheats, produces shock when touched, or if idle watts are excessive, test for a field ground with a test lamp across the field leads and frame. If grounded, remove the motor to the service shop for repairs.
(f) Open-circuited field	(f) These motors have a main and a phase (starting) winding. Apply current to each winding separately with a test lamp. Do not leave the windings connected too long while rotor is stationary. If either winding is open, remove the motor to the repair shop for repairs.
(g) Short-circuited field	(g) If the motor draws excessive watts and at the same time lacks torque, over-heats, or hums, a shorted field is indicated. Remove to the service shop for repairs.
(h) Incorrect end play	(h) Certain types of motors have steel-enclosed cork washers at each end to cushion the end thrust. Too great an end thrust, hammering on the shaft, or excessive heat may destroy the cork washers and interfere with the operation of the cutout switch mechanism. If necessary, install new end-thrust cushion bumper assemblies. End play should not exceed 0.01 in. (0.254 mm); if it does, install additional steel end-play washers. End play should be adjusted so that the cutout switch is closed at standstill and open when the motor is operating.
(i) Excessive load	(i) This may be approximately determined by checking the ampere input with the nameplate marking. Excessive load may prevent the motor from accelerating to the speed at which the governor acts and cause the phase winding to burn up.
(j) Tight bearings	(j) Test by turning armature by hand. If adding oil does not help, bearings must be replaced.

(continued)

Table 12-2 (continued)

Split-Phase Induction Motors

Symptom and Possible Cause	Possible Remedy
Motor Overheats	
(a) Grounded field	(a) Test for a field ground with a test lamp between the field and motor frame. If grounded, remove the motor to the service shop for repair.
(b) Short-circuited field	(b) Test for excessive current draw, lack of torque, and presence of hum. Any of these symptoms indicates a shorted field. Remove the motor to the service shop for repair.
(c) Tight bearings	(c) Test by turning armature by hand. If oiling does not help, new bearings must be installed.
(d) Low voltage	(d) Measure voltage at motor terminals with switch closed. Voltage should be within 10% of nameplate voltage. Overloaded transformers or power circuits may cause low voltage. Check with power company. Overloaded building circuits can be found by comparing the voltage at the meter with the voltage at the motor terminals with the switch closed.
(e) Faulty cutout switch	(e) See Paragraph (c) under *Failure to Start*.
(f) Excessive load	(f) See Paragraph (i) under *Failure to Start*.
Excessive Bearing Wear	
(a) Belt too tight	(a) Adjust belt to tension recommended by manufacturer.
(b) Pulleys out of alignment	(b) Align pulleys correctly.
(c) Dirty, incorrect, or insufficient oil	(c) Use type of oil recommended by manufacturer.
(d) Dirty bearings	(d) Clean thoroughly. Replace worn bearings.
Excessive Noise	
(a) Worn bearings	(a) See paragraphs (a), (b), (c), and (d) under *Excessive Bearing Wear*.
(b) Excessive end play	(b) If necessary, add additional end-play washers.

Table 12-2 *(continued)*

Split-Phase Induction Motors

Symptom and Possible Cause	Possible Remedy
(c) Loose parts	(c) Check for loose hold-down bolts, loose pulleys, etc.
(d) Misalignment	(d) Align pulleys correctly.
(e) Worn belts	(e) Replace belts.
(f) Bent shaft	(f) Straighten shaft, or replace armature or motor.
(g) Unbalanced rotor	(g) Balance rotor.
(h) Burrs on shaft	(h) Remove burrs.

Motor Produces Shock

(a) Grounded field	(a) See paragraph (e) under *Failure to Start.*
(b) Broken ground strap	(b) Replace ground strap.
(c) Poor ground connection	(c) Inspect and repair ground connection.

Rotor Rubs Stator

(a) Dirt in motor	(a) Thoroughly clean motor.
(b) Burrs on rotor or stator	(b) Remove burrs.
(c) Worn bearings	(c) Replace bearings and inspect shaft for scoring.
(d) Bent shaft	(d) Repair and replace shaft or rotor.

Radio Interference

(a) Poor ground connection	(a) Check and repair any defective grounds.
(b) Loose contacts or connections	(b) Check and repair any loose contacts on switches or fuses, and loose connections on terminals.

Repulsion-Start, Induction, Brush-Lifting Motors

Failure to Start

(a) Fuses blown	(a) Check capacity of fuses. They should not be greater in ampere capacity than recommended by the manufacturer, and in no case smaller

(continued)

Table 12-2 *(continued)*

Split-Phase Induction Motors

Symptom and Possible Cause	Possible Remedy
	than the full-load ampere rating of the motor or with a voltage capacity equal to or greater than the voltage of the supply circuit.
(b) No voltage or low voltage	(b) Measure voltage at motor terminals with switch closed. See that it is within 10% of the voltage stamped on the nameplate of the motor.
(c) Open-circuited field or armature	(c) Indicated by excessive sparking in starting, refusal to start at certain positions of the rotor, or by a humming sound when the switch is closed. Examine for broken wires, loose connections, or burned segments on the commutator at the point of loose or broken connections. Inspect the commutator for a foreign metallic substance that might cause a short between the commutator segments.
(d) Incorrect voltage or frequency	(d) Requires new motor built for operation on local power supply. DC motors will not operate on AC circuit, or vice versa.
(e) Worn or sticking brushes	(e) When brushes are not making proper contact with the commutator, the motor will have a weak starting torque. This can be caused by the worn brushes, brushes sticking in holders, weak brush springs, or a dirty commutator. The commutator should be polished with fine sandpaper (never use emery). The commutator should never be oiled or greased.
(f) Improper brush setting	(f) Unless a new armature has been installed, the brush holder or rocker arm should be opposite the index and locked in position. If a new armature has been installed, the position may be slightly off the original marking.
(g) Improper line connection	(g) See that the connections are made according to the connection diagram sent with the motor. The motor may, through error, be wired for a higher voltage.
(h) Excessive load	(h) If the motor starts with no load, and if all the foregoing conditions are satisfactory, then failure to start is most likely due to an excessive load.

Table 12-2 (continued)

Split-Phase Induction Motors

Symptom and Possible Cause	Possible Remedy
(i) Shorted field	(i) Take separate current readings on each of the two halves of the stator winding. Unequal readings indicate a short. Shorted coil may also feel much hotter than the normal coil. An increase in hum may also be caused by a shorted winding.
(j) Shorted rotor	(j) Remove the brushes from the commutator and impress full voltage on the stator. If there are one or more points at which the rotor "hangs" (fails to revolve easily when turned), the rotor is shorted. Forcing the rotor to the position where it is most difficult to hold will cause the shorted coil to become hot. Do not hold in position too long or the coil will burn out.
(k) Dirty bearings	(k) When bearings become clogged with dirt, the motor may need protection from excessive dust. The application may be such that a specially constructed motor should be used.
Motor Operates without Lifting Brushes	
(a) Dirty commutator	(a) Clean with fine sandpaper. Do not use emery.
(b) Governor mechanism or brushes sticking, or brushes worn too short for good contact	(b) See that brushes move freely in slots and that governor mechanism operates freely by hand. Replace worn brushes.
(c) Frequency of supply circuit incorrect	(c) Run motor idle. After brushes throw off, speed should be slightly in excess of full-load speed shown on nameplate. An idle speed varying more than 10% from nameplate speed indicates that motor is being used on an incorrect supply frequency. A different motor will be required.

(continued)

Table 12-2 *(continued)*

Split-Phase Induction Motors

Symptom and Possible Cause	Possible Remedy
(d) Low voltage	(d) See that voltage is within 10% of nameplate voltage with the switch closed.
(e) Line connection improperly or poorly made	(e) See that contacts are good and that connections correspond with diagram sent with motor.
(f) Incorrect brush setting	(f) Check to see that rocker-arm setting corresponds with index mark.
(g) Incorrect adjustment of governor	(g) The governor should operate and lift brushes at approximately 75% of speed stamped on nameplate. Below 65% or over 85% indicates incorrect spring tension.
(h) Excessive load	(h) An excessive load may be started but not carried to and held at full-load speed, which is beyond where the brushes lift. Tight motor bearings may contribute to overload. This is sometimes indicated by brushes lifting and returning to the commutator.
(i) Shorted field	(i) See Paragraph (i) under *Failure to Start*.
Excessive Bearing Wear	
(a) Belt too tight, or an unbalanced line coupling	(a) Correct the mechanical condition.
(b) Improper, dirty, or insufficient oil	(b) The lubrication system of most small motors provides for supplying the right amount of filtered oil to the bearings. It is necessary only for the user to keep the wool yarn saturated with a good grade of machine oil.
Motor Runs Hot	
(a) Bearing trouble	(a) See Paragraphs (a), (b), and (c) under *Excessive Bearing Wear*.
(b) Short-circuited coils in stator	(b) Make separate wattmeter readings on each of the two halves of the stator winding. Sometimes the shorted coil may be located by

Table 12-2 *(continued)*

Split-Phase Induction Motors

Symptom and Possible Cause	Possible Remedy
	the fact that one coil feels much hotter than the other. An increase over normal in the magnetic noise (hum) may also indicate a shorted stator.
(c) Rotor rubbing stator	(c) Extraneous matter may be between the rotor and the stator, or the bearings may be badly worn.
(d) Excessive load	(d) Be sure proper pulleys are on the motor and the machine. Driving the load at higher speeds requires more horsepower. Take an ammeter reading. If current draw exceeds the nameplate amperes for full load, the answer is evident.
(e) Low voltage	(e) Measure the voltage at the motor terminals with the switch closed. The reading should not vary more than 10% from the value stamped on the nameplate.
(f) High voltage	(f) See (e) above.
(g) Incorrect line connection to the motor	(g) Check the connection diagram sent with the motor.

Motor Burns Out

Symptom and Possible Cause	Possible Remedy
(a) Frozen bearing	(a) See Paragraphs (a), (b), and (c) under *Excessive Bearing Wear.*
(b) Some condition of prolonged excessive overload	(b) Before replacing the burned-out motor, locate and remove the cause of the overload. Certain jobs that present a heavy load will, under unusual conditions of operation, apply prolonged overloads that may destroy a motor and be difficult to locate unless examined carefully. On jobs where it is assumed that somewhat intermittent service will normally prevail and that consequently are closely monitored, the load cycle especially should be checked as a change in this feature will easily produce excessive overload on the motor.

Motor is Noisy

Symptom and Possible Cause	Possible Remedy
(a) Unbalanced rotor	(a) When transportation handling has been so rough as to damage the heavy shipping case,

(continued)

Table 12-2 *(continued)*

Split-Phase Induction Motors

Symptom and Possible Cause	Possible Remedy
	it is advised to test the motor for unbalanced conditions at once. It is even possible (though it rarely happens) that a shaft may be bent. In any event, the rotor should be rebalanced dynamically.
(b) Worn bearings	(b) See Paragraphs (a), (b), and (c) under *Excessive Bearing Wear*.
(c) Rough commutator or brushes not seating properly	(c) Noise from this cause occurs only during the starting period, but conditions should be corrected to avoid consequent trouble.
(d) Excessive end play	(d) Proper end play is as follows: ⅓ (248.7 W) and smaller—0.127 mm to 0.762 mm; ½ (373 W) to 1 hp (0.746 kW)—0.254 mm to 1.905 mm. Washers supplied by the factory should be used. Be sure to tell factory all figures involved. Remember that too little end play is as bad as too much.
(e) Motor not properly aligned with the driven machine	(e) Correct the mechanical condition.
(f) Motor not firmly fastened to mounting base	(f) All small motors have steel bases so they can be firmly bolted to their mounting without fear of breaking. It is, of course, not to be expected that the base should be strained out of shape in order to make up for roughness in the mounting base.
(g) Loose accessories in motor	(g) Such parts as oil covers, guards (if any), end plates, etc., should be checked, especially if they have been removed for inspection of any sort. The conduit box should be tightened when the top is fitted after the connections are made.
(h) Air gap not uniform	(h) This results from a bent shaft or an unbalanced rotor. See Paragraph (a).

Table 12-2   *(continued)*

Split-Phase Induction Motors

Symptom and Possible Cause	Possible Remedy
(i) Amplified motor noises	(i) When this condition is suspected, set the motor on a firm floor. If the motor is quiet, then the mounting is acting as an amplifier to bring about certain noises in the motor. Frequently the correction of slight details in the mounting will eliminate this, but rubber mounts are the surest cure.

Excessive Brush Wear

(a) Dirty commutator	(a) Clean with fine sandpaper (never use emery).
(b) Poor contact with commutator	(b) See that the brushes are long enough to reach the commutator, that they move freely in the slots, and that the spring tension gives firm but not excessive pressure.
(c) Excessive load	(c) If the brush wear is due to overload, it can usually be checked by noting the time required for the brushes to lift from the commutator. The proper time is less than 10 seconds.
(d) Failure to lift promptly and stay off during the running period	(d) Examine for conditions listed under *Motor Operates without Lifting Brushes.*
(e) High mica	(e) Examination will show this condition. Take a very light cut off the commutator face and polish with fine sandpaper. Undercut the mica.
(f) Rough commutator	(f) True up on lathe.

Brush-Holder or Rocker-Arm Wear

(a) Failure to lift properly and stay off during the running period	(a) No noticeable wear of this part should occur during the life of the motor. Troublesome wear indicates faulty operation. See *Motor Operates without Lifting Brushes.*

(continued)

Table 12-2 *(continued)*

Split-Phase Induction Motors

Symptom and Possible Cause	Possible Remedy

Radio Interference

(a) Faulty ground

(a) Check for poor ground connections and repair. Static electricity generated by the belts may cause radio noises if the motor frame is not thoroughly grounded. Check for loose connections or contacts in the switch, fuses, or starter.

Capacitor-Start Induction Motors
Failure to Start

(a) Blown fuses or overload device tripped

(a) Examine motor bearings. Be sure that they are in good condition and properly lubricated. Be sure the motor and driven machine both turn freely. Check the circuit voltage at the motor terminals against the voltage stamped on the motor nameplate. Examine the overload protection of the motor. Overload relays operating on either magnetic or thermal principles (or a combination of the two) offer adequate protection to the motor. Ordinary fuses of sufficient size to permit the motor to start do not protect against burnout. A combination fuse and thermal relay, such as *Buss Fusetron*, protects the motor and is inexpensive. If the motor does not have overload protection, the fuses should be replaced with overload relays or *Buss Fusetrons*. After installing suitable fuses and resetting the overload relays, allow the machine to go through its operating cycle. If the protective devices again operate, check the load. If the motor is excessively overloaded, take the matter up with the manufacturer.

(b) No voltage or low voltage

(b) Measure the voltage at the motor terminals with the switch closed. See that it is within 10% of the voltage stamped on the motor nameplate.

(c) Open-circuited field

(c) Indicated by a humming sound when the switch is closed. Examine for broken wires and connections.

Table 12-2 *(continued)*

Split-Phase Induction Motors

Symptom and Possible Cause	Possible Remedy
(d) Incorrect voltage or frequency	(d) Requires motor built for operation on power supply available. AC motors will not operate on DC circuit, or vice versa.
(e) Cutout switch faulty	(e) The operation of the cutout switch may be observed by removing the inspection plate in the end bracket. If the governor disc does not hold the switch closed, the motor cannot start. This may call for additional end-play washers between the shaft shoulder and the bearing. Dirty or corroded contact points may also keep the motor from starting. See that the contacts are clean. With the load disconnected from the motor, close the starting switch. If the motor does not start, start it by hand and listen for the characteristic click of the governor as the motor speeds up and also when the switch has been opened and the motor slows down. Absence of this click may indicate that the governor weights have become clogged, or that the spring is too strong. Continued operation under this condition may cause the phase winding to burn up. Remove the motor to the service shop for adjustment.
(f) Open field	(f) These motors have a main and phase winding in the stator. With the leads disconnected from the capacitor, apply current to the motor. If the main winding is all right, the motor will hum. If the main winding tests satisfactorily, connect a test lamp between the phase lead (the black lead) from the capacitor and the other capacitor lead. Close the starting switch. If the phase winding is all right, the lamp will glow and the motor may attempt to start. If either winding is open, remove the motor to the service shop for repairs.
(g) Faulty capacitor	(g) If the starting capacitor (electrolytic) is faulty, the motor starting torque will be weak and

(continued)

Table 12-2 *(continued)*

Split-Phase Induction Motors

Symptom and Possible Cause	Possible Remedy
	the motor may not start at all, but may run if started by hand. A capacitor can be tested for open circuit or short circuit as follows: Charge it with DC (if available), preferably through a resistance or test lamp. If no discharge is evident on immediate short circuit, an open or a short is indicated. If no DC is available, charge with AC. Try charging on AC several times to make certain that the capacitor has had a chance to become charged. If the capacitor is open, short-circuited, or weak, replace it. Replacement capacitors should not be of a lower capacity or voltage than the original. In soldering the connections, *do not use acid flux*.
	Note 1—Electrolytic capacitors, if exposed to temperatures of 20°F (-6.7°C) or lower, may temporarily lose enough capacity so that the motor will not start, and may cause the windings to burn up. The temperature of the capacitor should be raised by running the motor idle or by other means. Capacitors should not be operated in temperatures exceeding 165°F (74°C).
	Note 2—The frequency of operation of electrolytic capacitors should not exceed 2 starts per minute of 3 seconds' acceleration each, or 3 to 4 starts per minute at less than 2 seconds' acceleration, provided the total accelerating time (i.e., the time before the switch opens) does not exceed 1 to 2 minutes per hour. This may be approximately determined by checking the ampere input with the nameplate marking. Excessive load may prevent the motor from accelerating to the speed at which the governor acts, and thus cause the phase winding to burn up.
Radio Interference (a) Faulty ground	(a) Check for poor ground connections. Static electricity generated by the belts may cause radio noises if the motor frame is not thoroughly grounded.

Table 12-2   (continued)

Split-Phase Induction Motors

Symptom and Possible Cause	Possible Remedy
(b) Loose connections	(b) Check for loose connections or contacts in the switch, fuses, or starter. Capacitor motors ordinarily will not cause radio interference. Sometimes vibration may cause the capacitor to move so that it touches the metal container. This may cause radio interference. Open the container, move the capacitor, and replace the paper packing so that the capacitor cannot shift.

Summary

The fractional-horsepower motor is usually designed to operate on single-phase AC at standard frequencies, and is reliable, easy to repair, and comparatively low in cost. The various types of single-phase motors developed for use on alternating current are as follows: (1) split-phase; (2) capacitor-start; (3) permanent-capacitor; (4) repulsion; (5) shaded-pole; (6) universal; (7) reluctance-synchronous; and (8) hysteresis-synchronous.

The *split-phase* induction motor is one of the most popular of the fractional-horsepower motors. It is used in sizes ranging from $\frac{1}{30}$ hp (24.9 W) to $\frac{1}{2}$ hp (373 W) on fans, business machines, automatic musical instruments, buffing machines, grinders, etc. The *resistance-start* motor and the *reactor-start* motor are forms of the split-phase induction motor.

The *capacitor-start* motor is another form of split-phase motor that has a capacitor connected in series with the auxiliary winding. This circuit is opened when the motor has attained a predetermined speed.

The *permanent-capacitor* motor has its main winding connected directly to the power supply, and the auxiliary winding connected in series with a capacitor. Both the capacitor and the auxiliary winding remain in the circuit while the motor is in operation.

The *repulsion* motor has a stator winding arranged for connection to the source of power, and a rotor winding connected to a commutator. This type of motor has a varying speed characteristic. The *repulsion-start-induction* motor and the *repulsion-induction* motor are forms of the repulsion motor.

A *shaded-pole* motor is a single-phase induction motor equipped with an auxiliary winding displaced magnetically from, and connected in parallel with, the main winding. This type of motor is manufactured in fractional-horsepower sizes, and is used in a variety of household appliances, such as fans, blowers, hair driers, and other applications requiring a low starting torque. It operates only on alternating current, is usually nonreversible, is low in cost, and is extremely rugged and reliable.

A *universal* motor is a series-wound or compensated series-wound motor that may be operated either on direct current or on single-phase, alternating current at approximately the same speed and output. Universal motors are commonly manufactured in fractional-horsepower sizes, and are preferred because they can be used either on AC or on DC, especially in areas that supply both types of current. Common applications for the universal-type motor are vacuum cleaners, sewing machines, portable drills, saws, routers, home motion-picture projectors, and business machines of all types.

The *reluctance-synchronous motor* is really a variation of the class squirrel-cage rotor. The rotor is modified to provide equally spaced areas of high reluctance. The reluctance-synchronous rotor starts and accelerates like a regular squirrel-cage rotor. Reluctance-synchronous motors may be designed for operation on single- or three-phase power. In single-phase operation, they may take the split-phase, capacitor-start, or permanent-split capacitor configuration. The skew of the rotor aids in providing smoothness of operation when compared to the capacitor-start and split-phase motors.

The *hysteresis-synchronous motor* has a rotor that is made of a heat-treated, cast, permanent-magnet alloy cylinder. It has a nonmagnetic support securely mounted to the shaft. This type of motor starts on the hysteresis principle and accelerates at a fairly constant rate until it reaches the synchronous speed of the rotating field. This type is strictly dependent on the frequency of the power source for its speed.

Synchronous motors are known for their constant speed characteristic. They operate at a speed determined by the frequency of the power source. They are more expensive and heavier than their single-phase counterparts, so they are used only when speed is the absolute factor in the design of the machine being driven by the motor.

Review Questions

1. Describe the basic construction of the split-phase induction motor.
2. List the six types of split-phase motors.
3. What are the two types of repulsion motors?

4. What is the chief advantage of the universal-type motor?

5. What are the applications for the universal-type motor?

6. What factor determines the speed of a reluctance-synchronous motor?

7. What factor determines the speed of a hysteresis-synchronous motor?

8. What happens when a reluctance motor is overloaded?

9. What happens when a hysteresis motor is overloaded?

10. What is meant by "pull-in"?

11. What is the most popular of the fractional horsepower motors?

12. Which of the motors has its auxiliary winding connected in series with a capacitor?

13. How are shaded-pole motors constructed in order to have "shaded poles"?

14. Why can the universal type motor operate on both AC and DC?

15. What is the most desirable characteristic of a synchronous motor?

Part IV

Right Motor for the Job

Chapter 13

Motor Selection and Replacement

Selection of Motor Type

There are many factors involved in selecting a motor. The application of the motor to a specific job makes it easier. Motors have been designed for many special uses.

Selection of a motor affects its installation; it also affects its operation and its service requirements. Selection may be determined by the user- or the motor-driven-apparatus requirements. Facts, field tests, and analyses of operating conditions are criteria used for motor selection.

Some basic factors in the selection of a motor are power supply, horsepower rating, speed, motor type, and the enclosure. Other considerations include motor mounting, motor connection to a load, and mechanical accessories or modifications.

Note
The authors would like to express appreciation to the Bodine Electric Company and the General Electric Company for assistance in the preparation of this chapter.

Power Supply (Voltage)
The system voltage must be known in order to select the proper motor. The motor nameplate will normally be less than the nominal power system voltage. A joint committee of the Edison Electric Institute and NEMA (National Electrical Manufacturers' Association) has recommended standards for both power system voltage and motor nameplate voltages, which are described in Table 13-1.

Table 13-1   Power System Voltage Standards

Polyphase 60 Hz	
Nominal Power System Volts	*Motor Nameplate Volts*
208	200
240	230
480	440
600	575

(continued)

Table 13-1 *(continued)*

Single-Phase 60 Hz	
Nominal Power System Volts	*Motor Nameplate Volts*
120	115
240	230

Motor Description

Following is a list of the important items that make up a motor description:

Design Type—K, KF, P, PF, and so on

Frame—14T, 286T, and so on

Horsepower—through 30 horsepower

Synchronous Speed—1800 rpm, 900 rpm, and so on

Volts—230, 460, 575, and so on

Phases—One, three

Frequency—60 Hz, 50 Hz

Enclosure—drip-proof, TEFC, explosion-proof, and so on

Duty—continuous, 1-hour, 15-minute, and so on

Service Factor—1.15, 1.0, and so on

Ambient Temperature—40°C, 65°C, and so on

Mounting Assembly—F-1, F-2, C-2, W-4, and so on

Bearing Requirement—ball, sleeve, and so on

Direction of Rotation—cw (or ccw) facing drive end

Environmental or Operating Conditions—high ambient temperature, excessive moisture, low voltages, and so on

Special electrical or mechanical features

Voltage Variation

All motors rated 30 horsepower or less are designed to operate successfully at rated load with a voltage variation of plus or minus 10 percent when rated frequency is supplied. They will also operate successfully when the sum of the voltage and frequency variation does not exceed 10 percent, provided the variation of frequency does not exceed 5 percent above or below nominal ratings as stamped on the nameplate.

Frequency

In addition to operating successfully with a voltage variation, a motor rated 30 horsepower or less will operate successfully with a frequency variation that does not exceed 5 percent above or below its rated frequency.

The predominant frequency in North America is 60 Hz. However, 50-Hz systems are common in foreign countries. Systems such as 25 to 40 Hz are isolated and relatively small in number. However, motors can be designed for these frequencies.

Phases

In general, three-phase power supplies will be found in most industrial locations. However, single-phase alone is available for most residential and rural areas. Two-phase power supply is found infrequently.

Motor Design Types

The type of motor will determine the electrical characteristics of the design. The following designs are NEMA designations for three-phase motors:

- *Design B*—A Design B motor is a three-phase, squirrel-cage motor designed to withstand full-voltage starting and develop locked-rotor and breakdown torques adequate for general application and have a slip at rated load of less than 5 percent.

- *Design C*—A Design C motor is a three-phase, squirrel-cage motor designed to withstand full-voltage starting, develop high locked-rotor torque for special high-torque applications, and have a slip at rated load of less than 5 percent.

- *Design D*—A Design D motor is a three-phase, squirrel-cage motor designed to withstand full-voltage starting, develop 275 percent locked-rotor torque, and have a slip at rated load of 5 percent or more.

All percentages are in terms of percent of full-load torque with rated voltage and frequency supplied.

Horsepower Requirements

The horsepower required by the driven machine determines the motor rating. Where the load varies with time, a horsepower-versus-time curve will show the peak horsepower required and the calculation of the root-mean-square (RMS) horsepower, indicating the proper motor rating from a heating standpoint. In case of extremely

large variations in load, or where shutdown, accelerating, or decel-
erating periods constitute a large portion of the cycle, the RMS
horsepower may not give a true indication of the equivalent continuous load, and the motor manufacturer should be consulted.

Where the load is maintained at a constant value for an extended
period (varying from 15 minutes to 2 hours, depending on size), the
horsepower rating will not usually be less than this constant value,
regardless of other parts of the cycle.

If the driven machine is to operate at more than one speed, the
horsepower required at each speed must be determined.

Motor Mounting

The frame numbers of induction motors specifically identify mounting dimensions. All machines that have the same frame designation
will have identical essential mounting dimensions, regardless of
electrical characteristics, thereby providing interchangeability. For
the majority of applications, foot-mounted motors are utilized. The
motor is mounted on the driven equipment and secured by four
mounting bolts through holes in the feet (Figure 13-1).

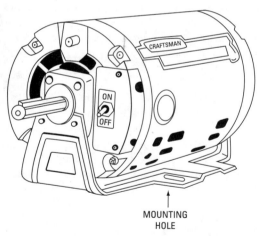

MOUNTING
HOLE

Figure 13-1 Note the mounting holes in the resilient mounting of the
motor. *(Courtesy Sears)*

Mounting Methods

There are a number of methods used to mount motors. The available space and the equipment location make it impossible in some

cases to mount the motor on the machine or on the floor next to the machine it is to power. Therefore, it becomes necessary to devise other means of mounting the motors. Some companies will furnish the brackets and others will make the motors according to your specifications if you order enough of them to make it worthwhile (Figure 13-2). In any case, the following methods of mounting a motor are given to indicate some of the possibilities (Figure 13-3).

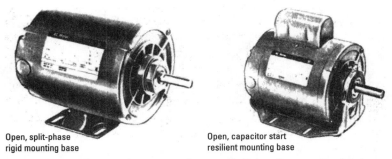

Open, split-phase
rigid mounting base

Open, capacitor start
resilient mounting base

Figure 13-2 The rigid mounting base and the resilient mounting base are shown here for comparison. *(Courtesy Westinghouse)*

Unless specified otherwise, most motors can be mounted in any position or any angle. However, drip-proof motors must be mounted in the normal horizontal position to meet the enclosure definition. They are mounted securely to the mounting base of the equipment or to a rigid, flat surface, preferably metallic.

For direct-coupled applications, align the shaft and coupling carefully, using shims as required under the motor base. Use a flexible coupling, if possible, but not as a substitute for good alignment practices.

For belted applications, align the pulleys and adjust the belt tension so that approximately ½ in. of the belt deflection occurs when the thumb force is applied midway between pulleys. With sleeve-bearing motors, position the motor so that the belt-pull is away from the oil hole in the bearing (approximately under the oiler of the motor).

End Shield Mountings
The industry has standardized three types of machined end shields, which have rabbets and bolt holes for mounting such items as pumps and gearboxes to the motor, or for overhanging the motor on a driven machine.

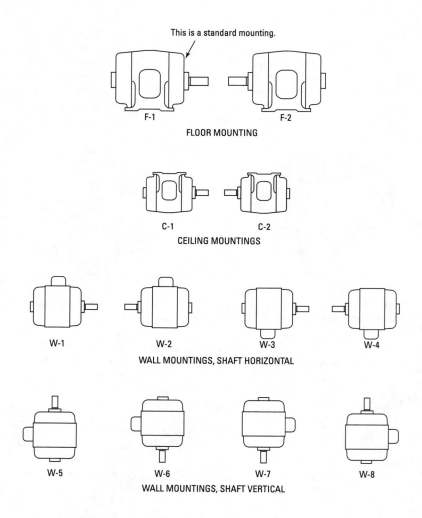

This is a standard mounting.

F-1   F-2

FLOOR MOUNTING

C-1   C-2

CEILING MOUNTINGS

W-1   W-2   W-3   W-4

WALL MOUNTINGS, SHAFT HORIZONTAL

W-5   W-6   W-7   W-8

WALL MOUNTINGS, SHAFT VERTICAL

Standard lead location – F-1, W-2, W-3, W-6, W-8, C-2
Lead location opposite standard – F-2, W-1, W-4, W-5, W-7, C-1

Figure 13-3   Mounting assembly symbols for standard mounting configurations for motors. *(Courtesy Doerr)*

The *Type C* face end shield provides a male rabbet and tapped holes for mounting bolts. This end shield is used for mounting a small apparatus to the motor.

The *Type D* flange has a male rabbet, with holes in the flange for through-bolts. This flange can be used on machine tool gearboxes where the motor is mounted to the apparatus.

The *Type P* base has a female rabbet, with holes for through-bolts in the flange, and is used for mounting vertical motors.

Part-Winding Starting

Part-winding starting is used to reduce the initial inrush of current. Power systems, especially in residential and commercial areas, are often limited in capacity. Consequently, when a large motor is started across the line, voltage fluctuation may cause annoying light flicker. To keep this under control, most power companies limit the amount and number of any sudden current demands on their system. Part-winding starting will meet these requirements by putting power into selected portions of the motor windings first and then energizing the rest of the windings a second or two later. The current inrush is limited to about 60 to 70 percent of the normal starting current and the starting torque is reduced to 40 to 50 percent of normal. This reduced current and torque condition exists for only one or two seconds. As the rest of the windings are energized on the second step, another inrush of current occurs, making the total current drawn equal to, or slightly less than, the normal across-the-line inrush current. As all the windings are energized, the motor displays standard characteristics.

Enclosures and Special Winding Treatment

To withstand special or extreme conditions, the following enclosures and/or special winding treatments are available.

Open, Drip-Proof

Normal insulation treatment that consists of one or more dips and bakes of varnish. The insulation system is composed of materials that will not absorb or retain moisture.

Open, Drip-Proof with Extra Varnish Treatments

Same as standard, open, drip-proof except with extra dips and bakes to increase the moisture resistance of the insulation system.

Totally Enclosed

The windings and internal parts are protected by the frame enclosure, which prevents free exchange of ambient air. Recommended for dirty or outdoor applications where high reliability and long life are prime considerations (Figure 13-4).

Figure 13-4 Totally enclosed, nonventilated, split-phase, resilient mounting base motor. *(Courtesy Westinghouse)*

Totally Enclosed, Severe Duty
Offers the same enclosure as above, but with special features; designed for use in chemical atmospheres or extremes of moisture and humidity. Table 13-2 is a guide to the proper selection of enclosure and/or winding treatments on applications where chemicals are encountered and/or where mechanical protection from dust is required.

Table 13-2   Rating Enclosures

Enclosure and/or Winding Treatment	Humidity Resistance	Mechanical Protection	Resistance to Chemicals
Open, Drip-Proof	4	5	4
Open, Drip-Proof with Extra Varnish Treatment	2	4	3
Moisture-Sealed	1	3	3
Totally Enclosed	3	2	2
Severe-Duty	2	1	1

Rank: 1 = High, 5 = Low

Each construction is ranked as to its ability to withstand the particular condition. Note that the drip-proof motor with extra varnish treatment is well suited to high humidity.

Mechanical Protection

Where clogging materials are present in severe proportions, the air gap of open motors may become clogged. Therefore, the recommendation is a totally enclosed motor.

Resistance to Chemicals

The first recommendation for any type of atmosphere containing chemicals, acids, bases, solvents, etc., should be severe-duty, enclosed construction.

Tropical Protection

Specifications for motors to be used in a hot and humid location may call for tropical protection or tropical insulation. This will be assumed to mean that the windings must be specially protected against moisture and fungus and able to operate with normal life expectancy in higher-than-normal ambient temperatures up to 65°C.

Motors specified for tropical use will be supplied with extra dips and bakes of insulating varnish and a 65°C, ambient insulation system. Special treatment for fungus proofing is not required, as the materials used in the insulation system are resistant to fungus growth.

Anti-Fungus Treatment

The anti-fungus requirement can be met by supplying either an insulation system that will not support fungus growth or a special anti-fungus varnish. Since either may be used, depending on the rating, do not specify anti-fungus varnish but request anti-fungus treatment.

Guide to Selecting the Right Motor

AC motors can be divided by power supply into two major electrical categories: polyphase and single-phase. The four important selection criteria to consider are horsepower, voltage, speed, and phase.

Over 90 percent of the motors used in the United States are single-phase. The most popular electrical types of single-phase motors for general use are the capacitor-start and the split-phase.

The split-phase motor has two windings, start and run, which are energized to start the motor. The start winding is cut out of the circuit at about 75 percent of operating speed. The capacitor-start motor is almost identical to the split-phase motor but delivers two or three times the starting torque per ampere of current.

Use

The lower-cost, split-phase motor is the logical motor to choose for applications where the starting load is light, such as fans and blowers, or where the load is applied after the motor has reached operating speed, for instance, saws and drill presses. Capacitor-start motors are necessary, of course, on applications such as conveyors where heavy loads must be started.

To make sure there is a correct motor for every job, single-phase motors are available in two types: general-purpose and special-service. Table 13-3 shows the differences between the four basic, single-phase motors.

Motor Speed

Almost all 60-Hz, split-phase and capacitor-start motors operate at one of these full-load speeds: 3450 rpm (two-pole), 1725 rpm (four-pole), 1140 rpm (six-pole), or 860 rpm (eight-pole).

Generally speaking, both motor price and physical size increase as rated rpm decreases. This means that savings can be achieved by selecting the highest-speed motor that will drive the device yet still remain within the practical limits of a 5:1 or 6:1 speed-reduction ratio. Obviously, such savings are most often available in belt-drive applications.

Bearings

All-angle, sleeve-bearing motors can be used over a wide range of applications where moderate thrusts are encountered. They can be mounted in any position and they are quieter and more economical than ball-bearing motors. The motor applications that demand ball-bearing motors are in powering devices that create high axial and/or radial thrust (Figure 13-5).

Installation (Power Supply/Connections)

Connect the motor for the desired voltage and rotation according to the connection diagram on the nameplate or in the terminal box (Figure 13-6).

Voltage, frequency, and phase of the power supply should be consistent with the motor nameplate rating. The motor will operate satisfactorily on voltage within 10 percent of nameplate value or frequency within 5 percent (combined variation not to exceed 10 percent) of nameplate value.

Table 13-3 Typical Characteristics of AC Motors

Motor Types	Split-Phase, General-Purpose	Split-Phase, Special Service	Capacitor-Start, Special Service	Capacitor-Start, General-Purpose	Polyphase, I HP and Below
Starting Torque (% Full Load Torque)	130%	175%	250%	350%	275%
Starting Current	Normal	High	Normal	Normal	Normal
Service Factor (% of Rated Load)	135%	100%	100%	135%	135%
Comparative Price Estimate (Based on 100% for Lowest Cost Motor)	110%	100%	135%	150%	150%
Remarks	Low starting torque. High service factor permits continuous loading—up to 35% over nameplate rating. Ideal for applications of medium starting duty.	Moderate starting torque, but has service factor of 1.0. Apply where load will not exceed nameplate rating for any extended duration of time. Because of higher starting current, use where starting is infrequent.	High starting torque but 1.0 service factor. Use only where load will not exceed nameplate rating for any extended duration of time. Starting current is normal.	Very high starting torque. High service factor permits continuous loading up to 35% over nameplate rating. Ideal for powering devices with heavy loads, such as conveyors.	Normal start current for polyphase is low compared to single-phase motors. High starting ability. High service factor permits continuous loading up to 35% over nameplate rating. Direct companion to general-purpose, capacitor-start motor.

Permanent-Split Capacitor Motors (KCP) are normally designed for direct-drive fan and blower applications.

Because of the uniqueness of design and applications, these types of motors are excluded from the above table. (Courtesy General Electric)

A. SLEEVE BEARING

High-density felt contactor wick

Large OD, cupped oil flingers for efficient oil return

Steel-backed bronze sleeve bearing for high-impact loads

Molydisulfide impregnated thrust washer

Locked thrust washer

Felt for thrust lubrication

Tempered steel thrust plate

Low-density felt to filter returned oil

Lubricating material (having high oil release rate)

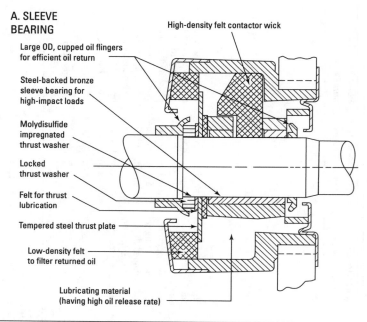

B. BALL BEARING

Steel insert

202 shielded ball bearing

Preload spring

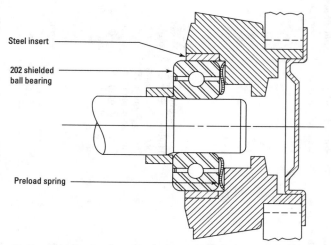

Figure 13-5 (A) shows the sleeve bearing and how it runs more quietly than the ball-bearing motor shown in (B).

Figure 13-6 The direction of rotation of this motor can be changed by slipping off the slip-on connectors and changing them to fit the directions given on the nameplate for the direction of rotation. *(Courtesy Westinghouse)*

Operation of 230-V Motors on 208-V Systems

Motors rated 230 V will operate satisfactorily on 208-Vt systems on most applications that require nominal starting torques. Starting and maximum running torques of a 230-V motor will be reduced approximately 25 percent when operated on 208-V systems. Fans, blowers, centrifugal pumps, and similar loads will normally operate satisfactorily at these reduced torques. Where the application torque requirements are high, it is recommended that a 230-V motor of the next highest horsepower or a 200-V motor be used. External motor controls for 230-V motors on 208-V systems should be selected from 230-V nameplate data.

Safety Precautions

1. Use safe practices when handling, lifting, installing, operating, and maintaining motors and motor-operated equipment.
2. Install motors and electrical equipment in accordance with the National Electrical Code (NEC) or sound local electrical

and safety codes and practices, and, when applicable, the Occupational Safety and Health Act (OSHA).

3. Ground motors securely. Make sure that grounding wires and devices are, in fact, properly grounded.

Caution
Failure to ground a motor properly may cause serious injury to personnel.

4. Before servicing or working on or near motors or motor-driven equipment, disconnect the power source from the motor and accessories (Figure 13-7).

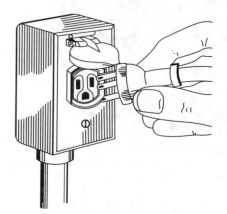

Figure 13-7   Remove the plug from the receptacle before working on a motor. If it is connected to a switchbox, make sure the switch is off. Note that there is a third hole here, which represents the grounding wire that connects to the outside frame of the motor.

(Courtesy Sears)

5. Remove shaft key from keyways of uninstalled motors before energizing the motor. Be sure the keys, pulleys, fans, and other attached parts are fully secured or installed on the motors before energizing the motor.

6. Make sure fans, pulleys, belts, and other parts are properly guarded if they are in a location that could be hazardous to personnel.

7. Provide proper safeguards against failure of motor-mounted brakes, particularly on applications involving overhauling loads.

8. Provide proper safeguards on applications where a motor is mounted on or through a gear-reducer to a holding or overhauling application. Do not depend on gear friction to hold the load.

9. Do not lift the motor-driven equipment with motor-lifting means. If eyebolts are used for lifting motors, they must be

securely tightened and the direction of the lift must not exceed a 15° angle with the shank of the eyebolt.

Caution

Motors subjected to overload, locked rotor, current surge, or inadequate ventilation conditions may experience rapid heat buildup, presenting risks of motor damage or fire. To minimize such risks, use of motors with proper overload protectors is advisable for most applications.

Do not use motors with automatic-reset protectors where automatic restarting might be hazardous to personnel or to equipment. Use motors with manual-reset protectors where such hazards exist. Such applications include conveyors, compressors, tools, and most farm equipment.

Location and Motor Enclosures

Open, drip-proof motors are designed for use in areas that are reasonably dry, clean, well ventilated, and usually indoors. If installed outdoors, it is recommended that the motor be protected with a cover that does not restrict flow of air to the motor.

Totally enclosed motors are suitable for use where exposed to dirt, moisture, and most outdoor locations, but not for very moist or for hazardous (atmosphere filled with explosive vapor dust) locations.

Severe-duty enclosed motors are suitable for use in corrosive or excessively moist locations.

Explosion-proof motors are made to meet Underwriters Laboratories Standards for use in hazardous (explosive) locations shown by the UL label on the motor (Figure 13-8).

Figure 13-8 Explosion-proof motor *(Courtesy Westinghouse)*

Certain locations are hazardous because the atmosphere does or may contain gas, vapor, or dust in explosive quantities. The National Electrical Code (NEC) divides these locations into classes and groups according to the type of explosive agent that may be present. Listed below are some of the agents in each classification. For a complete listing, see *Article 500* of the National Electrical Code. A newly revised code is published every three years.

Class I (Gases, Vapors):

Group A—Acetylene (*Note:* Motors are not available for this group.)

Group B—Butadiene, ethylene oxide, hydrogen, propylene oxide (*Note:* Motors are not available for this group.)

Group C—Acetaldehyde, cyclopropane, diethyl ether, ethylene, isoprene

Group D—Acetone, acrylonitrile, ammonia, benzene, butane, ethylene dichloride, gasoline, hexane, methane, methanol, naphtha, propane, propylene styrene, toluene, vinyl acetate, vinyl chloride, xylene

Class II (Combustible Dusts):

Group E—Aluminum, magnesium, and other metal dusts with similar characteristics

Group F—Carbon black, coke, or coal dust

Group G—Flour, starch, or grain dust

Ambient temperature around motors should not exceed 40°C unless the motor nameplate specifically permits a higher value.

Maintenance

Motors properly selected and installed are capable of operating for many years with a reasonably small amount of maintenance.

Before servicing motors and motor-operated equipment, disconnect the power supply from the motors and accessories. Use safe working practices during servicing of equipment.

Clean motor surfaces and ventilation openings periodically, preferably with a vacuum cleaner. Heavy accumulations of dust and lint will result in overheating and premature failure of motors.

Lubrication

Motors are lubricated at the factory to operate for long periods under normal service conditions without relubrication. Excessive or too frequent lubrication may damage the motor. Follow instructions furnished with the motor—usually on the nameplate or terminal box cover or on a separate instruction.

If instructions are not available, relubricate as follows: For sleeve-bearing motors, add electric motor oil (or SAE #10 or #20 nondetergent oil) after three years of normal or one year of heavy-duty service. For light-duty applications, add oil after 25,000 hours of operation. Ball-bearing motors of the newer types have sufficient lubrication for many years of operation. The time period between relubrication can vary from 10 years to 9 months, depending on the ambient temperature and the type of service. For specific regreasing recommendations, refer to the instruction provided with the motor or contact the manufacturer's service center locally.

Wet Motors

If a motor gets wet, disconnect and place in an oven set at its lowest temperature and leave it until it stops steaming (Figure 13-9).

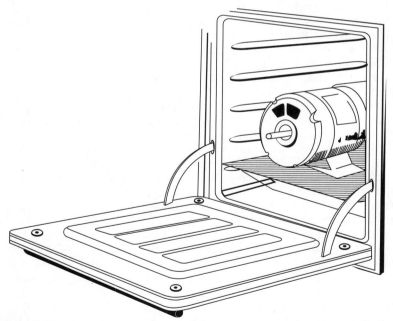

Figure 13-9 To dry wet motor, disconnect and place in oven that is set at its lowest temperature. Open door and leave in oven until steaming stops.

Summary

There are many factors involved in selecting a motor. Motors are available for a variety of purposes. Some basic factors are power supply availability, horsepower rating, speed, motor type, and the enclosure. Other considerations include motor mounting, motor connection to a load, and mechanical accessories or modifications.

The NEMA, National Electrical Manufacturers' Association, has recommended standards for both power system voltage and motor nameplate voltages. All motors rated 30 horsepower or less are designed to operate successfully at rated load with a voltage variation of plus or minus 10 percent when rated frequency is supplied. Motors are designed for 60 Hz in this country, but 50 Hz is used in many foreign countries. 25 Hz and 40 Hz power is still available in this country, but it is very limited today. Three-phase power is found in most industrial and business locations. Single-phase alone is available for most residential locations as well as rural areas.

Frame numbers of induction motors identify the mounting dimensions. There are a number of methods used to mount motors. Some come with brackets designed especially for a particular use. With direct-couple applications, align the shaft and coupling using shims if needed. End shield mountings have been standardized to three types. They are Type C, Type D, and Type P.

Part-winding starting is used to reduce the initial inrush of current. Power systems, especially those in residential and commercial areas are often limited in capacity. The starting and stopping of motors can cause spikes in voltage that can damage sensitive electronics equipment, especially computers.

To withstand special or extreme conditions, the enclosures are designed with the environment in mind. There are a number of enclosure types: open, drip-proof; open, drip-proof with extra varnish treatments; totally enclosed; and severe duty. Each construction is ranked as to its ability to withstand the particular condition under which the motor must operate.

Selecting a motor involves the intended use of the machine, the cost of the motor, speed, frequency of the power source, voltages available, service factor, starting current, and the bearing availability.

Safety precautions are necessary in the operation of any machine, especially an electric motor. Motors should be installed using equipment in accordance with the National Electrical Code, or NEC. Sound and local electrical and safety codes and practices should be followed as well as OSHA (Occupational Safety and Health Act) requirements. Certain safety procedures and noise level limits, as well as proper electrical installation procedures, should be followed.

Installation of motors should include an awareness of what the various types of special purpose motors were designed to do and locations where they may work. Keep in mind that ambient temperature should not exceed 40°C around the motor unless its nameplate specifically permits a higher value.

Motors properly selected and installed are capable of operating for many years with a reasonably small amount of maintenance. Motor surfaces should be cleaned and the ventilation openings periodically cleaned of lint and dirt. Motors are lubricated at the factory to operate for long periods under normal service conditions without relubrication. Excess or too frequent lubrication may damage a motor. Follow the instructions furnished with the motor. The time period between relubrication can vary from 10 years to 9 months, depending on the ambient temperature and the type of service. Wet motors may be dried out by placing them in an oven that is set to its lowest temperature setting.

Review Questions

1. What factors play an important role in selecting a motor?
2. What does NEMA stand for? What does it do?
3. What frequency is most commonly used for electrical power in the USA?
4. What frequency is most commonly used for electrical power in Europe?
5. How do you identify the mounting dimensions of a motor?
6. What are the three types of standardized end shield mountings for motors?
7. Why would you use or select a part-winding electrical motor?
8. What causes spikes of voltage in local power sources?
9. Where would you use an open, drip-proof motor?
10. What is the NEC? Where is it used?
11. What does OSHA mean? What is its function?
12. How often do motors usually need relubricating?
13. Why would the choice of a motor's bearings make a difference in your selection process?
14. Where could you install a severe-duty motor?
15. Where would you install a totally enclosed motor?

Chapter 14

Motor Calculations

The following simple calculations for motor circuits will serve to acquaint the reader with methods used in determining voltage drops, currents, and efficiency of DC and AC motors.

Problem A 115-V DC motor draws a current of 200 amperes and is located 1000 ft from the supply source. If the copper transmission wire has a diameter of 0.45 in., what must be the voltage of the supply source (Figure 14-1)?

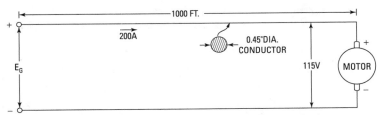

Figure 14-1 Voltage-drop calculation in a motor feeder line.

Solution The resistance of any conductor varies directly with its length and inversely as its cross-sectional area. Therefore, the formula is

$$R = \frac{P \times L}{A}$$

where

      R is the resistance in ohms

      P is the specific resistance of one mil ft of copper wire (10.4),

      L is the length in feet,

      A is the area in circular mils (diameter of wire in thousandths)

Note

The cross-sectional area of a circular conductor can be expressed in circular mils by squaring the diameter of the conductor expressed in thousandths. For example, a wire 0.625 in. in diameter has a circular-mil area of $625 \times 625 = 390{,}625$ circular mils.

Thus, the resistance of the feed lines is

$$R = \frac{10.4 \times 2000^*}{450^2} = 0.103 \text{ ohm}$$

where *indicates the length of 2 conductors and

$$E_C = E_R + I_L R = 115 + (200 \times 0.103) = 135.6 \text{ volts}$$

Problem A 10-hp, 230-V DC motor of 84 percent full-load efficiency is located 500 ft from the supply mains. If the motor starting current is 1.5 times the full-load current, what is the smallest cross-sectional area of copper wire required when the allowable voltage drop in the feeder at starting is 24 V (Figure 14-2)?

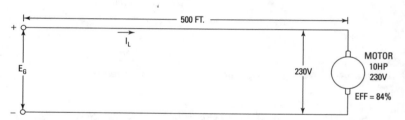

Figure 14-2   Cross-sectional area calculation for the feeder line to a 10-hp motor.

Solution The motor full-load current is

$$I_L = \frac{\text{hp} \times 746}{E \times \text{efficiency}} = \frac{10 \times 746}{230 \times 0.84} = 38.6 \text{ amperes}$$

where

> Efficiency is percent in decimal form
> E is voltage
> I_L is full current load

Motor starting current is

$$I_s = 38.6 \times 1.5 = 57.9 \text{ amperes}$$

where

I_S is starting current
R is resistance
A is cross-sectional area

Since the voltage drop in the feeder at starting is 24 V, then, according to Ohm's law, $R = E/I$

$$R = \frac{24\ V}{57.9A} = 0.415\ \text{ohm}$$

The minimum cross-sectional area is therefore

$$A = \frac{10.4 \times 2 \times 500}{0.415} = 25,060.241\ \text{circular mils}$$

Problem In the circuit of Figure 14-3, with loads L_1, L_2, L_3 drawing currents of 25, 8, and 40 amperes, respectively, calculate

1. The power supplied by each generator
2. The voltages E_1, E_2, and E_3

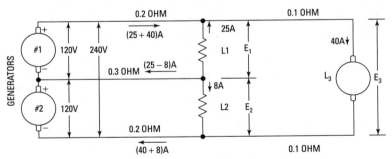

Figure 14-3 A three-wire system supplying three loads.

Solution By inspection, the currents supplied by generator Nos. 1 and 2 are 65 and 48 amperes, respectively.

$$P_{G1} = 120 \times 65 = 7800\ \text{watts}$$

$$P_{G2} = 120 \times 48 = 5760\ \text{watts}$$

where

P_{G1} is Power (Generator 1)
P_{G2} is Power (Generator 2)

According to Kirchhoff's law,

$E_1 = 120 - (65 \times 0.2) - (17 \times 0.3) = 101.9$ volts

$E_2 = 120 + (17 \times 0.3) - (48 \times 0.2) = 115.5$ volts

and

$E_1 + E_2 = 217.4$ volts

Similarly,

$E_3 = 217.4 - (0.2 \times 40) = 209.4$ volts

where

E_1 is Voltage (Generator 1)
E_2 is Voltage (Generator 2)
E_3 is Voltage output across both generators

Problem A three-wire system supplies the load shown in Figure 14-4. If the resistance of each lamp is 110 ohms and the motor takes a current of 25 amperes, calculate the voltage across each group of lamps

1. When the motor is disconnected

2. When the motor is operating

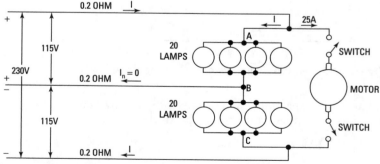

Figure 14-4 A three-wire system supplying a lamp and motor load.

Solution The combined resistance of a group of 20 lamps, each having a resistance of 110 ohms, is

$$R = \frac{110}{20} = 5.5 \text{ ohms}$$

Since the load is a balanced one, it is evident that the current in the neutral is zero.

The current through the lamps with the motor disconnected is

$$I = \frac{230}{11.4} = 20.2 \text{ amperes}$$

(a) Voltage $E_{AB} = E_{BC} = \dfrac{20.2 \times 11}{2} = 111.1 \text{ volts}$

The current through the lamps with the motor operating and drawing 25 amperes can be obtained if Kirchhoff's law is applied to the circuit, remembering that the current flowing in the line is now $(I + 25)$ amperes.

$$230 = 0.4(I + 25) + 11I = 11.4I + 10$$

and

$$I = \frac{220}{11.4} = 19.3 \text{ amperes}$$

(b) Voltage $E_{AB} = E_{BC} = \dfrac{19.3 \times 11}{2} = 106.15 \text{ volts}$

where

I is current

E_{AB} is voltage across points A to B

E_{BC} is voltage across points B to C

Thus, when the motor is thrown on the line, the voltage across the lamps will fall from 111.10 to 106.15 V.

Problem The field winding of a shunt motor has a resistance of 110 ohms, and the voltage applied to it is 220 V. What is the amount of power expended in the field excitation?

Solution The current through the field is

$$I_f = \frac{E_s}{R_f} = \frac{220}{110} = 2 \text{ amperes}$$

Power expended is

$E_s I_f = 220 \times 2 = 440$ watts

The same results will also be obtained by using the equation

$$\frac{E_s^2}{R_f} = P_f = \frac{220^2}{110} = 440 \text{ watts}$$

where

I_f is field current
E_s is source voltage
R_f is field resistance
P_f is power expended in field

Problem Assume that the DC motor shown in Figure 14-5 draws a current of 10 amperes from the line with a supply voltage of 100 V. If the total mechanical loss (friction, windage, etc.) is 90 W, calculate the

1. Copper losses in the field
2. Armature current
3. Copper losses in the armature
4. Total loss
5. Motor input
6. Efficiency

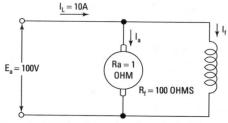

Figure 14-5 A DC-motor circuit.

Solution From the foregoing data we obtain the following results:

1. The copper losses in the field are

$$I_f^2 \times R_f = 1^2 \times 100 = 100 \text{ watts}$$

2. The armature current is

10 − 1 = 9 amperes

3. Copper losses in the armature are

$I_a^2 \times R_a = 9^2 \times 1 = 81$ watts

4. The total loss is

100 + 81 + 90 = 271 watts

5. Motor input is

$E_a \times I_L = 100 \times 10 = 1000$ watts

6. Motor output is equal to

input − losses = 1000 − 271 = 729 watts

7. The efficiency is equal to

$$\frac{\text{output}}{\text{input}} \times 100 = \frac{729}{1000} \times 100 = 72.9 \text{ percent}$$

where

I_f is field current
R_f is field resistance
I_a is armature current
I_L is load current

Problem A 50-hp, 500-V shunt motor draws a line current of 4.5 amperes at no load. The shunt field resistance is 250 ohms and the armature resistance, exclusive of brushes, is 0.3 ohm. The brush drop is 2 V. The full-load line current is 84 amperes. What is the horsepower output and efficiency?
Solution From the data supplied, we obtain the following:

Full-load armature current

$$I_a = I_L - I_f = 84 - \frac{500}{250} = 82 \text{ amperes}$$

No-load armature current

$$I_a = I_L - I_f = 4.5 - \frac{500}{250} = 2.5 \text{ amperes}$$

Stray power loss

$$P_{sp} = I_L \times E = 2.5 \times 500 = 1250 \text{ watts}$$

$$2 \times I_a = 2 \times 82 = 164 \text{ watts}$$

The efficiency of the motor

$$= \frac{500 \times 84 - [(82^2 \times 0.3) + (500 \times 2) + 164 + 1250]}{500 \times 84}$$

where

I_a is armature current
I_L is load current
I_a is field current
P_{sp} is stray power loss
E is voltage

$$\text{Motor efficiency} = \frac{42,000 - 4431}{42,000}$$

$$= \frac{37,569}{42,000} = 0.8945, \text{ or } 89.45\%$$

$$\text{Horsepower output} = \frac{37,569}{746} = 50.36 \text{ hp}$$

Problem Calculate the horsepower output, torque, and efficiency of a shunt motor from the following data:

$$I_1 = 19.8 \text{ amperes}$$

$$E_1 = 230 \text{ volts}$$

Balance reading $= 12$ lbs. corrected for zero reading

Brake arm $= 2$ ft.
Speed $= 1,100$ rpm

where

I_1 is current
E_1 is voltage

Solution The horsepower output of the motor is

$$\text{hp} = \frac{2\pi \times 2 \times 12 \times 1100}{33,000} = 5.03 \text{ hp}$$

$$\text{Torque} = FR = 12 \times 2 = 24 \text{ lb. ft.}$$

$$\text{Eff} = \frac{\text{output}}{\text{input}} = \frac{5.03 \times 746}{230 \times 19.8} = 0.824, \text{ or } 82.4\%$$

Problem A shunt motor with an armature and field resistance of 0.055 and 32 ohms, respectively, is to be tested for its mechanical efficiency by means of a rope brake. When the motor is running at 1400 rpm, the longitudinal pull on the 6-in.-diameter pulley is 57 lbs. Simultaneous readings of the line voltmeter and ammeter are 105 and 35, respectively. Calculate the

1. Counter emf
2. Copper losses
3. Efficiency

Solution From the foregoing data

$$I_a = I_L - I_F = 35 - \frac{105}{32} = 31.7 \text{ amperes}$$

where

I_a is current of armature
I_L is current of line
I_f is current of field

1. The counter emf is

$$(E_{armature} = E_{line} - (I_{arm} \times R_{arm}))$$
$$E_a = E_1 - (I_a R_a) = 105 - (31.7 \times 0.055) = 103.26 \text{ volts}$$

2. The copper losses are

$$\text{Power loss for copper} = (I^2 \times R_f) + (I_a^2 \times R_a)$$

$$P_c = I^2 R_f + I_a^2 R_a = (3.3^2 \times 32) + (31.7^2 \times 0.055)$$

$$= 404 \text{ watts}$$

3. The efficiency is

$$\text{Output} = \frac{2\pi \times \text{rpm} \times \text{radius of pulley in ft.} \times \text{pull on pulley in lbs.}}{33,000}$$

$$\text{Output} = \frac{2\pi \times 1400 \times 3/12 \times 57}{33,000} = 3.8 \text{ hp}$$

$$\text{Input} = \frac{\text{Voltage of line} \times \text{Current of line}}{746 \text{ watts}}$$

$$\text{Input} = \frac{105 \times 35}{746} = 4.93 \text{ hp}$$

$$\text{Efficiency} = \frac{\text{output}}{\text{input}}$$

or

$$\text{Eff.} = \frac{3.8}{4.93} = 0.771, \text{ or } 77\% \text{ (approx.)}$$

Problem A DC motor requires 10 kW to enable it to supply its full capacity of 10 horsepower to its pulley. What is its full-load efficiency?
Solution

$$\text{Efficiency} = \frac{\text{output}}{\text{input}} = \frac{10 \times 746}{10,000} + 0.746, \text{ or } 74.6\%$$

Problem A 7-hp motor takes 6.3 kW at full load. What is its efficiency?
Solution

$$\text{Efficiency} = \frac{\text{output}}{\text{input}}$$

$$\text{Output} = 7 \times 746 = 5,222 \text{ watts}$$

$$\text{Input} = 1000 \times 6.3 = 6,300 \text{ watts}$$

$$\text{Efficiency} = \frac{5,222}{6,300} = 0.829, \text{ or } 83\% \text{ (approx.)}$$

Problem A certain load to be driven at 1750 rpm requires a torque of 60 lb-ft. What horsepower will be required to drive the load?
Solution

$$\frac{2\pi \times T \times N}{33,000} = \frac{2\pi \times 60 \times 1,750}{33,000} = 20 \text{ hp}$$

Problem A 15-hp, 220-V, 1800 rpm shunt motor has an efficiency of 87 percent at full load. The resistance of the field is 440 ohms. Calculate the

1. Full-load armature current
2. Torque of the machine

Solution

1. The armature current is

$$I_L = \frac{15 \times 746}{0.87 \times 220}$$

$$I_f = \frac{220V}{440\Omega} = 0.5$$

$$I_a = I_L - I_f = \frac{15 \times 746}{0.87 \times 220} - \frac{220}{440}$$

$$= 58.46 - 0.5 = 57.96 \text{ amps}$$

where

I_L is current of line
I_f is current of field
N is speed of rotation in rpm

2. The torque is

$$T = \frac{\text{hp} \times 5,252}{N} = \frac{15 \times 5,252}{1,800} = 43.77 \text{ lb-ft.}$$

Problem When the field rheostat is cut out, a 230-volt shunt motor generates a counter emf of 220 V at no load. The resistance of the armature is 2.3 ohms and that of the field is 115 ohms. Calculate the

1. Current through the armature when the field rheostat is cut out
2. Current through the armature when sufficient external resistance has been inserted in the field circuit to make the field current one-half as great

Solution

1. The armature current when the field rheostat is cut out is

$$I_a = \frac{E_s - E_a}{R_a} = \frac{230 - 220}{2.3} = 4.35 \text{ amperes}$$

2. The current through the field without external resistance is

$$\frac{230}{115} = 2 \text{ amperes}$$

When the field current has been made half as great by inserting external resistance, the field flux and therefore the counter emf, will become half as great, or 110 V. The armature current in this particular case is therefore

$$I_a = \frac{230 - 110}{2.3} = 52.2 \text{ amperes}$$

Problem As shown in Figure 14-6, a resistance of 130 ohms and a capacitance of 30 microfarads are connected in parallel across a

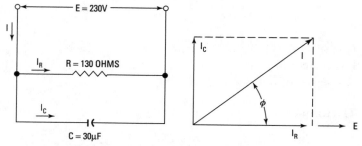

Figure 14-6   Resistance and capacitance in parallel.

230-V, 50-Hz supply. Find the following:

1. Current in each circuit
2. Total current
3. Phase difference between the total current and the applied voltage
4. Power consumed
5. Power factor

Solution The capacitive reactance (X_c) of the circuit is

$$X_C = \frac{10^6}{2\pi \times 50 \times 30} = 106 \text{ ohms}$$

where

X_C is capacitive reactance, $X_C = \dfrac{1}{2\pi FC}$

2π is 6.28; 10^6 converts μF to F

I_R is current through resistance

F is frequency (Hz)

C is capacitance (F)

1. Current through the resistance is

$$I_R = \frac{E}{R} = \frac{230}{130} = 1.77 \text{ amperes}$$

Current through the capacitance is

$$I_C = \frac{E}{X_C} = \frac{230}{106} = 2.17 \text{ amperes}$$

2. Total current is

$$I = \sqrt{I_R^2 + I_C^2} = \sqrt{1.77^2 + 2.17^2} = 2.8 \text{ amperes}$$

3. Phase difference is

$$\cos\angle\theta = \frac{I_R}{I} = \frac{1.77}{2.8} = 0.632$$

$$\angle\theta = 50.8° \text{ (angle of lead)}$$

4. Power consumed is

$$P = I_R^2 R = 1.77^2 \times 130 = 407 \text{ watts}$$

5. Power factor according to (3) is 0.632, or 63.2 percent.

Problem Two circuits, Figure 14-7, are connected in parallel as shown. If the voltage of the source is 120, calculate the:

1. Phase displacement
2. Power factor of the circuit
3. Total current

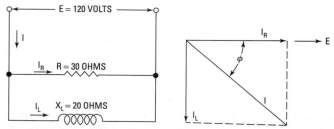

Figure 14-7   Resistance and inductance in parallel.

Solution With reference to the vector diagram in Figure14.7, the current through the ohmic resistance is

$$I_R = \frac{E}{R} = \frac{120}{30} = 4 \text{ amperes}$$

Current through the inductive reactance is

$$I_L = \frac{E}{X_L} = \frac{120}{20} = 6 \text{ amperes}$$

1. Phase displacement is

$$\tan\theta = \frac{I_L}{I_R} = \frac{6}{4} = 1.5$$

$$\theta = 56.3°$$

2. Power factor is

$$\cos\angle\theta = 0.555, \text{ or approximately } 56\% \text{ (lagging)}$$

3. Total current is

$$I = \sqrt{4^2 + 6^2} = 7.2 \text{ amperes}$$

Problem In two circuits (Figure 14-8) in parallel, one branch consists of a resistance of 15 ohms and the other of an inductive reactance of 10 ohms. When the impressed voltage is 110, find the

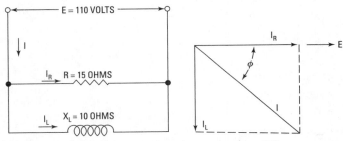

Figure 14-8   Resistance and inductance in parallel.

1. Current through the ohmic resistance (I_R)
2. Current through the inductive reactance (I_L)
3. Line current (I_T)
4. Power factor (cos $\angle\theta$)
5. Angle of lag of the line current ($\angle\theta$)

Solution With reference to the vector diagram in Figure 14-8, the following are true:

1. Current through the ohmic resistance is

$$I_R = \frac{E}{R} = \frac{110}{15} = 7.34 \text{ amperes}$$

2. Current through the inductive reactance is

$$I_L = \frac{E}{X_L} = \frac{110}{10} = 11 \text{ amperes}$$

3. Total line current (I_T) is

$$I = \sqrt{I_R^2 + I_L^2} = \sqrt{7.34^2 + 11^2} = 13.2 \text{ amperes}$$

4. Power factor is

$$\cos\angle\theta = \frac{I_R}{I_T} = \frac{7.34}{13.2} = 0.556$$

5. Angle of lag of the line current is

$$\angle\theta = 56.2°$$

where

   I_R is resistance current
   E is voltage applied
   R is resistance
   I_L is current through coil
   X_L is inductive reactance
   I_T is total current

Problem As shown in Figure 14-9, a resistance of 10 ohms, an inductance of 0.6 henry, and a capacitance of 300 microfarads are

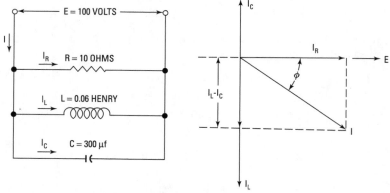

Figure 14-9   Resistance, inductance, and capacitance in parallel.

connected in parallel across a 100-V, 25-Hz supply. Find the

1. Current in each circuit
2. Total current
3. Power factor
4. Power consumption

Solution In this example, it is first necessary to find the inductive and capacitative reactances. They are, respectively,

$$X_L = 2\pi \, FL$$

$$X_L = 2\pi \times 25 \times 0.06 = 9.425 \text{ ohms}$$

$$X_C = \frac{1}{2\pi FC}$$

$$X_C = \frac{10^6}{2\pi \times 25 \times 300} = 21.2 \text{ ohms}$$

1. Current through the resistance is

$$I_R = \frac{E}{R} = \frac{100}{10} = 10 \text{ amperes}$$

Current through the inductive reactance is

$$I_L = \frac{E}{X_L} = \frac{100}{9.425} = 10.61 \text{ amperes}$$

Current through the capacitive reactance is

$$I_C = \frac{E}{X_C} = \frac{100}{21.2} = 4.72 \text{ amperes}$$

2. Total current is

$$I = \sqrt{I_R^2 + (L_L - I_C)^2}$$

$$= \sqrt{10^2 + (10.61 - 4.72)^2}$$

$$= 11.61 \text{ amperes}$$

3. Power factor is

$$\cos\angle\theta = \frac{I_R}{I_T} = \frac{10}{11.61} = 0.861 \text{ (lagging)}$$

$$\angle\theta = 30.53°$$

4. Power consumption is

$$P = EI\cos\angle\theta = 100 \times 11.61 \times 0.861 = 1000 \text{ watts}$$

where

    X_L is inductive reactance
    2π is constant 6.28
    F is frequency (Hz)
    L is inductance (H)
    X_C is capacitive reactance
    I_R is current through resistance
    E is applied voltage
    R is resistance

Problem A circuit connected as shown in Figure 14-10 contains a 10-ohm resistance and 0.5-henry inductance in parallel with a capacitor of 20 microfarads. The voltage and frequency of the source are 1000 and 60, respectively. Find the

  1. Current through the coil
  2. Phase angle between the current through the coil and the potential across it

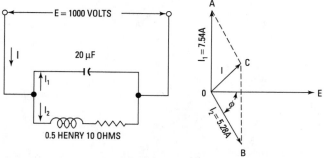

Figure 14-10 Impedance and capacitance in parallel.

3. Current through the capacitor
4. Total current

Solution

1. The current through the coil (I_L) is

$$I_2 = \frac{1000}{\sqrt{10^2 + (2\pi \times 60 \times 0.5)^2}} = 5.30 \text{ amperes}$$

2. Phase angle is

$$\tan\theta = \frac{2\pi 60 \times 0.5}{10} = 18.85$$

Use calculator and hit

[arc] [tan] = 86.9°

(tangent 18.85 is 86.9°)

3. Current through the capacitor (I_C) is

$$I_1 = \frac{1000}{X_C} = 20 \times 10^{-6} \times 2\pi \times 60 \times 1000$$

$$= 7.54 \text{ amps}$$

With reference to the vector diagram

$OA = I_1 = 7.54$ amps, and $OB = I_2 = 5.30$ amps

As the current, I, is the resultant of these two vectors, it is now possible to construct the parallelogram as indicated by the dotted lines. It follows from the construction that $\beta OBC = 90°$, and from the law of cosines,

$$I = \sqrt{(5.30^2 + 7.54^2) - (2 \times 5.30 \times 7.54 \sin\angle\theta}$$

$I = 2.26$ amperes (approximately)

Problem A series circuit consists of a 30-microfarad capacitance and a resistance of 50 ohms connected across a 110-V, 60-Hz supply. Calculate the

  1. Impedance of the circuit (Z)
  2. Current in the circuit (I_T)
  3. Voltage drop across the resistance (E_R)
  4. Voltage drop across the capacitance (E_C)
  5. Angle between the voltage and the current ($\angle\theta$)
  6. Power loss (P)
  7. Power factor of the circuit (PF or $\cos\angle\theta$)

Solution

  1. Impedance of the circuit is

$$X_C = \frac{1}{2\pi fC} = \frac{1}{2\pi \times 60 \times 0.000030} = 88.4 \text{ ohms}$$

$$Z = \sqrt{R^2 + X_C^2} = \sqrt{50^2 + 88.4^2} = 101.6 \text{ ohms}$$

  2. Current in the circuit is

$$I = \frac{E}{Z} = \frac{110}{101.6} = 1.08 \text{ amperes}$$

  3. Voltage drop across the resistance is

$$E_R = IR = 1.08 \times 50 = 54 \text{ volts}$$

  4. Voltage drop across the capacitance is

$$E_C = IX_C = 1.08 \times 88.4 = 95.5 \text{ volts}$$

  5. Angle between voltage and current is

$$\cos\theta = \frac{R}{Z} = \frac{50}{101.6} = 0.492$$

$$\angle\theta = 60.5°$$

6. Power loss is

$$P = I^2R = 1.08^2 \times 50 = 58.3 \text{ watts}$$

7. Power factor is cos, therefore

$$\cos\angle\theta = 0.492, \text{ or } 49.2 \text{ percent}$$

where

X_C is capacitive reactance
2π is constant 6.28
F is frequency (Hz)
C is capacitance (F)
E is voltage
Z is impedance
I is current
E_R is voltage across resistor
E_C is voltage across capacitor

Problem A single-phase motor is taking 20 amperes from a 400-V, 50-Hz supply, the power factor being 80 percent lagging. What value of capacitor connected across the circuit will be necessary to raise the power factor to unity?

Solution With reference to the vector diagram in Figure 14-11, it is evident that

$$I_C = I_m \sin\theta$$

Since

$$I_m = 20, \text{ and } \sin\theta = \sqrt{1 - 0.8^2} = 0.6$$

then

$$I_C = 20 \times 0.6 = 12 \text{ amperes}$$

and

$$X_C = \frac{E}{I_C} = \frac{400}{12} = 33.3 \text{ ohms}$$

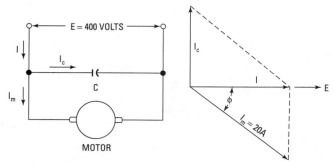

Figure 14-11 A motor circuit and vector diagram.

The equation for the capacitive reactance is

$$X_C = \frac{1}{2\pi fC}$$

or

$$C = \frac{1}{2\pi \, fXc}$$

Inserting values and simplifying,

$$C = \frac{1}{6.28 \times 50 \times 33.3} = .0000956F \text{ or } 95.6 \ \mu F$$

$$C = 95.6 \ \mu F$$

Problem A 220-V, 60-Hz, 10-hp, single-phase induction motor operates at an efficiency of 86 percent and a power factor of 90 percent. What capacity should be placed in parallel with the motor so that the feeder supplying the motor will operate at unity power factor?

Solution The current taken by the motor is

$$I = \frac{10 \times 746}{220 \times 0.9 \times 0.86} = 43.81 \text{ amperes (lagging)}$$

Current taken by the capacitor is

$$Ic = I \times \sin\angle\theta$$

$Ic = 43.81 \times 0.4358 = 19$ amperes

$PF = \cos\angle\theta$

$PF = 90\%$

$PF = .9$ or use calculator to obtain:

[arc] [cos] $= 25.84°$

[sin] of $25.84° = 0.4358$

The capacitive reactance of the capacitor is

$$Xc = \frac{220}{19} = 11.58 \text{ ohms}$$

Capacity necessary is

$$C = \frac{10^6}{2\pi \times 60 \times 11.58} = 229 \text{ microfarads}$$

Problem In a balanced, three-phase, 208-V circuit, the line current is 100 amperes. The power is measured by the two-wattmeter method; one meter reads 18 kW and the other 0. What is the power factor of the load? If the power factor were unity and the line current the same, what would each wattmeter read?
Solution The expression for power is

$$P = EI \cos\phi\sqrt{3}$$

Since one wattmeter reads zero, then

$$18,000 = 208 \times 100 \times \cos\phi\sqrt{3}$$

so

$$\cos\phi = \frac{18,000}{208 \times 100 \times \sqrt{3}} = 0.5$$

With the power factor unity and with the same line current, then

$$\tan\phi = 3\left[\frac{W_1 - W_2}{W_1 + W_2}\right] = 0, \text{ or } W_1 = W_2$$

Also

$$W_1 + W_2 = 208 \times 100 \times \sqrt{3} = 36 \text{ kW}$$

That is, each wattmeter reads 36/2, or 18 kW.
where

P is power
E is voltage
I is current
W_1, W_2 are wattmeters
ϕ is one phase of 3ϕ power

Problem In a certain manufacturing plant, a single-phase alternating current of 110 V, 60 Hz is applied to a parallel circuit having two branches. Branch 1 consumes 27 kW with a lagging power factor of 0.8, while Branch 2 consumes 12 kW with a leading power factor of 0.9. Calculate the

1. Total power supplied in kW and kVA
2. Power factor of the total load

Solution

1. Total power supplied is

$$P = P_1 + P_2 = 27 + 12 = 39 \text{ kW}$$

The reactive-power component supplied by Branch 1 is

$$P_1\left[\frac{\sin\phi_1}{\cos\phi_1}\right] = P_1\tan\phi_1 = 27 \times 0.7508 = 20.27 \text{ kVA}$$

Similarly, the reactive-power component supplied by Branch 2 is

$$P_1\left[\frac{\sin\phi_2}{\cos\phi_2}\right] = P_2\tan\phi_2 = 12 \times 0.4834 = 5.80 \text{ kVA}$$

The sum of the reactive-power components is

$$P_1 \tan\phi_1 - P_2 \tan\phi_2 = 20.27 - 5.80 = 14.47 \text{ kVA}$$

Total power supplied, in kVA, is

$$\text{kVA} = \sqrt{39^2 + 14.47^2} = 41.6$$

2. Power factor of the total load is

$$\cos\phi = \frac{39}{41.6} = 0.94$$

Problem A certain load takes 40 kVA at 50-percent lagging power factor, while another load connected to the same source takes 80 kVA at 86.7-percent lagging power factor. Find the

1. Total effective power
2. Reactive power
3. Power factor
4. Apparent power

Solution The apparent power taken by each load, expressed with reference to the voltage, is given in Figure 14-12. From a trigonometric table, the angle corresponding to a 50-percent power factor is 60°, and the angle corresponding to an 86.7 percent power factor is 30°. To find the resultant, it is necessary to complete the parallelogram. As shown in Figure 14-12, the sum of the effective power is indicated along the horizontal, and is as follows:

1. $(40 \times 0.5) + (80 \times 0.867) = 89.36 \text{ kW}$

The total reactive power is indicated along the vertical, and is

2. $(40 \times 0.867) + (80 \times 0.5) = 74.68 \text{ kVA}$

$$\tan\theta = \frac{74.68}{89.36} = 0.836$$

3. $\cos\theta = 0.767$

382 Chapter 14

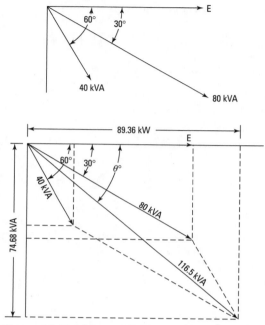

Figure 14-12 Vector diagrams of true, reactive, and apparent power in an AC circuit.

4. $\theta = 39.9°$

The apparent power is

5. $\dfrac{89.36}{\cos\theta} = \dfrac{89.36}{0.767} = 116.5 \text{ kVA}$

Problem A group of induction motors takes 100 kVA at 84-percent lagging power factor. A synchronous motor connected to the same line takes 60 kVA at 70.7-percent leading power factor. Determine the

1. Total effective power

2. Reactive power

3. Power factor

4. Apparent power

Solution This is a simple problem on power-factor correction. Remember here that, since the synchronous motor has a leading power factor, this vector must be laid out in the proper direction or 45° above the reference or voltage vector. Thus, from Figure 14-13

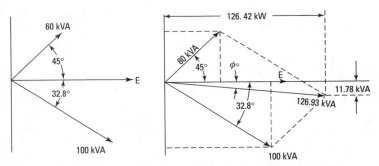

Figure 14-13 Vector diagrams of leading and lagging loads.

1. $(100 \times 0.84) + (60 \times 0.707) = 126.42$ kVA

The total reactive power is the difference between the two, since one is leading and the other is lagging. Thus

2. $(100 \times 0.542) - (60 \times 0.707) = 11.78$ kVA

3. $\tan\theta = \dfrac{11.78}{126.42} = 0.093$

$\cos\theta = 0.996$

The apparent power is

4. $\dfrac{126.42}{\cos\theta} = \dfrac{126.42}{0.996} = 126.93$ kVA

Table 14-1 shows NEMA resistor classifications for wound-rotor motors.

Table 14-1 NEMA Resistor Classification for Wound-Rotor Motors

Percent Full-Load Current on First Point	Starting Torque % of Full Load			Wound-Rotor Induction Motors		Resistor Class Number						
	Series Motors	Compound Motors	Shunt Motors	1 Ph. Stg.	3 Ph. Stg.	5 Sec. on Out of 80 Sec.	10 Sec. on Out of 80 Sec.	15 Sec. on Out of 90 Sec.	15 Sec. on Out of 60 Sec.	15 Sec. on Out of 45 Sec.	15 Sec. on Out of 30 Sec.	Cont.
25	8	12	25	15	25	111	131	141	151	161	171	91
30	30	40	50	30	50	112	132	142	152	162	172	92
70	50	60	70	40	70	113	133	143	153	163	173	93
100	100	100	100	55	100	114	134	144	154	164	174	94
150	170	160	150	85	150	115	135	145	155	165	175	95
200	250	230	200		200	116	136	146	156	166	176	96

Summary

It is often advantageous for anyone concerned with electric motors to be able to determine certain facts about the circuits by means of simple calculations. For example, it may be necessary to provide the proper size of wire to install a motor. This can be easily done by substituting the known factors in the proper formula. This chapter offers many examples of the solving of similar problems.

Review Questions

1. What is the efficiency of a 10-hp motor that requires 8 kW at full load?
2. What is the formula for finding capacitive reactance?
3. What is the formula for finding inductive reactance?
4. How much current flows through a 10-henry inductance when connected across a 220-V, 60-Hz, alternating current?
5. List the three forms of Ohm's Law for DC circuits.
6. What is the circular-mil area of a conductor, the diameter of which is 0.125 in.?
7. What is the formula for finding the resistance of any conductor?
8. How can the cross-sectional area of a circular conductor be expressed?
9. What is Kirchhoff's Law? How is it used with motors?
10. How do you find copper losses in a conductor?
11. How do you find copper losses in an armature?
12. How do you find copper losses in a field coil?
13. What is meant by full-load armature current?
14. How do you find the efficiency of a motor?
15. How do you find the horsepower of a motor? What is the formula?

Chapter 15

Modifying Motors

In order to modify a motor it is necessary to understand its operating characteristics. The characteristics can be changed in certain ways. The method used to change the operational characteristics is chosen with the limitations of the motor in mind. For instance, the most popular electrical types of single-phase motors for general use are capacitor-start and split-phase.

The split-phase motor has two windings, start and run, which are activated to start the motor. The start winding is cut out of the circuit at about 75 percent of operating speed.

The capacitor-start motor is virtually identical to the split-phase motor, but it delivers two to three times the starting torque per ampere of current. This is done by modifying the start-winding circuit and adding a capacitor.

Since the basic difference in these two types of motors is in their load-starting ability, the lower-cost, split-phase motor is the logical choice where the starting load is light or where load is applied after the motor has reached operating speed. Capacitor-start motors are required, of course, where heavy loads must be started.

Permanent-split capacitor and shaded-pole motors are designed specifically for light starting loads such as time-sequence switches, dampers, and fans.

Characteristics of the different types of motors are shown in Table 15-1. Keep in mind that, as a rule, both motor price and physical size increase as the rated rpm decreases for a given horsepower rating. This means that you may save money by selecting the highest-speed motor that will drive your device.

Motor Modifications for Shafts

A number of variations in adapters for motor shafts are available. This type of modification makes motors usable for different pulley sizes or drive mechanisms. Modification E of Figure 15-1 shows a shaft adapter for changing a ¼-in. shaft to ⁵⁄₁₆-in. Modification F shows how to change a ¼-in. shaft by coupling it with an extension 4 in. long. A flat spot is available along the entire 4-in. extension. This will make the motor work in an area where a longer shaft is needed to fit outside an enclosure in most cases.

Table 15-1 Typical Characteristics of AC Motors

	H.P. Ratings	Full-Load Speeds, rpm (60 Hz)	Starting Torque	Breakdown Torque	Starting Current	Comparative Cost (100 = lowest)	Guidelines
Shaded pole	1/65 to 1/20	1650	very low	low	low	100	Low-cost motor for light-duty applications. Compact, rugged, easy to maintain. R&M motors have higher starting torques than most.
Permanent-split capacitor	1/50 to 1/3 1/60 to 1/6	3250 1625	low		low	140	Very compact, easy to maintain. High efficiency, high power factor. Can operate at several speeds with simple control devices.
Split-phase	1/40 to 1/3 1/50 to 1/6	3450 1725	moderate		high	120	For constant-speed operation, varying loads. Where moderate torques are desirable, may be preferable to more expensive capacitor start.
Capacitor-start	1/40 to 1/3 1/50 to 1/6	3450 1725	high		moderate	150	Suitable for constant speed under varying load, high torques, high overload capacity.
Polyphase	1/30 to 1/3 1/75 to 1/6	3450 1725	high		moderate	150	Generally suited to same applications as capacitor-start motors if polyphase power is available. Gets to operating speed smoothly and quickly.

(Courtesy Robbins & Myers)

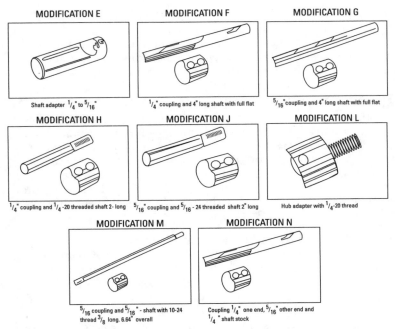

Figure 15-1 Modifications on shafts of AC motors. *(Courtesy General Electric)*

Modification G changes the ⁵⁄₁₆-in. shaft with a coupling to one with an extra 4 in. Modification H is a ¼-in. coupling with a ¼–20 threaded extension for a 2-in. addition. The setscrew in the coupler attaches to the existing motor shaft, and the other setscrew in the coupler attaches to the extension so that the ¼–20 thread is used to screw on a pulley or drive device or place the device on the shaft with a ¼–20 nut pulled up tightly to hold it in place on the shaft extension.

A larger shaft coupling and extension is shown in Modification J. This ⁵⁄₁₆-in. coupling has a ⁵⁄₁₆–24 threaded end so that the 2-in. extension can be used to handle larger devices. In some instances a shorter extension with screw threads is needed. Modification L shows a hub adapter with ¼–20 threads.

Modification M is a ⁵⁄₁₆-in. coupling and a ⁵⁄₁₆-in. shaft with a 10–24 threaded end that is ³⁄₈ in. long and 6.64 in. long overall. Note the flat spot on one end that fits into the coupling and is held from slipping on the shaft extension by a setscrew that makes contact with the flat spot.

Modification N shows a coupling with ¼ in. on one end and ⁵⁄₁₆ in. on the other end and a ¼-in. shaft stock.

These shaft adapters are available from the motor manufacturer or from a local supply house. Many variations can be made in the original motor shafts with these extensions and adapters.

Speed Controls

Another way to modify a motor is to change its speed. A number of devices are available for speed changing—in most instances, speed reduction. Both mechanical and electrical modifications can be made to utilize a motor of one speed at another desired speed. However, there are changes in the motor performance that must be taken into consideration when the speed is varied. Different motors react differently. A brief description of some of these differences follows.

Speed Control of Series-Wound Motors

Series motors can be used on AC or DC. They are capable of supplying high starting torque, high speeds, and high outputs.

A series motor does not inherently provide good speed regulation. It is classified as having a varying-speed characteristic. This means that the speed will decrease with an increase in load and increase with a decrease in load. The amount of speed change will depend upon the slope of the motor's speed-torque curve at the load point or throughout the load range. Furthermore, if a speed control is used, the regulation and the amount of speed adjustment will be influenced by, or will depend upon, the controlling means. Basically, the speed of a series motor can be changed by varying the voltage input to the motor. This can be done in a number of ways. The series motor is very versatile because it can yield a great variety of speed-torque characteristics, depending on its speed-controlling means (the power supply).

Series Resistance Control

Figure 15-2 shows the resistance method of voltage control, which in turn affects the speed of the motor. Figure 15-3 illustrates the principle using a DC current supplied to the motor by way of a full-wave bridge rectifier. A variable resistor or rheostat in series with the motor at any load will decrease the speed as the resistance is increased. Actually, the speed can be adjusted until the motor stops. Due to starting-torque limitations, armature cogging, and reduced ventilation, which causes overheating, the minimum speed is usually limited to some higher value.

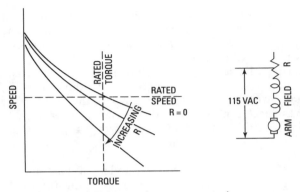

Figure 15-2   AC series rheostat control. *(Courtesy Bodine)*

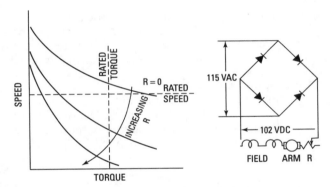

Figure 15-3   DC series rheostat control. *(Courtesy Bodine)*

In the running mode, a series resistor introduces a voltage drop in the circuit directly proportional to the current. The voltage drop across the resistor will increase as the motor is loaded since the motor current will increase with the load. It follows, therefore, that the voltage across the motor will decrease with an increase in load and the speed will drop more rapidly with load whenever a series resistor is used. The higher the resistance value, the greater the drop in speed as the load is increased.

When motor speeds are adjusted by resistance, they become much more sensitive to load variations. A given load increase on a series-wound motor with resistance in series will cause a larger drop in speed than it would without the added resistance. Also, the value of resistance in a resistor will change with temperature. To maintain speed, as the normal operating temperature is approached, the resistance will usually require adjustment.

The starting torque of a series motor will be noticeably affected by the presence of resistance in series with the winding. This is noticeable especially with a sewing machine, where the motor has to be started by hand each time a universal-type series motor is used to drive the sewing machine. This is due to the fact that at starting maximum current wants to flow, which will limit the motor voltage to its lowest value.

A series motor will sometimes fail to start under full load at the lowest speed or highest-resistance setting. The minimum full-load speed at which a series motor will operate with a series resistor is usually limited by the starting torque available to start the load with that value of resistance.

Typically on AC, the speed range of a series motor using a variable series resistor will be from 1.5:1 to 3:1, depending on the motor. On DC, the speed range will be increased because of the improved regulation and corresponding increase in starting torque.

Shunt-Resistance Speed Control

Figure 15-4 shows this shunt resistance method of speed control using AC. Figure 15-5 illustrates the principle using DC supplied to the motor by way of full-wave rectifier bridge. For purposes of comparison, Figures 15-4 and 15-5 were obtained on the same motor used in Figures 15-2 and 15-3.

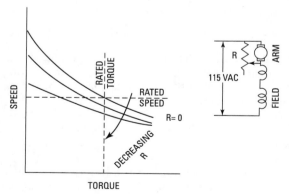

Figure 15-4   AC shunt rheostat control. *(Courtesy Bodine)*

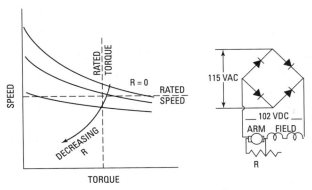

Figure 15-5   DC shunt rheostat control. *(Courtesy Bodine)*

The speed range is usually limited by this method because of increased current passing through the field coils and the corresponding heating effect. Care must be taken in choosing the minimum resistance across the armature because the high field current created may burn out the field coils. A wide speed range may be used only if the application has a very limited duty cycle.

Compared to the series resistance method, the shunt resistance method has a narrow or more limited speed range. However, the shunt method of speed improves the speed regulation of the motor and maintains good starting torque characteristics. It is an excellent method for matching motor speeds.

Series and Shunt Resistance

Another method used to modify a motor for speed control is by using a combination of the two systems previously discussed. Varying the rheostats singly or in tandem will give speed-control characteristics of each of these systems as well as those in between. Again, caution must be exercised to avoid overheating the field coils when an armature shunting-resistor is used.

Variable Transformer Control

Figures 15-6 and 15-7 show typical characteristic curves for the transformer control method. The motor used in Figure 15-5 is again employed to make the comparison meaningful.

By using a variable transformer to vary the voltage across a series motor, a speed range of 4:1 to 7:1 is typical, depending on the motor. If a full-wave bridge is used to convert the output of the transformer

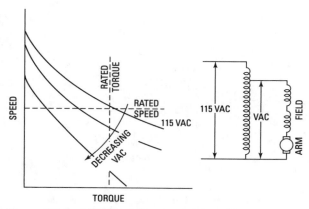

Figure 15-6 Variable AC voltage control. *(Courtesy Bodine)*

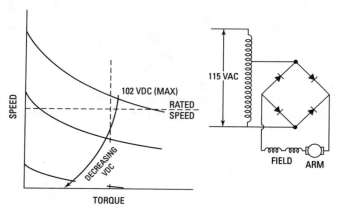

Figure 15-7 Variable DC voltage control. *(Courtesy Bodine)*

to DC, the speed range will be increased because of the improved regulation and starting torque.

Tapped Field Windings

Another way to modify a motor for speed control is by using tapped field windings and a switching arrangement connected to the taps. Such motors provide discrete speed steps rather than a continuously adjustable output.

Governors for Speed Control

To limit the operating speed of a series-wound motor to a constant value, governors can be used. Such governors are usually electromechanical devices having a set of contacts that periodically insert a fixed resistor into the circuit to limit the motor speed to a predetermined value.

Electronic Speed Control

Electronic methods of speed control are many and varied. Older methods such as thyratrons or grid-controlled rectifiers, saturable reactors, or magnetic amplifiers have fallen from popularity in favor of solid-state components because of size, weight, and cost. They all have one thing in common: to provide an adjustable voltage to control the speed of the series motor.

Today, a typical electronic series motor control is a half-wave SCR (silicon-controlled rectifier) device with feedback. Since these devices are half-wave, the maximum voltage to the motor is substantially less than 115 V, resulting in a relatively low top speed from the standard motor used in our comparison in Figure 15-8. However, the feedback feature, which corrects for drop in speed due to loading, usually allows an extension of the speed range on the low end due to improved starting torque and speed-regulation characteristics.

Prolonged operation of series-wound motors by SCR controls can reduce brush life. Some adjustable-speed, series-wound applications obtain best brush life with triac-type electronic controls.

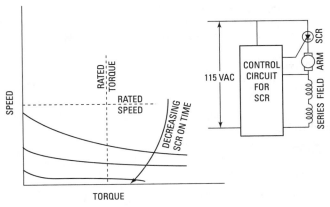

Figure 15-8 SCR control. *(Courtesy Bodine)*

Where excellent speed regulation, wide speed range, and continuous duty operation are required, shunt-wound or PM (permanent-magnet) motors with full-wave, feedback-type SCR controllers have become popular in recent years.

For specific applications, motor design factors can sometimes be changed to aid speed-control characteristics. For example, armature slots can be skewed to provide less cogging at the low end of the speed range. This normally will extend the usable speed range.

Extended operation of series-wound motors at low speeds causes overheating unless the motor is specially designed for such operation or external cooling is used.

Many methods are available to modify a motor for the control of speed. The final choice must be made based on the performance desired, the temperature limitations of the motor, size restrictions, and overall system cost.

Modifying AC Motors for Speed Control

Modifying AC motors for speed control is not easy. Most of the AC motors are made for a specific speed. The number of poles makes a difference in the speed. The universal motor (AC/DC) is controlled easily by placing any number of speed-control devices in the circuit with the motor. This type of speed control has already been discussed.

AC Motors Designed for Constant Speed

One of the characteristics of the AC induction motor is its ability to maintain a constant speed under normal voltage and load variations. Therefore, as would be expected, this type of motor does not lend itself to a simple method of speed control over a wide range.

There are, however, some variations of conventional induction motors that are designed for the express purpose of improved speed control. These motors may use wound rotors with variable resistance, brush shifting means, and other special modifications.

Keep in mind that the speed of an induction motor is controlled by two factors: the number of poles in the stator winding and the frequency of the power supply.

$$\text{rpm} = \frac{F \times 60}{P/2}$$

where

rpm = revolutions per minute

F = frequency (in Hz)

P = number of poles

The synchronous AC motor rotates at the exact speed defined by the formula. The nonsynchronous motor never operates at synchronous speed. The difference between the synchronous speed and the actual speed is known as rotor slip.

$$\text{Percent of slip} = \frac{\text{synchronous speed} - \text{actual speed}}{\text{synchronous speed}} \times 100$$

The amount of slip depends on the motor design, power input, and motor load. The nonsynchronous motor speed can be adjusted by changing the amount of rotor slip.

AC Motor Speed-Control Methods

Among the methods used to control AC motor speed are the following:

- Change in frequency
- Change in number of poles in the stator
- Change in rotor slip

Change in Frequency

Change in frequency has the advantage of providing stepless speed changes over a wide range, and may be used with either synchronous or nonsynchronous motors. The major disadvantage encountered with this method is the relatively high cost of frequency of the changing power supply. The frequency-change method must also operate within narrow limits since a motor designed for 50 to 60 Hz usually cannot be satisfactorily operated at a radically different frequency. The motor must be designed for power supply limitations. These may be the laminations and winding design. The inductive reactance of the motor is determined by the frequency and inductance of the winding. The laminations or core and the number of turns determine the inductance and thus the inductive reactance. The frequency of the power source also determines the inductive reactance. The inductive reactance has a definite bearing on how much the current will be limited. Excess current causes excess heat. This heating effect can take place in a very short time. It is important, therefore, to make sure the motor is limited in frequency changes. It is difficult to use a 400-Hz motor on a 60-Hz line without excessive current and heat generation being the result. 60 Hz of horse power will not operate a 400-Hz aircraft motor since the iron in the core and the number of winding turns are definitely less.

Change in Number of Poles in the Stator

Both synchronous and nonsynchronous motors can use the pole-changing method for speed control (Figure 15-9). This method of change or modification has a limited number of speeds to offer since the speed is a function of the number of poles. Most motors have no more than four definite speeds using this method. A portion of the winding is idle during the operation of one or more speeds. This results in motor inefficiency and a considerable reduction in the output rating for any given frame size. Switching methods for pole changing are also expensive and complicated, making the method useful in relatively few applications.

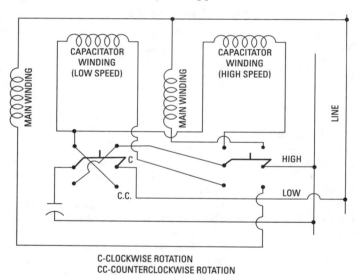

C-CLOCKWISE ROTATION
CC-COUNTERCLOCKWISE ROTATION

Figure 15-9 Simplified pole-changing circuit. *(Courtesy Bodine)*

Change in Rotor Slip

The changing of rotor slip is simpler, less costly, and the most widely used technique for varying the speed of an AC induction motor. There are three types of nonsynchronous motors to which this method is best suited. They are the shaded-pole, permanent-split capacitor, and polyphase.

Due to the sensitivity of the centrifugal or relay starting switches, the rotor-slip method should not be applied to split-phase-start and capacitor-start motors unless the speed will never go low enough to engage the starting switch. If the motor runs at the reduced speed with the starting switch closed, the auxiliary winding or the switch contacts will soon burn out.

There are several ways to change the power input to an induction motor. This will increase or decrease the amount of slip. Six modifications are shown here to show how the method works.

Tapped Winding

This method is most widely used for shaded-pole fan motors. The change in input is obtained by changing the motor impedance through the use of various portions of the total winding (Figure 15-10). The number of speeds is determined by the number of taps introduced into the winding. In addition to shaded-pole motors, the tapped-winding technique can be used with permanent-split capacitor motors.

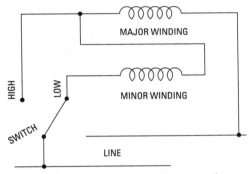

Figure 15-10   Tapped winding circuit for speed control. *(Courtesy Bodine)*

Series Resistance

A variable resistor can be used to vary the voltage across the winding of an induction motor (Figure 15-11). Series resistance can be used with either shaded-pole or permanent-split capacitor motors.

Variable Voltage Transformer

This method may be used in place of a series resistor to reduce voltage across the winding. It has the advantage of maintaining substantially the same voltage under the starting condition when the current is higher than during

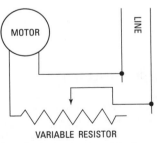

Figure 15-11   Simplified series resistance circuit.

(Courtesy Bodine)

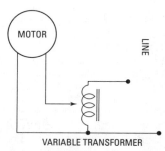

VARIABLE TRANSFORMER

Figure 15-12 Variable-voltage transformer method. *(Courtesy Bodine)*

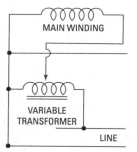

Figure 15-13 Variable-voltage transformer in a PSC motor circuit. *(Courtesy Bodine)*

the running mode. There is also much less power lost as heat than with a resistor (Figure 15-12). By reducing the voltage across the main winding of a permanent-split capacitor motor, full voltage is maintained across the capacitor winding, providing more stable operation at lower speeds (Figure 15-13).

Winding Change
The winding change is applicable to only the permanent-split capacitor motor (Figures 15-14 and 15-15). The functions of the main and the capacitor (starting) windings can be switched to provide a "high" and a "low" speed. High speed is obtained when the winding with fewer turns functions as the main, while the lower speed is achieved when the winding with more turns functions as the main. This is an extremely efficient technique, but it does require that the motor winding be exactly tailored to the load in order to provide the desired two speeds.

Shunt Resistor
Also confined to the permanent-split capacitor motor, this method has been found to provide stable speed in four-pole, 60-Hz motors up to $\frac{1}{100}$ horsepower (7.5 W) from 1500 rpm down to 900 rpm with constant torque output (Figure 15-16). With this method it is necessary to use a high-slip type of rotor.

Solid-State Semiconductor
A very popular and practical speed control means in home-installed heating and cooling equipment, semiconductor controls are usually used with induction motors in fractional-horsepower sizes and general-purpose construction using standard slip rotors. In the smaller sizes, shaded-pole and permanent-split capacitor motors may be used. For applications that require larger drives ($\frac{1}{8}$ horsepower and up), split-phase-start or capacitor-start motors may be specified

Figure 15-14   Permanent-split capacitor winding changes for high and low speeds. *(Courtesy Westinghouse)*

if the speed range is narrow and the lowest speed is well above the opening point of the starting switch. The high-inertial-blower–type load typical in heating and cooling equipment plus the relatively narrow speed-range requirement make possible the use of a standard motor and the simple series speed-control circuit. More sophisticated solid-state semiconductor controls have been developed that are capable of satisfactorily controlling the speed of

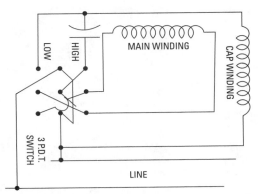

Figure 15-15   Winding function-change method. *(Courtesy Bodine)*

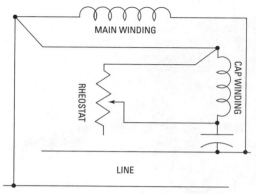

Figure 15-16 Shunt-resistor method.

induction motor driving loads with relatively small inertia. Such applications are generally confined to motors in the subfractional range (less than $\frac{1}{20}$ horsepower).

Reversing Direction of Rotation

Some motors come with quick snap connectors that can be switched easily to change the motor rotation. Take a look under the removable cover over the box located on the motor. The box for power input is usually located on the side of the motor (Figure 15-17).

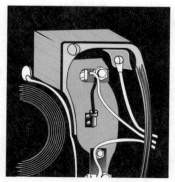

Figure 15-17 By moving the two leads it is possible to change the motor's direction of rotation. *(Courtesy General Electric)*

Other motors also have terminals that can be switched easily to change from a 120-V motor to one that will operate on 240 volts AC. The stator windings are in a series of the 240-V operation and in parallel for the 120-V operation. If a motor has the capability, its terminals will probably be marked on the motor itself or inside the junction box on the side of the motor.

Summary

In order to modify a motor it is necessary to understand its operating characteristics. The characteristics can be changed in certain ways. The method used to change the operational characteristics is chosen with the limitations of the motor in mind.

Permanent-split capacitor and shaded-pole motors are designed specifically for light starting loads, such as time-sequence switches, dampers, and fans.

A number of adapters are available for motor shafts. This type of modification makes a motor usable for different pulley sizes and drive mechanisms. They are available from the motor manufacturer or local parts supply houses. Another way to modify a motor is to change its speed. A number of devices are available for speed control or changing. Both mechanical and electrical modifications can be used for this purpose.

Another way to modify a motor for speed control is to use tapped field windings and a switching arrangement connected to the taps. Mechanical as well as electromechanical governors can be used with series-wound motors to keep them at a constant value. Electronic speed controls are many and varied. They vary from thyratrons to saturable reactors, and magnetic amplifiers to solid-state components. There are some problems with electronic control; the series motor with a prolonged use of an SCR control, for instance, will have reduced brush life. Extended operation of series-wound motors at low speeds causes overheating unless the motor is specially designed for such operation or external cooling is used.

Modifying AC motors for speed control is not easy. Most AC motors are made for a specific speed. The nonsynchronous motor can have its speed adjusted by changing the slip. Adjusting the frequency is not always possible. Using the 400-Hz, aircraft-type AC motor on 60 Hz can cause damage to the motor. Even variations in the 50/60 Hz motors when switched from one frequency power to the other can cause excessive heat and loss of speed and stability in operation.

The changing of rotor slip is less costly and the most widely used technique for varying the speed of an AC induction motor. There are three types of nonsynchronous motors to which this method is best suited. They are the shaded-pole, permanent-split capacitor, and polyphase motors.

There are several ways to change the power input to an induction motor. This will increase or decrease the amount of slip. Six modifications are tapped winding, series resistance, variable voltage transformer, winding change, shunt resistor, and solid-state semiconductor.

Some motors come with quick-snap connectors that can be switched easily to change the motor rotation. Other motors can be easily switched from 120- to 240-V operation.

Review Questions

1. What does it mean to modify a motor?
2. What do you need to know about the motor before modifying it?
3. How can a pulley change the speed of a motor?
4. How do you change the speed of a motor?
5. Name four types of electronic motor control.
6. What type of motor can have the speed adjusted by changing the slip?
7. Where are 400-Hz motors used? Why?
8. Why does a 60-Hz motor heat up on 50 Hz?
9. What is an SCR? How does it control motor speed?
10. What are the three types of motors best suited to have the slip adjusted for speed control?
11. List six modifications to motors that decrease or increase the motor slip.
12. How is the direction of rotation of an AC motor changed?
13. What is the advantage of the capacitor-start motor?
14. For what kind of applications are the permanent-split capacitor and shaded-pole motors designed?
15. How are shaft adapters classified?

Part V

Motor Controls

Chapter 16

Switches and Relays

One of the most troublesome things that happens to a motor is overheating and burnout of the insulation materials. Although motors have been designed to operate under various temperature conditions, it is always possible for a motor to overheat when a number of things take place:

1. Low line voltage increases the current drawn from the line or power source. Since an increase in current causes an increase in temperature, it is important to make sure that the motor windings are properly protected from such an increase in temperature.

2. Jamming the motor and causing it to stop when the power is on will also cause temperature increases. This takes place almost immediately and a good sensing device is needed to protect the windings.

3. Large increases in voltage can cause excess current to flow through the windings. The increased current will cause increased heat to be generated.

Overload Protection

The effect of an overload is a rise in temperature in the motor windings. The larger the overload, the more quickly the temperature will increase to a point that is damaging to the insulation and lubrication of the motor. An inverse relationship, therefore, exists between current and time. The higher the current, the shorter the time before motor damage, or burnout, can occur.

All overloads shorten motor life by causing deterioration of the motor insulation. Relatively small overloads of short duration cause little damage, but if sustained, they could be just as harmful as overloads of greater magnitude. The relationship between the magnitude (percent of full load) and duration (time in minutes) of an overload is shown in the graph of Figure 16-1.

The ideal overload protection for a motor is an element with current-sensing properties very similar to the heating curve of the motor, which could act to open the motor circuit when *full-load current* (FLC) is exceeded. The operation of the protective device should be such that the motor is allowed to carry harmless

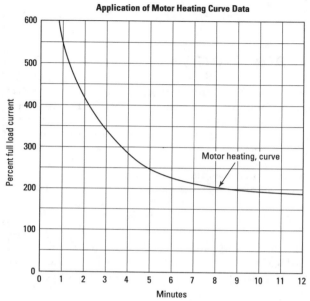

Figure 16-1 The relationship between the magnitude and duration of an overload.

overloads but is quickly removed from the line when an overload has persisted too long.

Fuses are not designed to provide overload protection. Their basic function is to protect against short circuits. A fuse chosen on the basis of motor FLC would blow every time the motor started. On the other hand, if a fuse were chosen large enough to pass the starting current or inrush current, it would not protect the motor against small, harmful overloads that might occur later (Figure 16-2).

Dual-element or time-delay fuses can provide motor overload protection. However, they suffer the disadvantage of being nonrenewable and must be replaced (Figure 16-3).

Overload relays consist of a current-sensing unit connected in-line to the

Figure 16-2 Screw-in type of fuse (top) and one-time knife type of fuse.

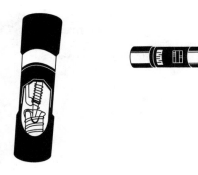

Figure 16-3   Cutaway view
of a dual-element fuse.

motor, plus a mechanism actuated by the sensing unit. This unit
serves to break the circuit directly or indirectly. In a manual starter,
an overload trips a mechanical latch, causing the starter contacts
to open and disconnect the motor from the line. In magnetic
starters, an overload opens a set of contacts within the overload
relay itself. These contacts are wired in series with the starter coil
in the control circuit of the magnetic starter. Breaking the coil circuit causes the starter contacts to open, disconnecting the motor
from the line.

There are two classifications for overload relays—*thermal* and
magnetic. Magnetic relays react only to current excesses and are
not affected by temperature. Thermal relays can be further divided
into two types—*melting alloy* and *bimetallic*.

Figure 16-4 shows the location of the overload in the circuit with
the motor windings. In this case it is a two-terminal external overload device. There is a run capacitor but no start capacitor or relay

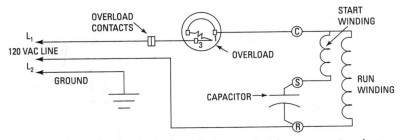

Figure 16-4   Two-terminal external overload with run capacitor but
no start capacitor or relay.

in this circuit. Figure 16-5 shows a motor overload device firmly attached to the motor housing. It quickly senses any unusual temperature rise or excess current draw. The bimetallic disc reacts to excess temperature and/or excess current draw and flexes downward, thereby disconnecting the motor windings from the power source.

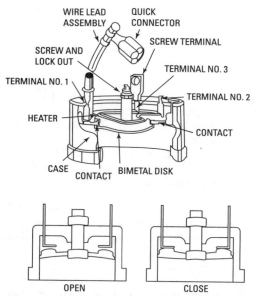

Figure 16-5   External line break overload.

Figure 16-6 shows how the overload device is connected in a motor control circuit with a potential relay to remove the start winding from the circuit as soon as the motor has come up to speed. Note the two capacitors.

Most small motors are protected from overheating by a bimetallic disc that has a red button sticking out of the motor frame. Figure 16-7 shows how the circuit breaker operates inside the motor. It stays off until the red button is pressed. If the motor is still too hot, it will open again. It usually takes a couple of minutes to cool off before being able to complete the circuit again.

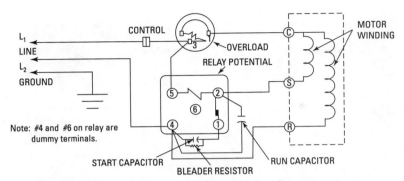

Figure 16-6 Two-terminal external overload with start components and potential relay to remove the start winding from the circuit.

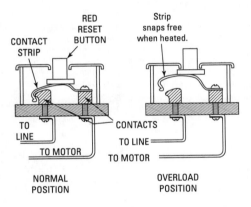

Figure 16-7 Push-button reset. This is located on most small-horsepower motors. The button is usually red and mounted in the motor frame.

Internal (Line Break) Motor Protector

Another safety feature for a motor is the internal line break overload, which is completely internal and tamper-proof; it cannot be bypassed. This type of circuit breaker is located precisely in the center of the "heat sink" portion of the motor windings. It protects the motor by detecting excessive motor winding temperature and protects the circuit source of power as well (Figure 16-8).

This type of circuit breaker is wired in series with the contactor holding coil and the pilot circuit contacts of the external supplementary current-sensitive overloads.

If the motor temperature rises above safe limits, the thermostat opens the holding coil circuit. The contactor disconnects the motor

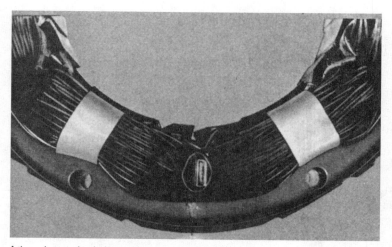

A thermal protective device guards motor windings and insulation from damaging heat, which can seriously decrease motor life expectancy. The thermally protected motor has an automatic reset device, connected in series with the power line, which acts to break the circuit when overheating occurs in the motor. Possible causes of overheating may be rotor jamming or loss of normal cooling air. The protector is "nested" for firm contact with the winding to insure uniform protector operation.

Figure 16-8   Internal line break overload. This is wedged inside the motor windings. It automatically resets when the windings cool down.

from the power source. Supplementary overloads do not break line current but instead interrupt current flow to the contactor holding coil. The contactor then disconnects the motor from the power source (Figure 16-9).

Manual Starters

Fractional-horsepower manual starters are designed to control and provide overload protection for motors of 1 horsepower or less on 115 or 230 V single-phase. They are available in single- and two-pole versions and are operated by a toggle handle on the front.

Manual motor starting switches provide on-off control of single-phase or three-phase AC motors where overload protection is not required or is separately provided (Figure 16-10).

Motor Controls

Figure 16-11 illustrates some of the various types of switches used for motor control. These are very much a part of the control circuits for fractional-horsepower motors. The toggle switches have a pole sticking up and can be operated in the on-off position by flipping the switches. A few of the different types of switches used to control motors are explained. Switches A, B, C, F, G, and H are toggle switches. Switches D and E are rocker types. A rocking action on the part of the switch control is required to operate the on-off portion. Switches I and J are push-button types. They can be momentarily on and then off, or they can be pushed for on and pushed for off. Switches K and L are the rotating type. They are rotated to on and off positions.

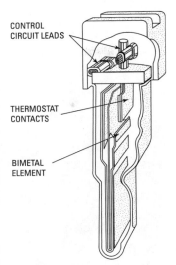

CONTROL
CIRCUIT LEADS

THERMOSTAT
CONTACTS

BIMETAL
ELEMENT

Removing Switches

In most cases it is best to remove the defective switch and replace it with an equal or better type. Make sure you check to see whether the switch will operate with an inductive load. The motor presents an arc after the switch is turned off and the magnetic field of

Figure 16-9 Internal pilot circuit thermostat and supplementary overload embedded in motor winding.

the motor coils collapses. This means that the motor control switch must be able to take a greater current surge than a switch you'd use for turning on and off a regular light bulb or resistive load.

Checking Switches

The quickest way to check a switch is to remove it from the circuit and test it with a continuity checker or with an ohmmeter. When the switch is off, it should read "infinity." When the switch is on, it should read a direct short on the ohmmeter. Any deviation from this indicates that the contacts of the switch are not working properly. It is very difficult to test for proper operation of a switch with an ohmmeter and be 100 percent sure that the switch is good. The low voltage of the ohmmeter indicates that the switch is not being tested under the same conditions as when it is in the circuit with power applied at its full voltage. This check with an ohmmeter will, however, give you some indication of whether the switch is good or bad. After checking the power source to make sure power is available and checking the fuse or circuit breaker on the motor, the next thing to check when a motor does not operate is the switch. If you have the proper insulated alligator clips on the ends of a piece of jumper wire,

Figure 16-10   Enclosed switch with handle guard and pilot light.
(Courtesy Square D)

you can short across the switch contacts to see whether the switch is operational. If the switch is defective, it will not operate the motor. However, if the jumper is placed across the contacts of the switch and the motor operates normally, the switch is defective. *Replace it!*

High-Resistance Contacts

Some switches develop high-resistance contacts after prolonged use. This means they could draw too much current at the contact point and arc, thus creating a hot switch. If the switch is hot to the touch, it should be checked to make sure that the line voltage is reaching the motor terminals. Use your voltmeter to check the line voltage and then check the voltage available at the motor terminals.

If the voltage isn't the same at both the checked points, you should replace the switch with a new one of equal or better current

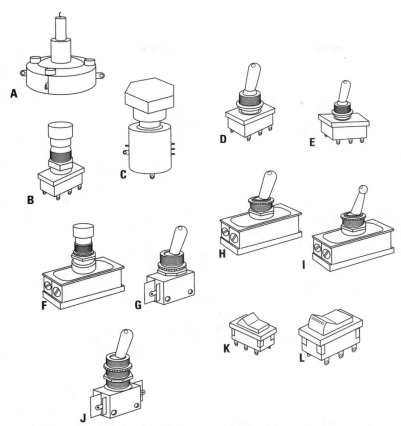

Figure 16-11 Different types of switches used for motor control.

ratings. Also, make sure the voltage rating of the switch is sufficient for the motor circuit.

Microswitches

Microswitches are used in many switch devices. They are usually enclosed in a box or some other type of enclosure. There could be any number of ways to operate the switch contacts. Levers, rollers, and direct force are used to operate the contacts inside the microswitch. The dimensions on the units will give you some idea as to size. These small switch contacts can handle large currents (Figure 16-12).

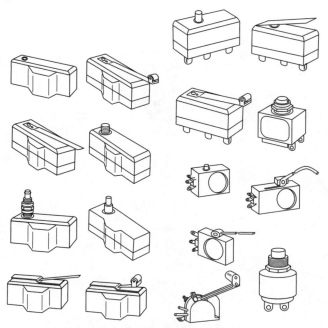

Figure 16-12 Microswitches. Note the different methods by which the switch can be actuated.

When checking a microswitch, look for NO, C, and NC marked on the side. The C is "common." NO is a "normally open" contact that will read "infinity" on an ohmmeter when checked between NO and C. When checked with an ohmmeter, it should read "zero" on NC ("normally closed") and C. Now, if you press the lever, the contacts will switch their arrangement—that is, C to NO will read "zero" and C to NC will read "infinity." Of course, some microswitches have only two contacts. This means that the switch is normally open if not otherwise stated.

Relays for Motor Control

With manual control, the starter must be mounted so that it is easily accessible to the operator. With magnetic control, the push-button stations or other pilot devices can be mounted anywhere on the machine and connected by control wiring into the coil circuit of the remotely mounted starter.

In the construction of a magnetic controller, the armature is mechanically connected to a set of contacts so that the contacts close when the armature moves to its closed position. Figure 16-13 shows several magnetic and armature assemblies in elementary form. Figure 16-14 shows what a motor control relay looks like in operational form. Figure 16-15 shows the principle on which the magnetic controller (relay) works. The electromagnet is the difference between a remote control possibility and a manual starter control. The electromagnet consists of a coil of wire placed on an iron core. When current flows through the coil, the iron bar, called the *armature*, is attracted by the magnetic field created by the current in the coil. To this extent, both will attract the iron bar (the arms of the core). The electromagnet can be compared to the permanent magnet shown in Figure 16-15.

The field of the permanent magnet, however, will hold the armature against the pole faces of the magnet indefinitely. The armature

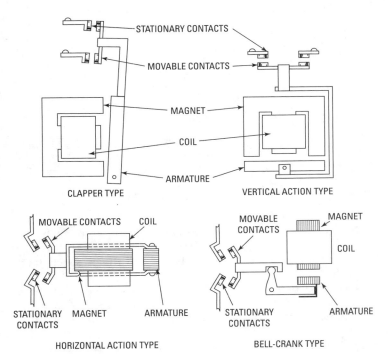

Figure 16-13   Magnetic frame and armature assemblies.

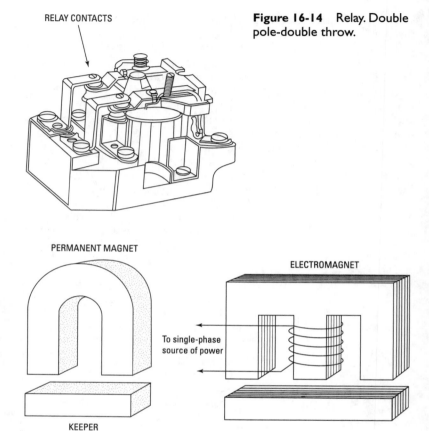

RELAY CONTACTS

Figure 16-14 Relay. Double pole-double throw.

PERMANENT MAGNET

ELECTROMAGNET

To single-phase source of power

KEEPER

Figure 16-15 Basic principles of a relay operation.

cannot be dropped out except by physically pulling it away. In the electromagnet, interrupting the current flow through the coil of wire causes the armature to drop out because of the presence of an air gap in the magnetic circuit.

In the other part of Figure 16-15, the coil, made up of the wire wound around the core, is controlled by a switch located near the power source and any distance from the electromagnet switch it is part of. The contacts can be attached as shown in Figure 16-13. When the coil is energized by closing a switch at some location and completing the circuit to the coil, it attracts the armature, which has the contacts attached to it. When the armature is near the iron core of the coil, it closes the contacts, completing the circuit. The contacts are the switch points that complete the circuit to the motor coils.

Maintenance and Repair of Relays

If the relay contacts do not close, a number of things could have caused the condition. If the contacts do not close and there is no magnetic field around the core of the electromagnet, then there is no power being applied to the coil. Check to see whether you have power. This means checking the fuses and circuit breakers associated with the line current. If you manually (carefully, with an insulated piece of fiber or similar material) close the contacts and the motor starts and runs while you have the contacts closed, it means that the coil or the switch controlling the coil circuit could be defective. Remove all power and check the resistance of the coil. It should give you a reading on the ohmmeter. If it reads "zero," it is shorted and useless. If it reads "infinity," it is open and must also be replaced. If you get the proper reading (this will vary with the manufacturer), it means the coil is okay but perhaps the contacts are not closing when the coil is energized by pushing its control switch. Check the switch in the control part of the coil circuit. If the switch is okay and the coil produces a hum and will not pull in, check for low voltage to the coil. Use your voltmeter across the terminals of the coil.

Contacts

If the contacts of the relay are pitted or burned, it could mean that you need to burnish (polish) them so they will close tightly against one another and complete the circuit. This can be done (with the power off, of course) by placing a piece of sandpaper (very fine grain) between the contacts and holding the contacts closed. You should then move the sandpaper until it sands down the high points on the contacts. Use an even finer grade and polish the points further. Make sure you don't allow the contacts to get to the point where they do not "mate" properly. It may be sufficient to use a tool such as that shown in Figure 16-16.

Figure 16-16
Relay contact burnisher also used for cleaning switch contacts for start windings in split-phase and capacitor-start motors.

Contact Burnisher

The relay contact burnisher is used to polish contacts on switches and relays. It looks like a large ballpoint pen. The cap comes off to reveal a small, very thin piece of serrated metal (file). The file is used to get between the contact points and clean them until they are polished. Different degrees of surface roughness are available for the tool. Just make sure you don't try to polish relay contacts when the power is on.

Maintenance of a relay involves checking it occasionally for the condition of the contacts and polishing them for good contact. It also means keeping the contacts free of dirt, dust, and, in some cases, extreme humidity and water. It is a good idea to enclose the relay in a waterproof container. Many types are made; they are available at any electrical supply store.

When replacing a relay, it is very important to make sure that the coil voltage is the same as the one you're replacing. There are hundreds of voltage possibilities in relays. Also, make sure that the relay contacts will handle the motor current.

Electronic Speed Control

The universal series-wound motor is easily controlled by a circuit such as the one shown in Figures 16-17 and 16-18. The SCR (silicon-controlled rectifier) is the heart of the control circuit. It is possible to obtain low speed and good torque with this type of control. Usually this type of circuitry (or something similar) is used to control the speed of a hand drill or other small appliance, such as a kitchen blender or mixer. There is a smoother change of speed with this than with switches moving from one position to another. In this type of circuitry, the volume-control type of resistor (R_1) shown

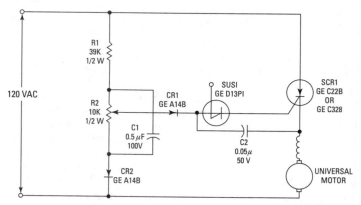

Figure 16-17   Triggered, universal-series, motor-speed control circuit with feedback. *(Courtesy General Electric)*

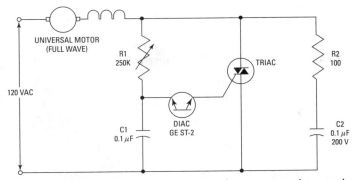

Figure 16-18   Another type of universal motor-speed control.

(Courtesy General Electric)

here is used to actually control the amount of current allowed to flow through the armature and field coil of the motor. If the motor reaches full speed when the full position is reached on the control but does not run when the other positions are in the circuit, then it is time to replace the entire electronic unit unless you have the proper experience and equipment to check the SCR and associated capacitors in the circuit. If the speed control works and causes the motor to vary in speed but not reach full speed, then the problem is the switch. This can, in most cases, be removed and replaced to make the control operate properly again. In most cases of this type, the speed control is located inside the cover of the motor or appliance. Care must be exercised in removing the control section.

Control Box

In some instances, you may want to control the speed of a motor you already have but find that it does not have control built in. You can do this with a control box (Figure 16-19). The SCR and its associated circuit parts are encased in a box where the power line comes out one end and the AC socket is adjacent to it. Just plug the box into the wall socket and plug the motor into the AC outlet. The control knob will change the speed of the motor as it is rotated from "0" to the "Full" position. At "Full" position, it should place the full line voltage across the coils and armature of the motor and full speed should be obtained.

Keep in mind that speed control of this type is usable only on universal motors and DC motors. This type of control cannot be used with split-phase or capacitor-start motors. The SCR actually

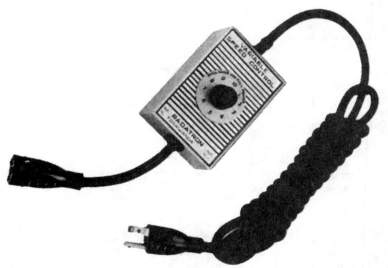

Figure 16-19   Packaged speed control for universal and DC motors.

changes the AC line voltage to DC. This DC will work with a motor with a wound armature such as the universal motor. Remember that the universal motor will work on AC or DC, but if it doesn't have brushes, it won't work with this type of speed control.

Solid-State Relays
Solid-state relays are made with silicon or germanium materials that operate on the same basic principles as transistors and diodes. In most instances, the relay is nothing more than a transistor. In other instances, the solid-state relay is a silicon-controlled rectifier (SCR). The circuit arrangements are such as to allow the switching needed for the relay action. Features not readily available in electromechanical relays are available in this type of relay. There are, of course, advantages and some disadvantages. In most instances, the manufacturer will point to advantages, but you have to become aware of the limitations by closely examining the information provided by the manufacturer of the device.

Note that the first difference between electromechanical and solid-state relays is the absence of a coil and contacts. The solid-state relay needs very low voltage and current to cause it to do its switching. The transistor or SCR does the actual switching, and the change in control voltage causes the semiconductor device to conduct or not conduct according to the control voltage applied to its elements. Figure 16-20 is an example of a solid-state relay.

Transistors
The transistor is a device used to do the switching in a solid-state relay, so let's take a closer look at its operation (Figure 16-21). The

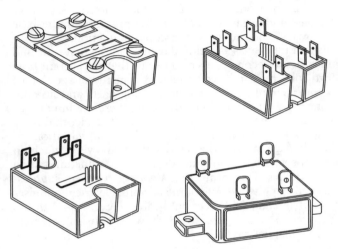

Figure 16-20   Various types of encapsulated solid-state relays.

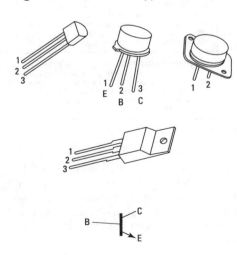

Figure 16-21   Various case configurations for transistors.

transistor is the key element in this type of relay. There are three elements in a transistor: base, collector, and emitter. The base-to-emitter voltage of the transistor can control the current flow between the emitter and the collector. In this PNP type of transistor, a negative voltage on the base allows emitter-base current to flow. This is due to the properties of the silicon-doped material at the junction of the emitter and base. The emitter-base voltage can then cause the transistor to conduct current from the emitter to the collector. A positive voltage on the base and negative on the emitter prevents emitter-base current from flowing, and the transistor stops conduction.

This means that it behaves as a closed contact in the first state and an open contact in the second. This also means that the current flow from emitter to collector can be controlled by a small voltage change in the base-emitter circuit. There are no moving parts and no contacts to be concerned with this time. However, there are limitations on how much current the transistor will conduct. The fact that there are no moving contacts, wear or acing, deterioration, vibration, or dust and dirt damage makes the solid-state relay very much in demand. Surge protection solid-state devices are sometimes used with magnetic switches. This means that a *voltage transient suppressor* may be needed to prevent some of the harmful electrical impulses generated by electromagnetic devices on the line. Figure 16-22 shows a transient suppressor, and Figure 16-23 shows a transient suppressor mounted on a magnetic relay.

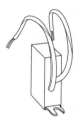 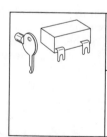

MANUAL TEST TOOL—Provides a means of manually switching the contacts of a basic relay or timing relay and holding all contacts in their switched state until the tool is removed. This simplifies the checking of control circuits without power on the coil or contacts.

TRANSIENT SUPPRESSOR—Consists of an R-C circuit designed to suppress coil 1=1 generated transients to approximately 200 percent of peak voltage. It is particularly useful when switching the TYPE X relay near solid state equipment. It is designed for use on 120 VAC coils only.

Figure 16-22   Transient suppressor.

Triacs

Another thyrister-type device is the triac. It is also used in solid-state switching. The SCR is limited in its application in that it can control current in only one direction. However, the triac can conduct in both directions. The triac has the same characteristics as the SCR and is often thought of as two SCRs placed in parallel but connected in the opposite direction.

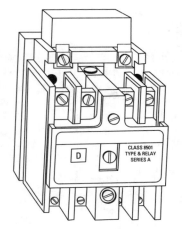

Figure 16-23 Transient suppressor mounted on top of a magnetic type relay (Square D).

CLASS 8501
TYPE & RELAY
SERIES A

D

Triac Construction Figure 16-24 shows the triac schematic and its construction. It has three terminals: main terminal 1 (MT_1), main terminal 2 (MT_2) and gate (G) as shown in Figure 16-25. A quick examination of the PN structure of the triac shows that it can pass current through the PNPN layer or through the NPNP layer. The device can be described as having an NPNP layer in parallel with a PNPN layer. This arrangement of four-layer material gives the triac a connection similar to two SCRs in parallel. The connection in Figure 16-25 is not how the triac operates. This is because the triac gate voltage responds differently from the SCR gate voltage. Figure 16-26 shows the schematic diagram of the triac as well as examples of what the devices look like in their cases. Because the triac can conduct current in both directions, the schematic diagram shows two diodes facing in opposite directions.

Triac Applications Because the triac conducts in both directions, it performs best when used to control AC power. Take a look at Figure 16-27. In this circuit, full power is applied to the load when the gate is triggered on. When S_1 is open, the triac cannot conduct. This is because the voltage applied to the triac is below the breakover point. When S_1 is closed, the triac is triggered on, and both halves of the AC power

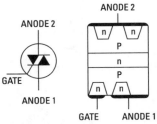

ANODE 2

ANODE 2

GATE

ANODE 1

n   n
P
n
P
n   n

GATE   ANODE 1

Figure 16-24 Schematic diagram and semiconductor materials stacked to make a triac.

are applied to the load. This differs from SCR operation. The SCR can apply only half of the line power to the load because it conducts in only one direction. The advantage of all thyristors is that small gate currents can control large load currents.

The triac conducts in both directions and requires a small current to operate. It does, however, have some disadvantages when compared with the SCR. The SCR has higher current ratings than the triac. The triac can handle currents up to 25 A, but the SCR can safely handle currents of around 800 A. That means the SCR is the better choice when large currents are required.

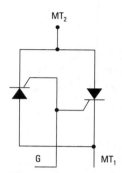

Figure 16-25 Two SCRs connected in parallel.

There are also some differences in the frequency-handling abilities of the SCR and triac. The triac is usually slower in turning on when used with an inductive load, such as when it is used to control a motor. The triac is also designed to operate mainly in the low-frequency range of 30 to 400 Hz. The SCR can safely handle frequencies up to 30 kHz.

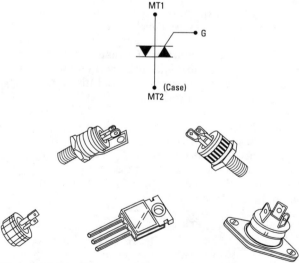

Figure 16-26 Triac symbol and various case configurations for the device.

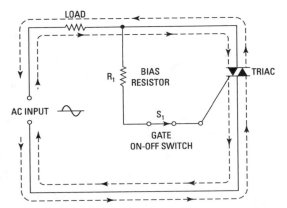

Figure 16-27   Note how a small gate current controls a larger current by using a triac.

Silicon-Controlled Rectifier

The silicon-controlled rectifiers (SCRs) are three-junction semiconductor devices that can operate as a switch (see Figure 16-28). It is basically a rectifier. The SCR will conduct current in only one direction. The best part of SCR operation is its ability to be turned on and off. The on-off action makes the SCR very useful in controlling current.

Construction of the SCR   The construction details of a component provide information about how the component will operate. Solid-state devices are made by joining P and N material into a junction or junctions. Bipolar transistors, diodes, and FETs are all constructed this way. The SCR is made by joining four alternating layers of P and N material. Most SCRs are made of silicon, but germanium is also used (see Figure 16-29).

Note how the layers of P and N material in Figure 16-29 are sandwiched together. Also, note the three junctions. Leads for external connections are attached to these layers. These three connections are called the anode (A), cathode (K), and gate (G).

Now take a look at the schematic of the SCR shown in Figure 16-24. The schematic symbol is about the same as that of a rectifier diode. The main difference is the gate. In some cases the circle around the symbol is not shown. The leads might not be identified on a schematic drawing. When the areas are marked, they are identified with the letters A, K, and G.

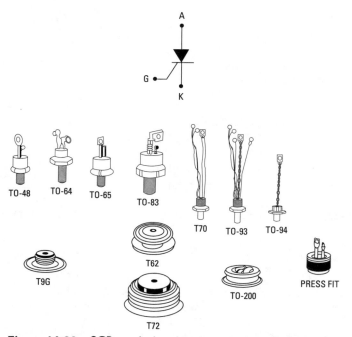

Figure 16-28 SCR symbol and various case configurations.

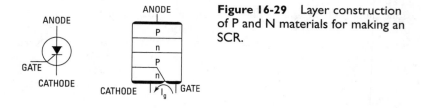

Figure 16-29 Layer construction of P and N materials for making an SCR.

Operation of the SCR Inasmuch as the SCR is a semiconductor device, it requires a biasing voltage to cause it to turn on. Figure 16-30 shows a simplified arrangement that causes the SCR to operate. A switch is used in the gate circuit to apply voltage to the gate. Resistor R_1 is used to limit the current flow in the gate circuit. A second voltage source supplies the needed forward bias to the anode and cathode. A resistor is in series with the anode-cathode circuit. This resistor is also used as a current-limiting resistor. It prevents high currents from causing damage to the SCR. Without the

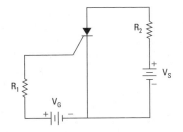

Figure 16-30 Biasing circuit for an SCR.

resistor, the SCR conducts hard in forward bias and burns out after a short operating period. A specified gate current must be reached before the SCR will become a conductor. Each SCR has its own breakover voltage. This means that each SCR must have the proper forward bias applied and the proper gate current in order to operate effectively as a switch.

AC Operation of the SCR The SCR can be used to control DC or AC. Because it is a rectifier, it operates on only one AC alternation. The SCR conducts only when the input cycle makes the anode positive and the cathode negative (see Figure 16-31). By closing S_1, a positive voltage is developed that turns the SCR on. The series resistor is in the gate circuit for current-limiting purposes. The diode is in the circuit to protect the anode and cathode during reverse voltage operation.

If S_1 is closed, the SCR conducts when the proper polarity appears at the anode. If the gate switch is opened, the SCR continues to conduct until the voltage between the anode and cathode falls below the breakover voltage. When the voltage falls below this level, the SCR remains off until S_1 is closed again.

There are many advantages to using an SCR instead of an electrical switch. For instance, the SCR will not wear out. It will not develop

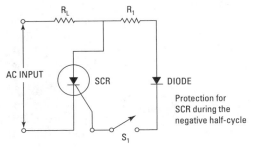

Figure 16-31 SCR used to control voltage to R_L (load resistor).

contact arcing, nor will it stick in one position. This means that the SCR is a more reliable component than the mechanical switch, especially in high-current applications. The SCR can be controlled by a switch or by an electrical pulse from a computer. The most important characteristic is that a small amount of power applied to the gate controls large amounts of current to a load.

Foot Pedal

Another type of speed control is the foot-pedal control used with sewing machine motors (Figure 16-32). This is nothing more than a resistor that has its resistance altered when the foot pedal is depressed. One thing to remember with this type of control and universal motors is that a resistor in the circuit causes the universal motor to have almost no torque when the motor is started from 0 rpm. That is why most sewing machine motors have to be started by hand each time the machine is stopped and started again.

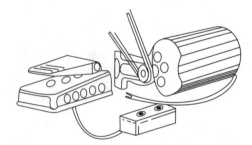

Figure 16-32 Foot-operated control for a sewing machine or any small universal motor.

About the only trouble you have with this control is when it is used on a motor of greater than $\frac{1}{15}$ horsepower. The contacts of the switch become pitted and cause more resistance in the circuit than is called for in proper operation. It can be easily disassembled and checked. Remove the screws in the bottom plate of the control and check for dirt, dust, or any corrosion resulting from humid conditions. Clean the contact points with sandpaper and blow out the sand dust. Check with an ohmmeter to see if it will alter the resistance once reassembled.

The best care can be that of *not* going from "0" to "Full" in one quick motion. This can create sparking or arcing of the contact surfaces between the slide and the resistance wire. Arcing causes damage to any surface because the temperature approaches 2200°F at the source of an arc.

Summary

Motors can overheat if there is low line voltage, jamming, and large increases in voltage that cause increased current and, as such, an increase in heat. The larger the overload, the more quickly the temperature rises. Overloads shorten the life of a motor. Ideal overload protection is an element with current-sensing properties similar to the heating curve of the motor.

Fuses are not designed to provide overload protection. They protect against short circuits. Dual-element or time-delay fuses can provide motor overload protection. However, they are nonrenewable and have to be replaced.

There are two classifications of overload relays: magnetic and thermal. Most small motors are protected from overheating by a metallic disc that has a red button sticking out of the motor frame. A thermal protective device guards motor windings and insulation from damaging heat, which can seriously decrease motor life expectancy.

Another safety feature for a motor is the internal line break overload. If the temperature rises, the thermostat opens the holding coil circuit. The contactor disconnects the motor from the line.

Fractional-horsepower manual starters are designed to control and provide overload protection for motors of 1 horsepower or less on 115- or 230-volt single-phase motors. Manual motor starting switches provide on-off control of single-phase or three-phase AC motors where overload protection is not required or is separately provided.

Many types of switches are used for motor control. Switches should be checked to see if they will operate properly with an inductive load similar to that provided by a motor. If they are defective, they should be replaced. The quickest way to check the switch is to check it with an ohmmeter when removed from the circuit.

Some switches develop high-resistance contacts after long use and should be replaced. This can be detected by checking the temperature of the switch. Microswitches can be activated by levers, rollers, and direct force. These small switch contacts can handle large currents.

In the construction of a magnetic controller, the armature is mechanically connected to a set of contacts so that when the armature moves to its closed position the contacts also close. The electromagnet consists of a coil of wire placed on an iron core.

If the relay contacts do not close, a number of things could have caused the condition. If the contacts of the relay are pitted or burned, it could mean that you need to burnish them so that they will close tightly against one another and complete the circuit. The relay contact burnisher is used to polish contacts on switches and relays. Maintenance of a relay involves checking it occasionally for the condition of the contacts and polishing for good contact. It must also be kept free of dirt, dust, and, in some cases, extreme humidity and water. The relay should be enclosed in a waterproof container. Make sure the coil voltage is the same as the one you are replacing. Make sure the contacts will handle the motor current.

The universal series-wound motor is easily controlled by a circuit with an SCR in it. It is possible to obtain low speed and good torque with this type of control. Control boxes with an SCR in the circuit are available so that the motor can be plugged into the socket on the box and the speed control inserted in the motor circuit.

Foot-controlled speed control for a sewing machine or DC motor is common. This is nothing more than a resistor that has its resistance altered when the foot pedal is depressed. It is a design that works well with universal or series-wound motors.

Review Questions

1. What causes motors to overheat?
2. Why are standard fuses not suitable for the protection of a motor from damage?
3. How does overheating damage an electric motor?
4. What are the two classifications for overload relays?
5. How are most small motors protected from overheating?
6. Where is a manual starter used?
7. Name three types of switches used for controlling motors.
8. What is the quickest way to check a switch to see if it is properly operating?
9. Why do switches develop high-resistance contacts?
10. What can cause a relay contact not to close properly?
11. How is maintenance accomplished on relay contacts?
12. What is a burnisher? How does it work?
13. How is the universal series-wound motor easily controlled?
14. What component does a variable-speed control box for small motors contain in its circuitry?
15. Where are foot-controlled motor speed devices used?

Chapter 17

Motion Control

Terminology is important when working with or working on a motor or any other device. Disc magnet stepper terminology is presented here in an effort to provide persons working with this type of motor with common ground on which to base their discussion, be it for the purpose of repair, selection, or operation of the motor.

Disc Magnet Stepper Terminology

You should remember that a stepper motor is fundamentally a single- or multi-phase synchronous motor designed for step-by-step operation. Driving such a motor with pulses will result in a rotation that is more or less uneven, depending on the pulse rate and the load characteristics. However, the drive signal could also be continuous, producing a rotating field; the motor then works as a normal synchronous motor. The difference between these and large synchronous motors is that the excitation field is created by a permanent magnet (disc magnet motor).

Table 17-1 shows common terms and their definitions.

Table 17-1   Disc Magnet Stepper Terminology

Term	Description
Two-Phase Motor	Motor with two phases, electrically shifted by 90°. Each phase may comprise one or more windings, which may be interconnected externally.
Step (or Step Angle)	Angle of rotation for one increment in the elementary step sequence (see four-step sequence). By definition, one step in a two-phase stepper corresponds to 90 electrical degrees.
Driving Sequence	Sequence of the increments applied to the motor phases in order to establish rotation.
Four-Step Sequence	Elementary sequence for two-phase steppers. One cycle corresponds to four full steps and 360 electrical degrees. A four-step sequence can be achieved with either one or two phases energized.

(continued)

Table 17-1 (continued)

Term	Description
Half-Step	Angle of rotation for one increment of an eight-step sequence (two-phase stepper).
Eight-Step Sequence	Sequence of the drive signals energizing either one or both phases to get eight stable positions in a cycle of 360 electrical degrees.
Microstepping	Dividing a full step into a number of intermediate stable positions achieved by using drive voltages or currents of variable levels.
Power Rate	The power rate is a basic figure used mainly in robotics applications. It shows the speed of increase in mechanical power. The power rate is directly related to the time necessary to move a load from one position to another.
Unipolar Drive	The drive voltage is applied between one end and the common point of a center-tapped winding. Two switches per phase are needed, and only 50 percent of the coils are used at any time. The maximum performances of the stepper cannot be obtained.
Bipolar Drive	Four switches per phase are needed. Use of 100 percent of the copper volume results in better performance than in unipolar mode.
Holding Torque	The maximum steady torque that can be applied to the shaft of an energized motor without causing continuous rotation.
Detent Torque	The maximum torque that can be applied to the shaft of a non-energized motor without causing continuous rotation.
Pull-In Torque	The maximum torque that can be applied to a motor shaft when starting at the pull-in rate.
Pull-Out Torque	The maximum torque that can be applied when running at the pull-out rate.
Pull-In Frequency	The maximum switching rate at which a motor can start or stop without losing steps. This rate is load-dependent.
Pull-Out Frequency	The maximum switching rate at which a motor can operate without losing steps. This rate is load-dependent.
Detent Torque and Rest Position	The remaining torque without excitation is due to the permanent magnet; the friction torque is to be added. The rest positions without current are the same as those with one phase energized.

Table 17-1   *(continued)*

Term	Description
Dynamic Measurements	Measuring dynamic torque is difficult because the measuring method itself could influence the system performance. Therefore, the accuracy of such measurements is difficult to guarantee, and it is not always possible to compare measurements made with a composite load (friction plus torque opposed to motor torque). For escap®, dynamic measurements are made using pure friction load.
System Performance	A stepper motor is as good as the driver allows it to be. A good motor can give its best performance only with an appropriate driver. By the way, it is always possible to economize on the driving circuit if the maximum performances of the motor are not needed.

Dual-Voltage Control (Bi-Level Drive)

A dual-voltage power driver can be utilized to reduce power dissipation without the loss of motor performance. Essentially this scheme uses a high voltage output to initially energize a motor phase. This ensures maximum current rise in minimum time. When the current reaches its predetermined maximum value, the high voltage is reduced to a lower "maintenance" voltage for the remainder of the step.

Chopper Drive

The chopper drive system limits the motor current by means of voltage modulation. It is a highly efficient means of motor control that permits an initial high voltage (10 to 20 times the rated voltage) to energize the motor until the desired current is obtained. The voltage is then switched off until the current decays to a predetermined level, after which the voltage is again turned on. This cycle of on-off voltage (chopping) is continued through the drive pulse time. This method of motor drive is well suited for stop-start operation.

Disc Magnet Steppers

The elementary sequence of a two-phase stepper has four states corresponding to four steps. Half-steps can be obtained with an eight-step sequence. The possible sequences are shown in Table 17-2.

Table 17-2 Driving Sequences

Step No.	2 phases energized Phase A	B	Direction of rotation C.W.	C.C.W.
1	+	+		
2	–	+		
3	–	–		
4	+	–		

Step No.	1 phase energized Phase A	B	Direction of rotation C.W.	C.C.W.
1	+	0		
2	0	+		
3	–	0		
4	0	–		

Half-step No.	Phase A	B	Direction of rotation C.W.	C.C.W.
1	+	0		
2	+	+		
3	0	+		
4	–	+		
5	–	0		
6	–	–		
7	0	–		
8	+	–		

Full-step and half-step switching sequences control voltage or current signals of constant amplitude, whereas mini- or micro-step driving sequences imply variable current levels in both phases.

Note
The number of phases, two in a two-phase stepper, should not be confused with the number of coils, which is very often four. In this respect, permanent-magnet and hybrid steppers differ from variable-reluctance steppers, which must have three or four phases.

Using Steppers at Their Upper Limits

In continuous operation, coil heating happens to be the primary limiting parameter. The winding temperature must not exceed 155°C (310°F). Temporarily, the phase current may be increased without any risk of demagnetization.

The chopper drive mode—constant current level—leads to additional iron losses, which increase with the chopping frequency. This higher temperature risk should not be omitted in thermal calculations. Drive circuit configurations are shown in Figures 17-1 to 17-3.

Disc Magnet, Two-Phase, Stepper Motor

This is a motor that weighs only 40 grams (Figure 17-4). The rotor consists of a thin axially magnetized disc. The stator design is a short, low-loss magnetic circuit resulting in the conservation of the motor characteristics at high speeds. The microstepping motor is designed for applications for high efficiency, a wide speed range, and high power-to-volume and power-to-weight ratios.

DRIVE CIRCUIT CONFIGURATION

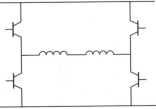

BIPOLAR

Figure 17-1   Disc magnet stepper, bipolar drive circuit.
(Courtesy Stock Drive Products, Div. of Designatronics, Inc.)

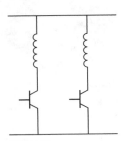

UNIPOLAR

Figure 17-2   Disc magnet stepper, unipolar drive circuit.
(Courtesy Stock Drive Products, Div. of Designatronics, Inc.)

MOTOR LEADS COLOR CODE

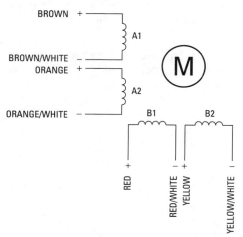

Figure 17-3 Disc magnet stepper motor leads color code.
(Courtesy Stock Drive Products, Div. of Designatronics, Inc.)

Figure 17-4 Disc magnet, two-phase, stepper motor.

(Courtesy Stock Drive Products, Div. of Designatronics, Inc.)

With a single step, the motor can run at 1000 rpm and features a maximum pullout rate of 10,000 rpm. Theoretically, it can accelerate under maximum current to 465,000 rad/sec^2. This dynamic speed and acceleration allows it to be utilized with a more favorable coupling ratio and thus eliminate low-speed jitters at the load.

Figure 17-5 is a stepper that has been designed for microstepping and, therefore, has a sinusoidal torque curve and reduced detent torque among its features. It weighs 250 grams with a peak holding torque of 35.7 oz-in.

A larger motor, 600 grams in weight, is shown in Figure 17-6. It has a peak holding torque of 139.1 oz-in.

PM, Two- and Four-Phase Stepper Motors

A permanent-magnet (PM), rotor-type stepper motor is shown in Figure 17-7. The stator is pressed steel and plated for corrosion resistance. The coil is wound on a bobbin molded of high-

Figure 17-5 Disc magnet, two-phase stepper motor.
(Courtesy Stock Drive Products, Div. of Designatronics, Inc.)

Figure 17-6 Disc magnet, two-phase stepper motor.
(Courtesy Stock Drive Products, Div. of Designatronics, Inc.)

Figure 17-7 PM, rotor-type stepper **motor.** *(Courtesy Stock Drive Products, Div. of Designatronics, Inc.)*

temperature nylon and fully shrouded. The rotor spindle is hardened and polished and runs in nylon bearings requiring no maintenance.

Figure 17-8 shows the wiring diagram for this type of motor. The step angle is 7.5° with a five-percent accuracy. It weighs 68 grams and uses 12 V for its power source. Figure 17-9 shows the typical characteristics.

The four-phase motor in Figure 17-10 is two-part, with a 48-pole stator that surrounds the toroidal coil with two separate *double windings* for *unipolar* operation.

The two-phase motor is two-part, with a 48-pole stator that surrounds the torpidly toroidal coil with two separate *single windings* for *bipolar* operation.

The rotor shaft that carries the high coercivity 24-pole permanent magnet is hardened, ground, and polished. The motor is fitted with maintenance-free sintered bronze bearings housed in plastic. This protects the bearings from dirt and prevents seepage of the

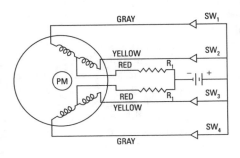

Figure 17-8 Wiring diagram for PM, rotor-type stepper motor. *(Courtesy Stock Drive Products, Div. of Designatronics, Inc.)*

TYPICAL CHARACTERISTICS AT 20°C.

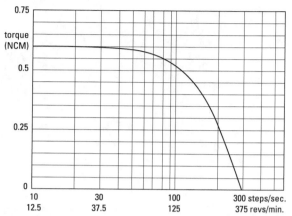

Figure 17-9 Typical characteristics curve for PM, rotor-type stepper motor. *(Courtesy Stock Drive Products, Div. of Designatronics, Inc.)*

lubricating oil by capillary effect. The step angle is 7.5°. Different motors are made to operate on 3, 6, 12, or 24 V, and the weight of the motor is 65 grams. This same type of motor is also available with a 15° step angle and a weight of up to 400 grams.

Another design (Figure 17-11) for a 7.5° permanent-magnet stepper motor can run clockwise or counterclockwise, depending on the input. Holding torque is measured with two adjacent phases at the rated current and can be as much as 16 oz-in. with a phase current of 0.47. This is accomplished with a weight of 250 grams or 0.55 lbs.

The hybrid 1.8° stepper motor is shown in Figures 17-12 to 17-14. It can run clockwise or counterclockwise, depending on the input, and can weigh 340 grams (0.75 lbs) up to 2.2 lbs or 1 kg. The motors are also available with double-ended shafts.

The motor shown in Figure 17-13 weighs 10 lbs or 4.6 kg with a holding torque of 781 oz-in. It, too, is a 1.8° hybrid stepper. The motor in Figure 17-14 is somewhat larger and weighs 23.1 lbs or 10.5 kg. The holding torque is 2130 oz-in.

As can be seen from this information, stepper motors are available in a wide variety of sizes, shapes, and holding torques.

Wiring data is important when checking for proper motor operation. Figure 17-15 shows the unipolar and bipolar connections

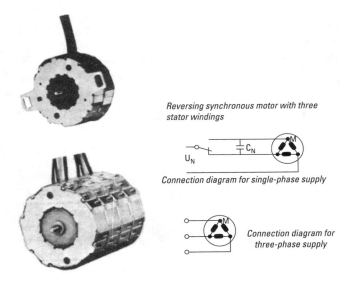

Reversing synchronous motor with three stator windings

Connection diagram for single-phase supply

Connection diagram for three-phase supply

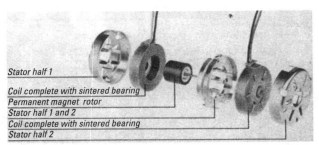

Stator half 1
Coil complete with sintered bearing
Permanent magnet rotor
Stator half 1 and 2
Coil complete with sintered bearing
Stator half 2

Figure 17-10   Two- or four-phase, two-part, 48-pole stator motor.
(Courtesy of Stock Drive Products, Div. of Designatronics, Inc.)

with 6-, 8-, and 12-lead motor designs. Sequencing for half-step and full-step is also given.

Two- and Four-Phase Linear Stepper Motors

The linear motor consists of a moving element called a *forcer*, which travels along a special ferromagnetic track called a *platen* (Figure 17-16). As the electromagnets that make up the forcer respond to varying currents, embedded teeth in the forcer move to successive positions relative to similar teeth in the platen to produce

Figure 17-11 7.5°
permanent-magnet
stepper motor. *(Courtesy of
Stock Drive Products, Div. of
Designatronics, Inc.)*

Figure 17-12 1.8° hybrid
stepper motor. *(Courtesy of Stock
Drive Products, Div. of Designatronics, Inc.)*

the commanded motion. The motor shown in Figure 17-16 produces a static force of 4.0 to 4.8 pounds, whereas the motor in Figure 17-17 is somewhat larger and can produce a static force of 7.5 to 10 pounds with the two-phase unit and 8 to 11 pounds with the four-phase. This unit is suited for small- to moderate-size, high-performance positioning systems where it can replace rotary motors and the hardware associated with converting rotary to linear motion. The complete absence of moving contact parts provides the motor with design simplification, size reduction, unlimited life expectancy, and reliability.

The four-phase has an advantage over two-phase operation in that it reduces cyclic error and offers slightly higher static force

Figure 17-13 1.8° hybrid
stepper motor. *(Courtesy of
Stock Drive Products, Div. of
Designatronics, Inc.)*

Figure 17-14 1.8° hybrid
stepper motor. *(Courtesy of Stock
Drive Products, Div. of Designatronics, Inc.)*

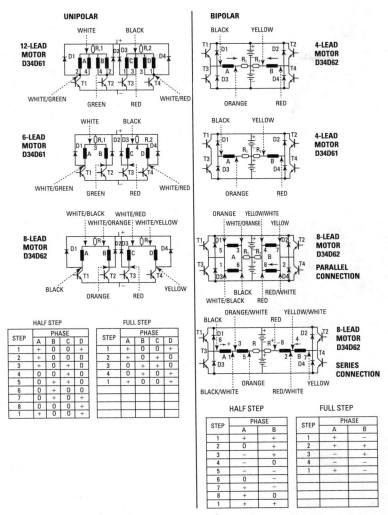

Figure 17-15 **Stepper motor wiring data.** *(Courtesy of Stock Drive Products, Div. of Designatronics, Inc.)*

because of its lower force ripple. In high-speed operation, the benefits of four-phase operation are available with no increase in electronics hardware. Features of this type of motor are fine resolution, repeatability of 0.00005π (open loop), and a velocity of 100 in. per second. It also has air bearings.

Figure 17-16 Two-phase linear stepper motor.
(Courtesy of Stock Drive Products, Div. of Designatronics, Inc.)

Figure 17-17 Two- and four-phase linear stepper motor. *(Courtesy of Stock Drive Products, Div. of Designatronics, Inc.)*

Figure 17-18 Two- and four-phase linear stepper motor. *(Courtesy Stock Drive Products, Div. of Designatronics, Inc.)*

For larger-size systems, the motor in Figure 17-18 will do the job; it can replace many of the rotary motors and their associated hardware. This type has a static force of 44 pounds and can move at 100 ips with a repeatability of 0.00005π (open loop).

The linear motor platen (the flat part of the motor) is shown in Figure 17-19. There are two types—the bar, or flat, type and the tubular type. The tubular type weighs 3.75 lbs, and the bar type is available in three different sizes: 3.5, 7, and 14 lbs.

If larger sizes are required, the 7300 series (Figure 17-20) is available and can provide much larger platens, up to 72 in. long and weighing in at 167 lbs.

Brushless DC Motors

The brushless DC motor (Figure 17-21) is a combination of the unirotational synchronous motor and an electronic circuit that enables it to be connected to DC. The electronics converts the DC into the pulse form required for the synchronous motor by means of an RC oscillator. The result is a brushless DC motor having the advantages of a synchronous motor—no brush wear, long life, and speed that is not load dependent.

The three-phase brushless DC motor with Hall sensors has 12 poles and 1.4 oz-in. of torque per ampere. It is small in size, only 16 mm thick, and is manufactured with an internal series resistor, which significantly reduces temperature-tracking errors.

Ironless-Rotor DC Motors

The construction of this small motor (Figure 17-22), with its permanent

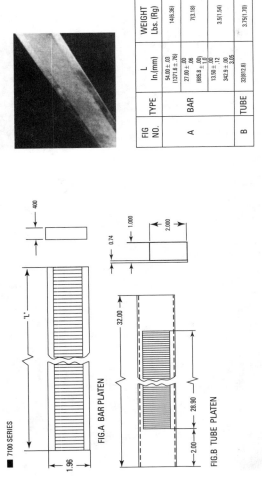

FIG NO.	TYPE	L In.(mm)	WEIGHT Lbs. (Rg)
A	BAR	54.00 ± .03 (1371.6 ± .76)	14(6.36)
		27.00 ± .00 .06 (685.8 ± 1.0 .00)	7(3.18)
		13.50 ± .00 .12 342.9 ± .00 3.05	3.5(1.54)
B	TUBE	32(812.8)	3.75(1.70)

■ 7100 SERIES

FIG.A BAR PLATEN

FIG.B TUBE PLATEN

Figure 17-19 Linear motor platens. (Courtesy Stock Drive Products, Div. of Designatronics, Inc.)

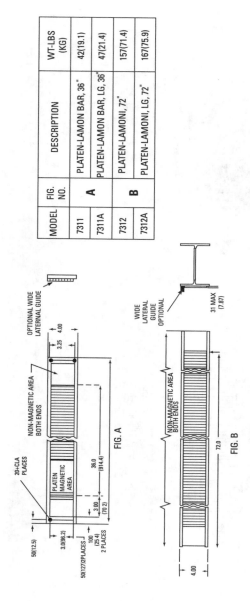

MODEL	FIG. NO.	DESCRIPTION	WT-LBS (KG)
7311	A	PLATEN-LAMON BAR, 36"	42(19.1)
7311A		PLATEN-LAMON BAR, LG, 36"	47(21.4)
7312	B	PLATEN-LAMONI, 72"	157(71.4)
7312A		PLATEN-LAMONI, LG, 72"	167(75.9)

Figure 17-20 Larger linear motor platens. *(Courtesy Stock Drive Products, Div. of Designatronics, Inc.)*

magnet and ironless rotor, results in a high power-to-volume ratio, smooth torque with almost no ripple, very rapid starting, and very low electrical noise at commutation. The low inertia rotor exhibits no cogging.

The precious metal commutator (silver alloy) and brushes (gold alloy), as well as the ferrite magnet with 12 poles, result in a small (8.2 mm thick) motor with very low inductance and low moment of inertia. It can be obtained in 5- and 12-V models. Torque of 1.05 oz-in. is provided by the 5-V motor, and the 12-V size produces 2.08 oz-in. per ampere of torque. A series resistor is built in for precise speed control. No-load speed is 6100 rpm and 7200 rpm for the 12-V model.

Not all ironless rotor DC motors resemble Figure 17-22; many resemble Figure 17-23. These are made for operation on 15 24 V with no-load speeds of 4300 or 5400 rpm, depending on the voltage applied. They pull 20 mA for the 15-V model; the 24-V pulls 25 mA. The weight of the motor is 230 grams.

One of the main advantages of this type of motor over some conventional types is its ability to deliver several times the nominal stall torque without the danger of demagnetization by high current pulses.

Summary

Disc magnet stepper terminology is used to describe what is fundamentally a single- or multi-phase synchronous motor, designed for step-by-step operation. Driving such a motor with pulses will result in a more or less uneven rotation, depending on the pulse rate and the load characteristics.

Figure 17-21 Brushless DC motor. *(Courtesy of Stock Drive Products, Div. of Designatronics, Inc.)*

Figure 17-22 Ironless-rotor DC motor. *(Courtesy of Stock Drive Products, Div. of Designatronics, Inc.)*

Figure 17-23 Larger ironless-rotor DC motor. *(Courtesy Stock Drive Products, Div. of Designatronics, Inc.)*

Driving sequence is the term used to describe the increments applied to the motor phases in order to establish rotation. *Microstepping* is the dividing of a full step into a number of intermediate stable positions achieved by using drive voltages or currents of variable levels.

Power rate is a basic figure used mainly in robotics applications. It shows the speed of increase in mechanical power. Unipolar and bipolar drives are possible with the proper number of switches.

In this type of small motor, you will come across the following terms to add to your motor vocabulary: holding torque, detent torque, pull-in torque, pull-out torque, pull-in frequency, detent torque and rest position, dynamic measurements, and system performance. Stepper motors have many uses, from robots to timing devices.

A dual-voltage power driver can be used to reduce power dissipation without the loss of motor performance. The chopper drive system limits the motor current by means of voltage modulation.

In disc magnet steppers, the permanent-magnet and hybrid steppers differ from variable-reluctance steppers. When using steppers, the coil-heating problem is the primary limiting parameter when in continuous operation. Temporarily, the phase current may be increased without any risk of demagnetization.

The disc magnet, two-phase stepper motor weighs only 40 grams. The rotor is nothing more than a thin axially magnetized disc. The stator design is a short, low-loss magnetic circuit resulting in the conservation of the motor characteristics at high speeds. The unipolar disc magnet stepper motor weighs about 250 grams with a peak holding torque of 35.7 oz-in. It can also be made in the 600-gram weight class with a holding torque of 139.1 oz-in.

A permanent-magnet, rotor-type stepper motor is made with a stator of pressed steel and plated for corrosion resistance. The coil is wound on a bobbin molded of high temperature nylon and fully shrouded. The rotor spindle is hardened and polished and runs in nylon bearings requiring no maintenance. Stepper motors are available in a wide variety of sizes, shapes, and holding torques.

The linear motor consists of a moving element called the forcer, which travels along a special ferromagnetic track called a platen. This type of unit is suited for small- to moderate-size, high-performance positioning systems where it can replace rotary motors and the hardware associated with converting rotary to linear motion.

The brushless DC motor is a combination of the unirotational synchronous motor and an electronic circuit that enables it to be connected to DC. The electronics converts the DC into the pulse form required for the synchronous motor.

The ironless motor has a high power-to-volume ratio, smooth torque with almost no ripple, very rapid starting, and very low electrical noise at commutation. The rotor exhibits no cogging. The precious metal commutator (silver alloy) and brushes (gold alloy), as well as the ferrite magnet with 12 poles, result in a small motor with very low inductance and low moment of inertia. It comes in 5- and 12-V models; other models are available in 15- and 24-V arrangements. One of the main advantages of this type of motor is its ability to deliver several times the nominal stall torque without the danger of demagnetization by high current pulses.

Review Questions

1. What type of motor is the stepper?
2. How is a motor driven by pulses?
3. What use can a pulse-driven motor have?
4. What is power rate?
5. What type of torque does the stepper have?
6. Why would you prefer to use a dual-voltage power driver in a stepper motor?
7. How do reluctance steppers differ from disc magnet steppers, permanent-magnet steppers, and hybrid steppers?
8. How much do the stepper motors weigh?
9. Of what is the stator made on a permanent-magnet rotor-type stepper motor?
10. Describe a linear motor.
11. Where is the linear motor used?
12. Describe the brushless DC motor.
13. What is the advantage of using a brushless DC motor? What type of power supply does it need?
14. Of what is the metal commutator of the ironless motor constructed?
15. How many poles does the ironless motor have?

Chapter 18

Motor Control Methods

Because over 90 percent of all motors are used on AC, DC motors and their controls will not be discussed in this chapter.

Wound-rotor motors and AC commutator motors have a limited application and are also excluded. Because the squirrel-cage induction motor is the most widely used, its control is the subject of this chapter. The use of higher voltages (2400, 4800, and higher) introduces requirements that are additional to those for 600-V equipment, and although the basic principles are unchanged, these additional requirements are not covered here.

Selection of Motor Control

The motor, machine, and motor controller are interrelated. They need to be considered as a package when choosing a specific device for a particular application. In general, five basic factors influence the selection of a controller:

- Electrical service
- Motor
- Operating characteristics of the controller
- Environment
- National codes and standards

Electrical Service

Establish whether the service is DC or AC. If AC, determine the number of phases and the frequency, in addition to the voltage.

Motor

The motor should be matched to the electrical service and correctly sized for the machine load (horsepower rating). Other considerations include the motor speed and torque. To select the proper protection for the motor, its full-load current rating, service factor, and time rating must be known.

Operating Characteristics of the Controller

The fundamental job of a motor controller is to start and stop the motor. It should also protect the motor, machine, and operator.

The controller might also be called upon to provide supplementary functions, which could include reversing, jogging or inching, plugging, and operating at several speeds or at reduced levels of current and motor torque.

Environment
Controller enclosures serve to provide protection for operating personnel by preventing accidental contact with live parts. In certain applications, the controller itself must be protected from a variety of environmental conditions, which might include the following:

* Water, rain, snow, or sleet
* Dirt or noncombustible dust
* Cutting oils, coolants, or lubricants

Both personnel and property require protection in environments made hazardous by the presence of explosive gases or combustible dusts.

National Codes and Standards
Motor control equipment is designed to meet the provisions of the National Electrical Code (NEC). Code sections applying to industrial control devices are *Article 430* on motors and motor controllers and *Article 500* on hazardous locations.

The 1970 Occupational Safety and Health Act (OSHA), as amended in 1972, requires that each employer furnish employment free from recognized hazards likely to cause serious harm. Provisions of the act are strictly enforced by inspection.

Standards established by the National Electrical Manufacturers Association (NEMA) assist users in the proper selection of control equipment. NEMA standards provide practical information concerning construction, test, performance, and manufacture of motor control devices such as starters, relays, and contactors.

One of the organizations that actually test for conformity to national codes and standards is Underwriter's Laboratories (UL). Equipment tested and approved by UL is listed in an annual publication, which is kept current by means of bimonthly supplements that reflect the latest additions and deletions.

Motor Controller
A motor controller will include some or all of the following functions: starting, stopping, overload protection, overcurrent protection, reversing, changing speed, jogging, plugging, sequence control,

and pilot light indication. The controller can also provide the control for auxiliary equipment such as brakes, clutches, solenoids, heaters, and signals. A motor controller may be used to control a single motor or a group of motors (Figure 18-1).

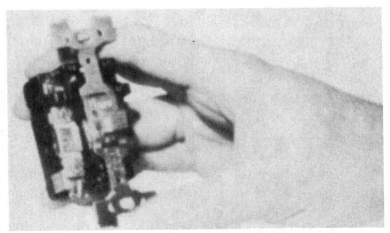

Figure 18-1 Motor controllers can be simple or complex. Both the small fractional horsepower manual starter above and the special control panel in Figure 18-2 qualify as motor controllers. *(Courtesy Square D Co.)*

Starter
The terms *starter* and *controller* mean practically the same thing. Strictly speaking, a starter is the simplest form of controller and is capable of starting and stopping the motor and providing it with overload protection (Figure 18-2).

Squirrel-Cage Motor
The workhorse of industry is the AC squirrel-cage motor. Of the thousands of motors used today in general applications, the vast majority are of the squirrel-cage type. Squirrel-cage motors are simple in construction and operation—merely connect power lines to the motor and it will run.

The squirrel-cage motor gets its name because of its rotor construction, which resembles a squirrel cage and has no wire winding.

Full-Load Current (FLC)
The current required to produce full-load torque at rated speed is called FLC.

Figure 18-2 A more complex motor controller. *(Courtesy Square D Co.)*

Locked-Rotor Control (LRC)

During the acceleration period at the moment a motor is started, it draws a high current called the *in-rush* current. The in-rush current when the motor is connected directly to the line (so that full line voltage is applied to the motor) is called the *locked-rotor* or *stalled-rotor* current. The locked-rotor current can be from four to ten times the motor full-load current. The vast majority of motors have an LRC of about six times FLC, and therefore this figure is generally used. The "six times" value is often expressed as 600 percent of FLC.

Motor Speed

The speed of a squirrel-cage motor depends on the number of poles on the motor's winding. On 60 Hz, a two-pole motor runs at about

3450 rpm, a four-pole at 1725 rpm, and a six-pole at 1150 rpm. Motor nameplates are usually marked with actual full-load speeds, but frequently motors are referred to by their synchronous speeds—3600, 1800, and 1200 rpm, respectively.

Torque
Torque is the turning or twisting force of the motor and is usually measured in pound-feet. Except when the motor is accelerating up to speed, the torque is related to the motor horsepower. If a motor is not able to furnish the proper amount of torque for a given load, it will draw an excess of current and overheat.

Ambient Temperature
The temperature of the air where a piece of equipment is situated is called the ambient temperature. Most controllers are of the enclosed type, and the ambient temperature is the temperature of the air outside the enclosure, not inside. Similarly, if a motor is said to be in an ambient temperature of 30°C (86°F), it is the temperature of the air outside the motor, not inside. Per NEMA standards, both controllers and motors are subject to an ambient temperature limit of 40°C (104°F).

Temperature Rise
Current passing through the windings of a motor results in an increase in the motor temperature. The difference between the winding temperature of the motor when running and the ambient temperature is called the temperature rise.

The temperature rise produced at full load is not harmful provided the motor ambient temperature does not exceed 40°C (104°F).

Higher temperature caused by increased current or higher ambient temperatures produces a deteriorating effect on motor insulation and lubrication. An old rule of thumb states that for each increase of 10°F above the rated temperature, motor life is cut in half.

Time (Duty) Rating
Most motors have a continuous duty rating permitting indefinite operation at rated load. Intermittent duty ratings are based on a fixed operating time (5, 15, 30, or 60 minutes), after which the motor must be allowed to cool.

Motor Service Factor
If the motor manufacturer has given a motor a service factor, it means that the motor can be allowed to develop more than its rated or nameplate horsepower without causing undue deterioration of the insulation. The service factor is a margin of safety. If, for example,

a 10-hp motor has a service factor of 1.15, the motor can be allowed to develop 11.5 hp. The service factor depends on the motor design.

Jogging

Jogging describes the repeated starting and stopping of a motor at frequent intervals for short periods of time. A motor would be jogged when a piece of driven equipment had to be positioned fairly closely—for example, when positioning the table of a horizontal boring mill during setup. If jogging is to occur more frequently than five times per minute, NEMA standards require that the starter be derated.

For example, NEMA Size 1 starter has a normal duty rating of 7.5 hp at 230 V, polyphase. On jogging applications, this same starter has a maximum rating of 3 hp.

Plugging

When a motor running in one direction is momentarily reconnected to reverse the direction, it will be brought to rest very rapidly. This is referred to as plugging. If a motor is plugged more than five times per minute, derating of the controller is necessary because of the heating of the contacts. Plugging can be used only if the driven machine and its load will not be damaged by the reversal of the motor torque.

Sequence (Interlocked) Control

Many processes require a number of separate motors that must be started and stopped in a definite sequence. This happens, for example, in a system of conveyors. When starting up, the delivery conveyor must start first, with the other conveyors starting the sequence. This is to avoid a pileup of material. When shutting down, the reverse sequence must be followed with time delays between the shut-downs (except for emergency stops) so that no material is left on the conveyors. This is an example of a simple sequence control. Separate starters could be used, but it is common to build a special controller that incorporates starters for each drive, timers, control relays, etc.

Enclosures

NEMA and other organizations have established standards of enclosure construction for control equipment. In general, equipment would be closed for one or more of the following reasons:

- To prevent accidental contact with live parts

- To protect the control from harmful environmental conditions
- To prevent explosion or fires that might result from the electrical arc caused by the control

The following identifies some common types of NEMA enclosures according to their numbers.

NEMA 1 General Purpose

The general-purpose enclosure is intended primarily to prevent accidental contact with the enclosed apparatus. It is suitable for general-purpose applications indoors where it is not exposed to unusual service conditions. A NEMA 1 enclosure serves as protection against dust and light, indirect splashing but is not dust-tight (Figure 18-3).

Figure 18-3   NEMA 1 general-purpose enclosure. *(Courtesy Square D Co.)*

NEMA 3 Dust-tight, Rain-tight

This enclosure is intended to provide suitable protection against specified weather hazards. A NEMA 3 enclosure is suitable for application outdoors, on ship docks, canal locks, and construction work, and for application in subways and tunnels. It is also sleet-resistant (Figure 18-4).

Figure 18-4   NEMA 3 dust-tight, rain-tight enclosure. *(Courtesy Square D Co.)*

NEMA 3R Rainproof, Sleet-Resistant

This enclosure protects against interference in operation of the contained equipment due to rain and resists damage from exposure to sleet. It is designed with conduit hubs and external mounting, as well as drainage provisions.

NEMA 4 Watertight

A watertight enclosure is designed to meet the following hose test. Enclosures are tested by subjection to a stream of water from a hose with a 1-in. nozzle that delivers at least 65 gallons per minute. The water is directed on the enclosure from a distance of not less than 10 ft and for a period of 5 minutes. During this period, it may be directed in any one or more directions as desired. There should be no leakage of water into the enclosure under these conditions.

A NEMA 4 enclosure is suitable for applications outdoors on ship docks and in dairies, breweries, etc. (Figure 18-5).

Figure 18-5 NEMA 4 watertight enclosure. *(Courtesy Square D Co.)*

NEMA 4X Watertight, Corrosion-Resistant
These enclosures are generally constructed along the lines of NEMA 4 enclosures but are made of a material that is highly resistant to corrosion. For this reason, they are ideal in applications such as paper mills, meatpacking, fertilizer, and plants, where contaminants would ordinarily destroy a steel enclosure over a period of time (Figure 18-6).

Figure 18-6 NEMA 4X watertight, corrosion-resistant enclosure.
(Courtesy Square D Co.)

NEMA 7 Hazardous Locations, Class I

Here the design is to meet the application requirements of the NEC for Class I hazardous locations. In this type of equipment, the circuit interruption occurs in air (Figure 18-7).

NEMA 9 Hazardous Locations, Class II

Class II locations are those that are hazardous because of the presence of combustible dust. This enclosure has been designed to handle the requirements of Class II (Figure 18-8).

NEMA 12 Industrial Use

When it is necessary to exclude materials such as dust, lint, fibers and flyings, oil seepage, or coolant seepage, the NEMA 12 is required. There are no conduit openings or knockouts in the

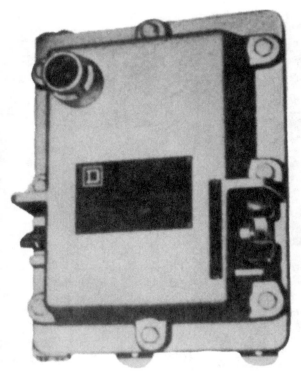

Figure 18-7   NEMA 7, Class I, Groups C and D, hazardous locations enclosure. Class I locations are those in which flammable gases or vapors are or may be present in the air in quantities sufficient to produce explosive or ignitable mixtures. *(Courtesy Square D Co.)*

enclosure, and mounting is achieved with flanges or mounting feet (Figure 18-9).

NEMA 13 Oil-tight, Dust-tight
This enclosure is usually made by casting. It has a gasket and can be used in the same atmospheres and locations as NEMA 12. The essential difference is that, due to its cast housing, a conduit entry is provided as an integral part of the enclosure. Mounting is accomplished with blind holes rather than mounting brackets (Figure 18-10).

Figure 18-8   NEMA 9, Class II, Groups E, F, and G, hazardous locations enclosure. The letter or letters following the type number indicate the particular group or groups of hazardous locations for which the enclosure is designed. The designation is incomplete without the suffix letter or letters. *(Courtesy Square D Co.)*

Overcurrent Protection

The function of the overcurrent protective device is to protect the motor branch-circuit conductors, control apparatus, and motor from short circuits or grounds. The protective devices commonly used to sense and clear overcurrents are thermal magnetic circuit breakers and fuses. The short-circuit device should be capable of carrying the starting current of the motor, but the device setting should not exceed 250 percent of full-load current with no code letter on the motor, or from 150 to 250 percent of full-load current depending on the code letter on the motor. If the value is not sufficient to carry the starting current, it may be increased, but it should never exceed 400 percent of the motor full-load current. The National Electrical Code requires (with few exceptions) a means to disconnect the motor and controller from the line, in addition to an overcurrent protective device to clear short-circuit faults. The circuit breaker in Figure 18-11 incorporates fault protection and disconnects in one basic device. When the overcurrent protection is provided by fuses, a disconnect switch is

Figure 18-9   NEMA 12 enclosure. *(Courtesy Square D Co.)*

required. The switch and fuses are generally combined as shown in Figure 18-12.

Overload Protection

The effect of an overload is a rise in temperature in the motor windings. The larger the overload, the more quickly the temperature will increase to a point damaging to the insulation and lubrication of

Figure 18-10 NEMA 13, oil-tight, dust-tight. *(Courtesy Square D Co.)*

the motor. An inverse relationship therefore exists between current and time—the higher the current, the shorter the time before motor damage or burnout can occur.

All overloads shorten motor life by deteriorating the insulation. Relatively small overloads of short duration cause little damage but, if sustained, could be just as harmful as overloads of greater magnitude. The relationship between the magnitude (percent of full load) and duration (time in minutes) of an overload is illustrated by the Motor Heating Curve shown in Figure 18-13.

The ideal overload protection for a motor is an element with current-sensing properties very similar to the heating curve of the motor, which could act to open the motor circuit when *full-load current* is exceeded. The operation of the protective device should be such that the motor is allowed to carry harmless overloads but is quickly removed from the line when an overload has persisted too long.

Figure 18-11   Circuit breaker, overcurrent protective device.
(Courtesy Square D Co.)

Fuses are not designed to provide overload protection. Their basic function is to protect against short circuits. A fuse chosen on the basis of motor FLC would blow every time the motor started. On the other hand, if a chosen fuse were large enough to pass the starting or in-rush current, it would not protect the motor against small, harmful overloads that might occur later.

Dual-element or time-delay fuses can provide motor overload protection. However, they suffer the disadvantage of being nonrenewable and must be replaced.

Overload relays consist of a current-sensing unit connected in the line to the motor, plus a mechanism actuated by the sensing unit. This unit serves to break the circuit directly or indirectly. In a manual starter, an overload trips a mechanical latch, causing the starter contacts to open and disconnect the motor from the line. In magnetic starters, an overload opens a set of contacts within the overload relay itself. These contacts are wired in series with the starter coil in the control circuit of the magnetic starter. Breaking

Figure 18-12   Fusible disconnect, overcurrent protective device.
(Courtesy Square D Co.)

the coil circuit causes the starter contacts to open, disconnecting the motor from the line.

There are two classifications of overload relays—*magnetic* and *thermal*. Magnetic relays react only to current excesses and are not affected by temperature. Thermal relays can be further divided into two types—*melting alloy* and *bimetallic*.

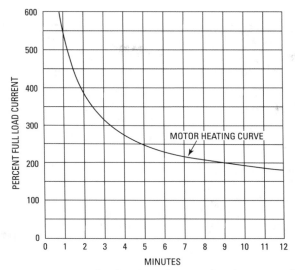

Figure 18-13 Application of motor heating curve data. Note: On 300 percent overload, the particular motor for which this curve is characteristic would reach its permissible temperature limit in three minutes. Overheating or motor damage would occur if the overload persisted beyond this time limit.

Melting-alloy, thermal-overload relays are also referred to as *solder-pot relays*. The motor current passes through a small heater winding (Figure 18-14). Under overload conditions, the heat causes a special solder to melt. This allows a ratchet wheel to spin free. The ratchet wheel then opens the contacts. When this occurs, the relay is said to trip. To obtain approximate tripping current for motors of different sizes or different full-load currents, a range of thermal units (heaters) is available. A reset button is usually mounted on the cover of enclosed starters. Thermal units are rated in amperes and are selected on the basis of motor full-load current, not horsepower (Figures 18-15, 18-16, and 18-17).

Bimetallic thermal-overload relays employ a U-shaped bimetal strip that is associated with a current-carrying heater element (Figure 18-18). When an overload occurs, the heat will cause the bimetal to deflect and open a contact. Different heaters give different trip points. In addition, most relays are adjustable over a range of 85 to 115 percent of the nominal heating rating.

ONE-PIECE THERMAL UNIT

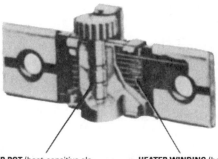

SOLDER POT (heat-sensitive element) is an integral part of the thermal unit. It provides accurate response to overload current, yet prevents nuisance tripping.

HEATER WINDING (heat-producing element) is permanently joined to the solder pot, so proper heat transfer is always insured. No chance of misalignment in the field.

Figure 18-14   One-piece thermal unit. Cutaway view of how it works.
(Courtesy Square D Co.)

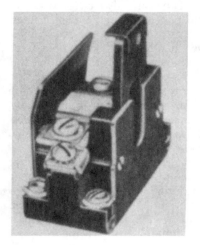

Figure 18-15   Single-pole, melting-alloy, thermal-overload relay.
(Courtesy Square D Co.)

Figure 18-16 Three-pole, melting-alloy, thermal-overload relay.
(Courtesy Square D Co.)

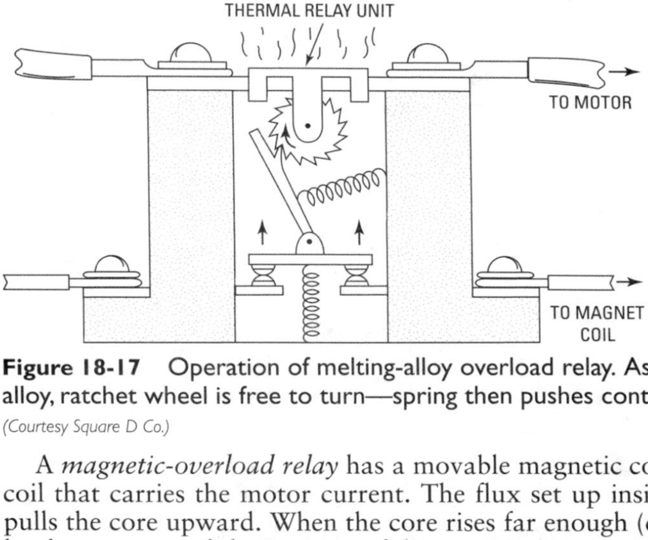

THERMAL RELAY UNIT

TO MOTOR

TO MAGNET
COIL

Figure 18-17 Operation of melting-alloy overload relay. As heat melts alloy, ratchet wheel is free to turn—spring then pushes contacts open.
(Courtesy Square D Co.)

A *magnetic-overload relay* has a movable magnetic core inside a coil that carries the motor current. The flux set up inside the coil pulls the core upward. When the core rises far enough (determined by the current and the position of the core), it trips a set of contacts on the top of the relay. The movement of the core is slowed by a piston working in an oil-filled dashpot (similar to a shock absorber) mounted below the coil. This produces an inverse-time characteristic (Figure 18-19). The effective tripping current is adjusted by moving the core on a threaded rod. The tripping time is altered by uncovering oil-bypass holes in the piston. Because of the time and current adjustments, the magnetic-overload relay is sometimes used to protect motors with long accelerating times or unusual duty cycles. The instantaneous-trip magnetic-overload relay is similar but has no oil-filled dashpot (Figures 18-20 and 18-21).

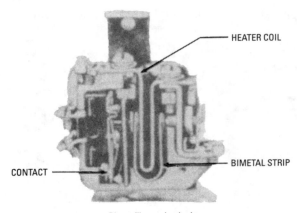

HEATER COIL

BIMETAL STRIP

CONTACT

Bimetallic overload relay
with side cover removed

Figure 18-18 Cutaway view of the bimetallic thermal-overload relay.
(Courtesy Square D Co.)

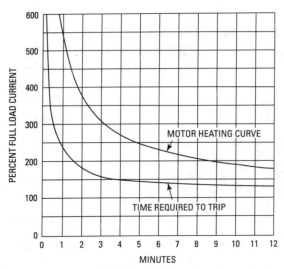

Figure 18-19 Motor heating curve and overload-relay trip curve. The
overload relay will always trip out at a safe value. *(Courtesy Square D Co.)*

Figure 18-20   Magnetic overload relay. Note the size of the wire that makes up the coil.
(Courtesy Square D Co.)

Figure 18-21   Magnetic overload relay. *(Courtesy Square D Co.)*

Manual Starter

A manual starter is a motor controller whose contact mechanism is operated by a mechanical linkage from a toggle handle or pushbutton, which is in turn operated by hand. A thermal unit and direct-acting overload mechanism provides motor-running overload protection. Basically, a manual starter is an "on-off" switch with overload relays.

Manual starters are generally used on small machine tools, fans, blowers, pumps, compressors, and conveyors. They are the least expensive of all motor starters, have a simple mechanism, and provide quiet operation with no AC magnet hum. Moving a handle or pushing the start button closes the contacts, which remain closed until the handle is moved to "off," the stop button is pushed, or the overload-relay thermal unit trips.

Fractional Horsepower (FHP) Manual Starter

FHP manual starters are designed to control and provide overload protection for motors of 1 hp or less on 115 or 230 V, single-phase. They are available in single- and two-pole versions and are operated by a toggle handle on the front.

Manual Motor-Starting Switches

Manual motor-starting switches provide on-off control of single-phase or three-phase AC motors where overload protection is not required or is separately provided. Two- or three-pole switches are available with ratings up to 5 hp, 600 V, three-phase (Figure 18-22). The continuous current rating is 30 amperes at 250 V maximum and 20 amperes at 600 V maximum.

The toggle operation of the manual switch is similar to the FHP starter, and typical applications of the switch include small machine tools, pumps, fans, conveyors, and other electrical machinery that have separate motor protection. They are particularly suited to switch nonmotor loads, such as resistance heaters.

Integral-Horsepower Manual Starter

The integral-horsepower manual starter is available in two- or three-pole versions. It is used to control single-phase motors up to 5 hp and polyphase motors up to 10 hp, respectively.

The two-pole starters have one overload relay. Three-pole starters usually have two overload relays but are available with three overload relays. When an overload relay trips, the starter mechanism unlatches, opening the contacts to stop the motor. The contacts cannot be re-closed until the starter mechanism has been reset by pressing the stop button or moving the handle to the reset position, after allowing time for the thermal unit to cool (Figure 18-23).

Figure 18-22   Enclosed switch with handle guard and pilot light.
(Courtesy Square D Co.)

Figure 18-23   Integral-HP manual starter in general-purpose
enclosure with pilot light. *(Courtesy Square D Co.)*

Magnetic Control

A high percentage of applications require the controller to be capable of operation from remote locations or to provide automatic operation in response to signals from pilot devices such as thermostats, pressure or float switches, limit switches, etc. Low-voltage release or protection might also be desired. Manual starters cannot provide this type of control, and therefore magnetic starters are used.

The operating principle that distinguishes a magnetic from a manual starter is the use of an electromagnet (Figure 18-24). The electromagnet consists of a coil of wire placed on an iron core. When current flows through the coil, the iron bar, called the armature, is attracted by the magnetic field created by the current in the coil. To this extent, both will attract the iron bar (the arms of the core). The electromagnet can be compared to the permanent magnet shown in Figure 18-24.

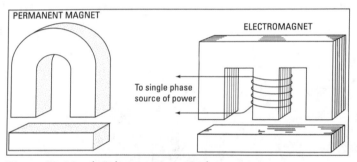

Figure 18-24   An electromagnet and a permanent magnet.
(Courtesy Square D Co.)

The field of the permanent magnet, however, will hold the armature against the pole faces of the magnet indefinitely. The armature could not be dropped out except by physically pulling it away. In the electromagnet, interrupting the current flow through the coil of wire causes the armature to drop out due to the presence of an air gap in the magnetic circuit.

With manual control, the starter must be mounted so that it is easily accessible to the operator. With magnetic control, the push-button stations or other pilot devices can be mounted anywhere on the machine and connected by control wiring into the coil circuit of the remotely mounted starter.

Magnet-Frame and Armature Assemblies

In the construction of a magnetic controller, the armature is mechanically connected to a set of contacts so that when the armature moves to its closed position, the contacts also close. The drawings in Figure 18-25 show several magnet and armature assemblies in elementary form. When the coil has been energized, the armature has moved to the closed position, the controller is said to be "picked up," and the armature is "seated" or sealed-in.

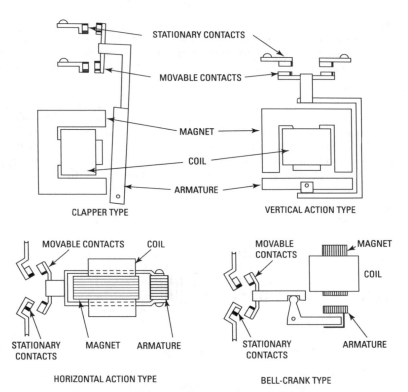

Figure 18-25   Magnet-frame and armature assemblies. *(Courtesy Square D Co.)*

Magnetic Circuit

The magnetic circuit of a controller consists of the magnet assembly, the coil, and the armature. It is so named from a comparison with an electrical circuit. The coil and the current flowing in it cause magnetic flux to be set up through the iron in a manner similar to a voltage causing current to flow through a system of conductors. The changing magnetic flux produced by alternating currents results in a temperature rise in the magnetic circuit. The heating effect is reduced by laminating the magnet assembly and armature.

Magnet Assembly
The magnet assembly is the stationary part of the magnetic circuit. The coil is supported by and surrounds part of the magnet assembly in order to induce magnetic flux into the magnetic circuit.

Armature
The armature is the moving part of the magnetic circuit. When it has been attracted into its sealed-in position, it completes the magnetic circuit.

Air Gap
When a controller's armature has sealed in, it is held closely against the magnet assembly. However, a small gap is always deliberately left in the iron circuit. When the coil is deenergized, some magnetic flux (residual magnetism) always remains; if it were not for the gap in the iron circuit, the residual magnetism might be sufficient to hold the armature in the sealed-in position (Figure 18-26).

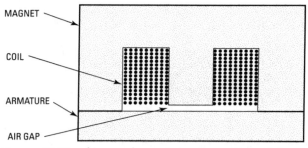

Figure 18-26 Air gap shown in center section of the magnet.
(Courtesy Square D Co.)

Shading Coil
A shading coil is a single turn of conducting material (generally copper or aluminum) mounted in the face of the magnet assembly or armature (Figure 18-27). The alternating main magnetic flux induces currents in the shading coil; these currents set up auxiliary magnetic flux, which is out of phase from the main flux. The auxiliary flux produces a magnetic pull that is out of phase from the pull due to

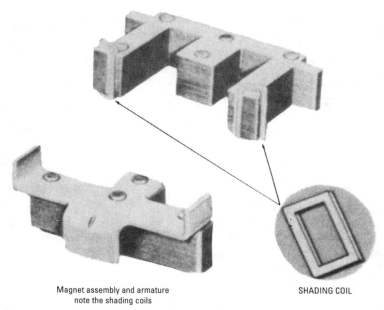

Magnet assembly and armature
note the shading coils

SHADING COIL

Figure 18-27   Magnet assembly and armature. Note the shading coils.
(Courtesy Square D Co.)

the main flux, and this keeps the armature sealed in when the main
flux falls to zero (which occurs 120 times per second with 60-Hz
AC). Without the shading coil, the armature would tend to open
each time the main flux went through zero. Excessive noise, wear
on the magnet faces, and heat would occur (Figure 18-28).

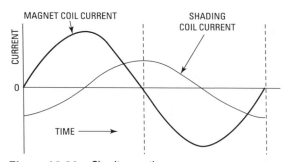

Figure 18-28   Shading coil current versus magnet coil current.
(Courtesy Square D Co.)

Effects of Voltage Variation (Voltage Too Low)

Low control voltage produces low coil currents and reduced magnetic pull. On devices with vertical action assemblies, if the voltage is greater than pick-up voltage but less than seal-in voltage, the controller may pick up but will not seal. With this condition, the coil current will not fall to the sealed value. As the coil is not designed to carry continuously a current greater than its sealed current, it will quickly get very hot and burn out. The armature will also chatter. In addition to the noise, wear on the magnet faces results (Figure 18-29).

AC Hum

All AC devices that incorporate a magnetic effect produce a characteristic hum. This hum or noise is due mainly to the changing magnetic pull (as the flux changes) inducing mechanical vibrations. Contactors, starters, and relays could become excessively noisy as a result of some of the following operating conditions:

- Broken shading coil
- Operating voltage too low
- Wrong coil
- Misalignment between the armature and magnet assembly— the armature is then unable to seat properly
- Dirt, rust, filings, and so forth, on the magnet faces—the armature is unable to seal in completely
- Jamming or binding of moving parts (contacts, springs, guides, yoke bars) so that full travel of the armature is prevented
- Incorrect mounting of the controller, as on a thin piece of plywood fastened to a wall, for example, so that a "sounding board" effect is produced

Holding Circuit Interlock

The holding circuit interlock is a normally open (NO) auxiliary contact provided on standard magnetic starters and contactors. It closes when the coil is energized to form a holding circuit for the starter after the *start* button has been released. As a matter of economics, vertical action contactors and starters in the smaller

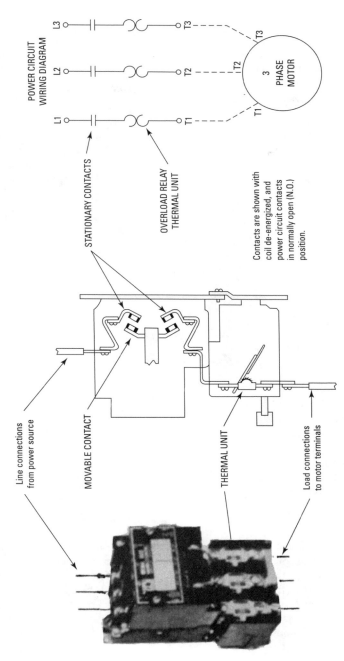

POWER CIRCUIT
WIRING DIAGRAM

L3 ○ T3

L2 ○ T2

L1 ○ T1

3 PHASE MOTOR

T1   T2   T3

STATIONARY CONTACTS

OVERLOAD RELAY
THERMAL UNIT

Contacts are shown with coil de-energized, and power circuit contacts in normally open (N.O.) position.

Line connections from power source

MOVABLE CONTACT

THERMAL UNIT

Load connections to motor terminals

Figure 18-29 Magnetic starter power circuit. *(Courtesy Square D Co.)*

479

NEMA sizes (size 0 and size 1) have a holding interlock that is physically the same size as the power contacts (Figure 18-30).

Figure 18-30   Electrical interlock, normally closed contact. *(Courtesy Square D Co.)*

Electrical Interlocks

In addition to the main or power contacts that carry the motor current and the holding circuit interlock, a starter can be provided with externally attached auxiliary contacts, commonly known as electrical interlocks. Interlocks are rated to carry only control-circuit currents, not motor currents. NO and NC versions are available.

Among a wide variety of applications, interlocks can be used to control other magnetic devices where sequence operation is desired, to electrically prevent another controller from being energized at the same time, and to make and break circuits indicating or alarm devices such as pilot lights, bells, or other signals.

Electrical interlocks are packaged in kit form and can be easily added to the field (Figure 18-31).

Figure 18-31   Magnetic contactor with externally attached electrical interlocks. *(Courtesy Square D Co.)*

Control Device (Pilot Device)

A device that is operated by some nonelectrical means (such as the movement of a lever) and has contacts in the control circuit of a starter is called a control device. Operation of the control device will control the starter and hence the motor. Typical control devices are control stations, limit switches, foot switches, pressure switches, and float switches. The control device may be of the maintained-contact or momentary-contact type. Some control devices have a horsepower rating and are used to directly control small motors through the operation of their contacts. When used in this way, separate overload protection (such as a manual starter) normally should be provided, as the control device does not usually incorporate overload protection.

Maintained Contact

A maintained-contact control device is one that, when operated, will cause a set of contacts to open (or close) and stay open (or closed) until a deliberate reverse operation occurs. A conventional thermostat is a typical maintained-contact device.

Momentary Contact

A standard pushbutton is a typical momentary-contact control device. Pushing the button will cause NO contacts to close and NC contacts to open. When the button is released, the contacts revert to their original states. Momentary-contact devices are used with three-wire control or jogging service.

Reversing Starter

Reversing the direction of motor shaft rotation is often required. Three-phase squirrel-cage motors can be reversed by reconnecting any two of the three line connections to the motor. By interwiring two contactors, an electromagnetic method of making the reconnection can be obtained.

As seen in the power circuit (Figure 18-32, part A), the contacts of the *forward* contactor (F), when closed, connect lines 1, 2, and 3

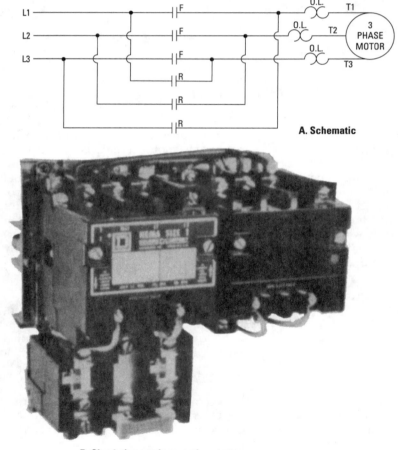

A. Schematic

B. Size 1, three-pole reversing starter.

Figure 18-32   Reversing starter circuit. *(Courtesy Square D Co.)*

to motor terminals T_1, T_2, and T_3, respectively. As long as the *forward* contacts are closed, mechanical and electrical interlocks prevent the *reverse* contactor from being energized.

When the *forward* contactor is de-energized, the second contactor can be picked up, closing its contacts R, which reconnect the lines to the motor. Note that by running through the *reverse* contacts, Line 1 is connected to motor terminal T_3 and line 3 is connected to terminal T_1. The motor will now run in the opposite direction.

Whether operating through either the forward or reverse contactor, the power connections are run through an overload relay assembly, which provides motor overload protection. A magnetic reversing starter, therefore, consists of a starter and contactor, suitably interwired, with electrical and mechanical interlocking to prevent the coils of both units from being energized at the same time. Figure 18-32, part B, shows a Size 1, three-pole reversing starter as it looks in operation.

Motor-Reversing Characteristics

Various reversing characteristics of the different motor types should be checked before attempting to reverse any motor. Table 18-1 shows the reversing characteristics of a number of different motors.

Timers and Timing Relays

A pneumatic timer or timing relay is similar to a control relay, except that certain portions of its contacts are designed to operate at a preset time interval after the coil is energized or de-energized. A delay on energization is also referred to as "on delay." A time delay on de-energization is called "off delay."

A timed function is useful in applications such as the lubricating system of a large machine, in which a small oil pump must deliver lubricant to the bearings of the main motor for a set period of time before the main motor starts (Figure 18-33).

In pneumatic timers, the timing is accomplished by the transfer of air through a restricted orifice. The amount of restriction is controlled by an adjustable needle valve, permitting changes to be made in the timing period.

Drum Switch

A drum switch is a manually operated, three-position, three-pole switch that carries a horsepower rating and is used for manual reversing of single- or three-phase motors. Drum switches are available in several sizes and can be spring-return-to-off (momentary-contact) or maintained-contact. Separate overload protection, by

Table 18-1 Motor-Reversing Characteristics

Motor Type	Power	Duty	Typical Reversibility	Speed Character	Typical Start Torque* Percent
Polyphase	AC	Continuous	Rest/Rot.	Relatively Constant	175 and up
Split-phase Synchronous	AC	Continuous	Rest Only	Relatively Constant	125–200
Split-phase Nonsynch-ronous	AC	Continuous	Rest Only	Relatively Constant	175 and up
PSC Nonsynch-ronous High Slip	AC	Continuous	Rest/Rot.[c]	Varying	175 and up
PSC Nonsynch-ronous Norm. Slip	AC	Continuous	Rest/Rot.[c]	Relatively Constant	75–150
PSC Reluctance Synch.	AC	Continuous	Rest/Rot.[c]	Constant	125–200
PSC Hysteresis Synch.	AC	Continuous	Rest/Rot.[c]	Constant	125–200
Shaded Pole	AC	Continuous	Uni-directional	Constant	75–150
Series	AC/DC	Int./Cont.	Uni-directional[b]	Varying[d]	175 and up
Permanent Magnet	DC	Continuous	Rest/Rot.[e]	Adjustable	175 and up
Shunt	DC	Continuous	Rest/Rot.	Adjustable	125–200
Compound	DC	Continuous	Rest/Rot.	Adjustable	175 and up
Shell Arm	DC	Continuous	Rest/Rot.	Adjustable	175 and up
Printed Circuit	DC	Continuous	Rest/Rot.	Adjustable	175 and up
Brushless DC	DC	Continuous	Rest/Rot.	Adjustable	75–150
DC Stepper	DC	Continuous	Rest/Rot.	Adjustable	[a]

* Percentages are relative to full-load rated torque. Categorizations are general and apply to small motors.
[a] Dependent upon load inertia and electronic driving circuitry.
[b] Usually unidirectional—can be manufactured as bidirectional.
[c] Reversible while rotating under favorable conditions (generally when inertia of the driven load is not excessive).
[d] Can be adjusted, but varies with load.
[e] Reversible down to 0°C after passing through rest.
(Courtesy Bodine)

Figure 18-33   Timing relay.
(Courtesy Square D Co.)

manual or magnetic starters, must usually be provided, as drum switches do not include this feature (Figure 18-34).

Control Station (Pushbutton Station)

A control station may contain pushbuttons, selector switches, and pilot lights. Pushbuttons may be momentary- or maintained-contact. Selector switches are usually maintained-contact, or can be spring-return to give momentary contact operation. Standard duty stations will handle the coil currents of contactors up to Size 4. Heavy-duty stations have higher contact ratings; they provide greater flexibility through a wider variety of operators and interchangeability of units (Figure 18-35, parts A, B, and C).

Limit Switch

A limit switch is a control device that converts mechanical motion into an electrical control signal. Its main function is to limit movement, usually by opening a control circuit when the limit of travel is reached. Limit switches may be momentary-contact

Figure 18-34   Drum switch. *(Courtesy Square D Co.)*

| A. Standard push-
button station. | B. Heavy-duty
push-button station. | C. Heavy-duty oil-tight
push-button station. |

Figure 18-35   Control pushbutton station. *(Courtesy Square D Co.)*

(spring-return) or maintained-contact types. Among other applications, limit switches can be used to start, stop, reverse, slow down, speed up, or recycle machine operations (Figure 18-36, parts A and B).

A. Heavy-duty limit switch with
   lever arm operator.

B. Turret head limit switch with
   roller push rod operator.

Figure 18-36   Limit switch. *(Courtesy Square D Co.)*

Snap Switch

Snap switches for motor control purposes are enclosed precision switches that require low operating forces and have a high repeat accuracy. They are used as interlocks and as the switch mechanisms for control devices such as precision limit switches and pressure switches. They are also available with integral operators for use as compact limit switches, door-operated interlocks, etc. Single-pole, double-throw and two-pole, double-throw switches are available (Figure 18-37, parts A and B).

A. Double-pole, double throw snap switch.

B. Single-pole, double throw.

Figure 18-37 Snap switches. *(Courtesy Square D Co.)*

Pressure Switch

The control of pumps, air compressors, welding machines, lube systems, and machine tools requires control devices that respond to the pressure of a medium such as water, air, or oil. The control device that does this is a pressure switch. It has a set of contacts that are operated by the movement of a piston, bellows, or diaphragm against a set of springs. The spring pressure determines the pressures at which the switch closes and opens its contacts.

Float Switch

When a pump motor must be started and stopped according to changes in water (or other liquid) level in a tank or sump, a float switch is used. This is a control device whose contacts are controlled by movement of a rod or chain and counterweight, fitted with a float. For closed-tank operation, the movement of a float arm is transmitted through a bellows seal to the contact mechanism (Figure 18-38).

Solid-State, Adjustable-Speed Controllers

Solid-state, adjustable-speed controllers are available to produce smooth starts and energy savings. They are, in most instances, maintenance-free and are easy to operate and make it simple to train new operators.

The reason for using solid-state, adjustable-speed controllers is because they provide stepless, smooth, adjustable-speed control of

Figure 18-38   Typical two-pole float switch. Float and float rod are not shown. *(Courtesy Square D Co.)*

the AC wound-rotor motor. This means the elimination of resistors, liquid rheostats, and reactors, as well as magnetic clutches. They all consume energy, which is not the case with the solid-state controller. Solid-state controller circuitry is used to provide excitation to the rotor. By controlling the rotor current it is possible to control the motor speed and thereby its torque.

Frequency Speed Control

Solid-state AC motor control is accomplished by changing the frequency of the power source. Westingthouse's ACCUTROL line is an adjustable-speed AC drive packaged in ratings from 1 to 5 hp at 230 V and 3 to 250 hp at 460 V, three-phase, 60 Hz.

Motor speed is adjustable by controlling the output voltage and frequency of the unit. This is accomplished by rectifying the incoming AC supply voltage and changing it to DC. The DC voltage is inverted by a three-phase inverter section to an adjustable frequency output whose voltage is adjustable proportionately to the frequency to provide constant volts per hertz excitation to the motor terminals up to 60 Hz. Above 60 Hz, the voltage may remain constant at rated volts. In this way energy-efficient, low-loss speed control is obtained in the range of 2 to 120 Hz.

This type of speed control does have advantages over DC machines inasmuch as the DC motors are hard to maintain and have problems in environments that are wet, corrosive, or explosive. These controls are found in food-packing plants, dairies, chemical plants, sand and gravel plants, paper mills, and cement plants. Centrifugal pumps and blowers are particularly suited for use with this type of control, as considerable reduction in energy consumption can be achieved by varying the speed to control the flow of gases or fluids instead of using throttling devices such as valves, dampers, or fluid recirculators.

Multispeed Starters
Multispeed starters are designed for the automatic control of two-speed, squirrel-cage motors of either the consequent pole or separate winding types. These starters are available for constant-horsepower, constant-torque, or variable-torque, three-phase motors. Multispeed motor starters are commonly used on machine tools, fans, blowers, refrigeration compressors, and many other types of equipment.

Speed Monitoring
A speed-sensing switch can be used to sequence conveyors where it is necessary for one conveyor to be running at nearly full speed before a second conveyor is started. The switch can also be used to indicate which direction materials on a conveyor are moving from the rotation of a suitable driven shaft.

The electronic speed switch is a rugged, self-contained, rotary-shaft speed detector, Figure 18-39. If the shaft speed exceeds or falls below an adjustable, pre-set value, the speed switch detects the

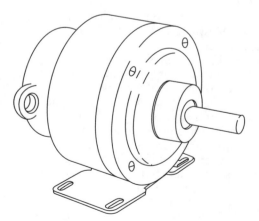

Figure 18-39
Electronic speed switch.

(Courtesy Reliance Electric Co.)

change and actuates external relays, audible alarms, or warning lamps. Output power is switched by a triac, solid-state switch. This model is available in the pictured foot-mounted model or the flange-mounted model. Table 18-2 shows the available speed ranges for the speed switch.

Table 18-2   Speed Ranges*

	Speed Range	
Range Dial Setting	**5–5000 rpm (Standard)**	**0.7–700 rpm (Option D)**
1	5–15	0.7–2
2	15–50	2–7
3	50–100	7–20
4	150–500	20–70
5	500–1500	70–100
6	1500–5000	200–700

* Speed switch range is field-adjustable by dial setting to the range limits shown. After the desired general speed range is selected, specific speed is set by turning an adjustable potentiometer.

A tachometer generator allows accurate monitoring of machine operating speeds. When this is tied into a closed-loop speed regulator, the tachometer generator can be used to control the machine speed, Figure 18-40.

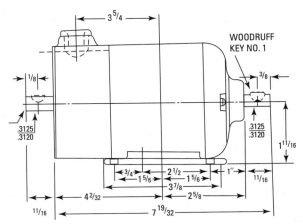

Figure 18-40   Tachometer generator. *(Courtesy Reliance Electric Co.)*

Summary

The squirrel-cage induction motor is the most widely used. It is the workhorse of industry. An electric motor should be matched to the electrical service.

The fundamental job of a controller is to start and stop the motor. It should also protect the motor, machine, and operator. Controllers are enclosed to provide protection for operating personnel by preventing accidental contact with live parts.

Motor controls are designed to meet the provisions of the National Electrical Code. Standards established by the National Electrical Manufacturers' Association assist in the proper selection of control equipment.

Current required to produce full-load torque at rated speed is called FLC. Locked-rotor current is nothing more than the inrush current that occurs when the motor is first energized by line voltage. Locked-rotor current can be from 4 to 10 times the motor full-load current.

Motor speed of a squirrel-cage motor depends on the number of poles on the motor winding. The torque is the turning or twisting force of the motor and is usually measured in pound-feet or in metric kilogrammeters or gram-centimeters.

Motor controllers are used for jogging, plugging, and sequence control steps. Derating of the controls is sometimes called for when the frequency of jogging, plugging, or sequencing operation is abnormal.

Overcurrent protection is provided for motors to protect the motor branch circuit conductors, control apparatus, and motor from short circuits or grounds.

Overloads shorten motor life by deteriorating the insulation. The ideal overload protection for a motor is an element with current-sensing properties very similar to the heating curve of the motor. Fuses are designed to provide overload protection. Overload relays are actuated by a sensing unit. They may be taken out of the circuit by breaking the coil circuit. This opens the starter contacts, disconnecting the motor from the line.

The holding circuit interlock is a normally open auxiliary contact provided on standard magnetic starters and contactors. It closes when the coil is energized to form a holding circuit for the starter after the *start* button has been released.

Reversing the direction of motor rotation is often required. Three-phase, squirrel-cage motors can be reversed by reconnecting any two of the three line connections to the motor. Other motors react differently to being reversed.

A drum switch is a manually operated, three-position, three-pole switch that carries a horsepower rating and is used for manual reversing of a single- or three-phase motor. Limit switches, snap switches, and float switches are examples of other types of control devices.

Review Questions

1. Which induction motor is the most widely used as the workhorse of industry?
2. What is the fundamental job of the controller?
3. What is the National Electrical Code?
4. What does the National Electrical Manufacturers' Association do?
5. What does the speed of a squirrel-cage motor depend on?
6. Why is derating of motor controllers called for when jogging and/or plugging?
7. How do overloads shorten the life of a motor?
8. How are overload relays actuated?
9. What is an interlock?
10. How can three-phase motors be reversed?
11. How is a drum switch operated?
12. What is the motor controller's main purpose?
13. What standards must the controller be designed to meet?
14. What is the Code that governs the use of controllers?
15. What is meant by derating a controller?

Note
The information and pictures herein have been furnished by Square D Company. Some of the material has been copyrighted by Square D and this publication does not claim copyright for such information. Square D assumes no obligations or liabilities arising from reproduction.

Part VI

Motor Repair and Maintenance

Chapter 19

Motor Repair Tools

Whenever you need to repair an electric motor, you will need standard tools, as well as some tools that are not ordinarily found in a home workshop. In some instances, you may need to locate or devise your own holders, pullers, or winders. This chapter will deal with some of the special tools needed, as well as some of the ordinary tools found in every workshop.

Screwdrivers

Most people have screwdrivers around the house or the shop. There are two types of screwdriver blade: the standard slot (flat-blade) type and the Phillips-head type, which has a crossed end. Screwdrivers come in thousands of variations. They may have wooden handles or they may have plastic handles; today most screwdrivers have plastic handles.

Figure 19-1 shows a flat-blade screwdriver with a plastic handle. The plastic handle is very helpful when working around electricity. Plastic is supposed to be shockproof if the handle is kept clean. The blade tip may vary in size from ⅛ to ¼ in. The shaft is from 4 to 8 in. long and is usually of nickel-plated, chrome-vanadium steel. The tips or points must withstand the force applied when a screw sticks or is hard to remove. The main thing to remember when using a screwdriver is to get a good fit between the tip of the screwdriver and the slot in the screw. This will prevent damage to the screw head and the screwdriver.

Phillips screwdrivers (Figure 19-2) have point sizes. The No. 0 point is ⅛ in., the No. 1 is ³⁄₁₆ in., the No. 2 is ¼ in., the No. 3 is ⁵⁄₁₆ in., and the No. 4 is ⅜ in. In this case it is very important to make sure that the most appropriately sized point is used for the screw head. That is why it is best to obtain a complete set, from No. 0 to No. 4, to fit all screw heads. Figure 19-3 shows an assortment of screw heads and matching screwdrivers.

Figure 19-1 Flat-blade screwdriver.

(Courtesy Stanley)

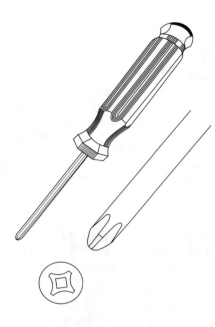

Figure 19-2   Phillips-head
screwdriver. *(Courtesy Stanley)*

Pliers

There are a number of pliers available for special jobs. The pliers in Figure 19-4 are indicative of the variety available for work in the electrical or electronics field. Each is designed for a particular job, as described here:

1. A 4-in. midget for close work.

2. 4-in. pliers for fast, clean tip cutting. It has a tapered nose and nearly flush cutting edges and it will cut to the tip. It produces burr-free cuts.

3. 7-in. diagonal pliers for heavy-duty cutting.

4. 4½-in., thin needle-nose pliers with cutter at the tip.

6. 5-in., thin chain-nose pliers whose smooth jaws are slightly beveled on inside edges.

7. 5½-in. pliers with the fine, serrated jaws for firm gripping or looping wire.

8. Slim serrated jaws (6 in. long) permit entry into areas that are inaccessible by regular long-nose pliers.

9. Long-nose pliers (6½ in.) with side cutter.

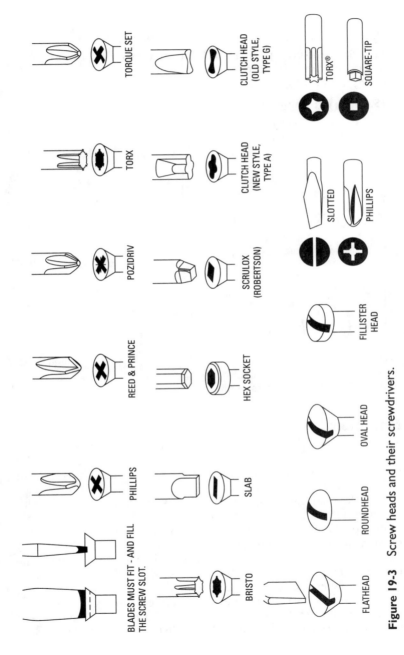

Figure 19-3 Screw heads and their screwdrivers.

TORQUE SET

CLUTCH HEAD
(OLD STYLE,
TYPE G)

TORX®

SQUARE-TIP

TORX

CLUTCH HEAD
(NEW STYLE,
TYPE A)

SLOTTED

PHILLIPS

POZIDRIV

SCRULOX
(ROBERTSON)

FILLISTER
HEAD

REED & PRINCE

HEX SOCKET

OVAL HEAD

PHILLIPS

SLAB

ROUNDHEAD

BLADES MUST FIT - AND FILL
THE SCREW SLOT.

BRISTO

FLATHEAD

499

10. Long-nose pliers (6½ in.) without side cutter.

11. Thin bent-nose pliers (5 in.) with fine serrated jaws and 60°-angle, thin bent nose for thin wire.

14. 8-in., serrated upper and lower jaws with side cutter.

15. 8-in., chrome-plated combination pliers for general use.

16. Four-position, 10-in. utility pliers with forged rib and lock design with serrated jaws.

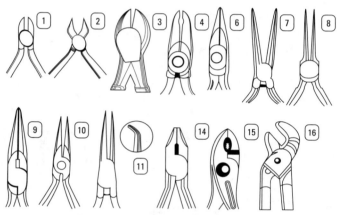

Figure 19-4   Various types of pliers.

Hammers

There are many different types of hammers. The three types mentioned here are best suited for motor repair work. Most people who have a basement or garage workshop will probably have one of these already.

The claw hammer is used to drive nails and to work mainly with wood; it has claws with which to pull out nails. It is the most common type of hammer, but it is used only occasionally in motor repair work when the ball-peen hammer is not available. Figure 19-5 shows the claw hammer most often found around the home. The ball-peen hammer (Figure 19-6) has a rounded top and a flat, larger-diameter bottom surface.

In some cases a mallet is needed to force a connection between the case and the stator or to move the armature gently into place. In most instances you will use a rubber mallet, but in more demanding

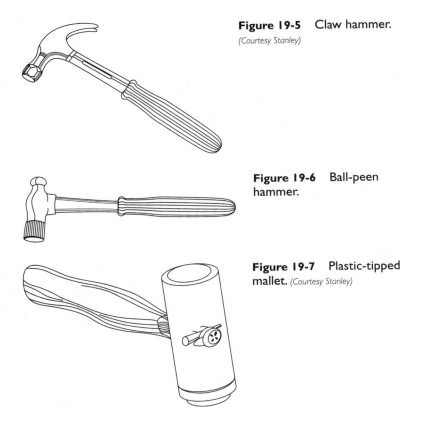

Figure 19-5 Claw hammer.
(Courtesy Stanley)

Figure 19-6 Ball-peen hammer.

Figure 19-7 Plastic-tipped mallet. *(Courtesy Stanley)*

situations you may need a plastic-tipped mallet. These will do the job without marring the surface of the metal. You should be careful with your hammer or mallet blows. Figure 19-7 shows the plastic-tipped mallet.

Hacksaws

Hacksaws are very useful for cutting metal. Figure 19-8 shows the hacksaw blade being installed and as it is used to cut metal. A blade will break if it is not properly inserted and tightened; therefore, tighten the blade as soon as it is pointing in the right direction—away from the handle. Note that the cutting takes place when the hacksaw is pushed away from the operator. Lift the saw when you bring it back to the starting position. Riding the blade as it is drawn back through the metal ruins it.

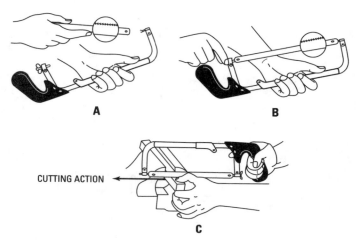

A   B

CUTTING ACTION ←

C

Figure 19-8 Inserting a hacksaw blade, and using the hacksaw.

Blades come in 8- , 10- , and 12-in. lengths. The hacksaw is usually adjustable to fit any of the three lengths of blades. Hacksaw
blades also come in a number of tooth sizes. Use 14-teeth-per-inch
blades for cutting 1-in. or thicker sections of cast iron, machine
steel, brass, copper, aluminum, bronze, or slate. Use the 18-teeth-
per-inch blades for cutting materials ¼- to 1-in. thick in sections of
annealed tool steel, high-speed steel, rail, bronze, aluminum, light
structural shapes, and copper. The 24-teeth-per-inch blade is used
for cutting material that is ⅛- to ¼-in. thick in sections; it is usually
best for iron, steel brass, and copper tubing, wrought-iron pipe,
drill rod, conduit, light structural shapes, and metal trim. The 32-
teeth-per-inch blade is used for cutting material similar to that recommended for 24-tooth blades.

A hacksaw blade is handy when you want to check for a short
in an armature or remove a wedge, as shown in Figure 19-9. A
power hacksaw blade is usually handy to have for removing
wedges.

Wrenches

Wrenches are used to tighten and loosen nuts and bolts. There are
two general types of wrenches: adjustable and nonadjustable.
Adjustable wrenches have one jaw that can be adjusted to accommodate different size nuts and bolts and may range from 4 to 18
in. in length for different types of work (Figure 19-10). When

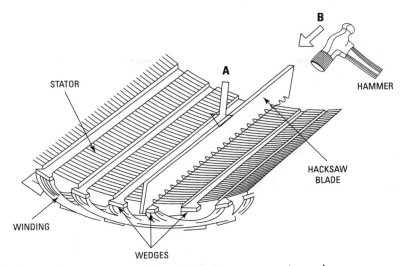

STATOR

A

B

HAMMER

HACKSAW
BLADE

WINDING

WEDGES

Figure 19-9   Using a hacksaw blade in motor repair work.

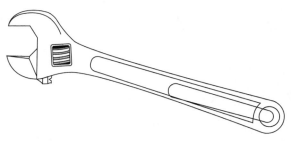

Figure 19-10   Adjustable wrench.

adjustable wrenches are used, there are two rules to remember:
(1) Place the wrench on the nut or bolt so that the force will be
placed on the fixed jaw, and (2) tighten the adjustable jaw so that
the wrench fits the nut or bolt snugly (Figure 19-11).

Nonadjustable wrenches have fixed openings to fit nuts and bolt
heads. Figure 19-12 shows a nonadjustable open-end wrench, while
Figure 19-13 shows a nonadjustable box-end wrench. These
wrenches are available in sets. They are also available in metric
sizes. Openings are actually 0.005 to 0.015 in. larger than the size

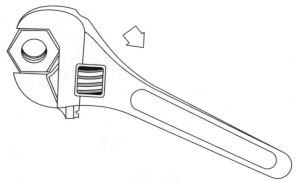

Figure 19-11 Note the direction of force with this type of wrench.

marked on the wrench to allow the wrench to be slipped over the nut or bolt head easily. Make sure, however, that the wrench fits the nut or bolt head properly (Figure 19-14). If it doesn't, you may damage the nut or the bolt head. It is safer to pull than to push a wrench. If you exert pressure on a wrench and the nut or bolt head suddenly breaks loose, you might injure your hand. If the wrench must be pushed rather than pulled, use the palm of the hand so that the knuckles will not be injured if there is a slip.

Allen Wrenches

Allen wrenches are designed to be used with headless screws, which are used in many devices as setscrews (Figure 19-15). Allen wrenches come in various sizes to fit any number of setscrews (Figure 19-16). A complete set is especially helpful to the motor repairperson.

A newer type of recessed setscrew head has an Allen-type hole, but ridges in each flat side make it difficult for an Allen wrench to fit. Newer hex-socket key sets are made in $3/32$- to $1/4$-in. sizes, with eight blades in a set (Figure 19-17).

Socket Wrenches

Socket wrenches may be used in some locations that are not easily accessible by the box-end wrench or the open-end wrench. Sockets are easily taken off the ratchet and replaced

Figure 19-12
Open-end wrench.

with another size. Sockets come in a 12-point or 6-point arrangement. Make sure you use the proper size for each nut or bolt head. This type of wrench is also available in metric sizes. Figure 19-18 shows some of the sockets, extensions, flexible handles, and universal joints that make sockets effective in almost any location.

Torque Wrenches

Torque wrenches are made so that you can apply the proper torque to various bolts and nuts, and are made to fit various socket drives. Two popular sizes of torque wrenches are the ⅜-in. drive and the ½-in. drive. Torque wrenches are made to measure in inch-pounds and also in foot-pounds. Use the properly sized wrench for the torque that has to be applied. Torque wrenches come in various handle lengths. Normally, the longer the wrench, the greater the torque

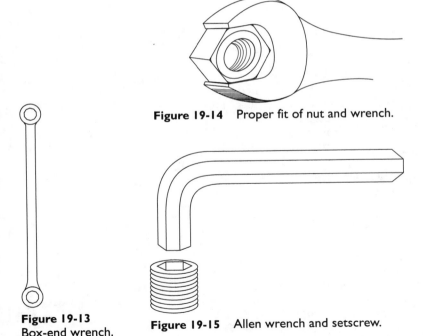

Figure 19-14   Proper fit of nut and wrench.

Figure 19-13
Box-end wrench.

Figure 19-15   Allen wrench and setscrew.

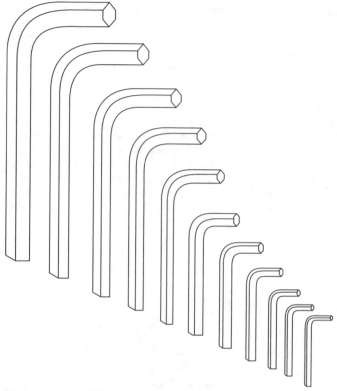

Figure 19-16 Allen wrench set.

the wrench will measure. A typical torque wrench is shown in Figure 19-19.

Nut Drivers

The nut driver is nothing more than a socket attached to a screwdriver handle. It is an excellent tool for most motor repair applications. Nut drivers come in a variety of sizes and usually have the size stamped on the plastic handle. Sometimes they are color-coded according to size. Figure 19-20 shows a set of nut drivers.

Figure 19-17 Socket key set.

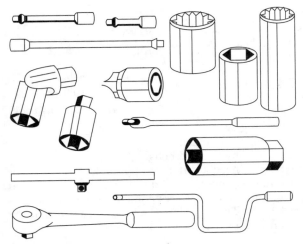

Figure 19-18   Socket wrench set.

Figure 19-19   Torque wrench.

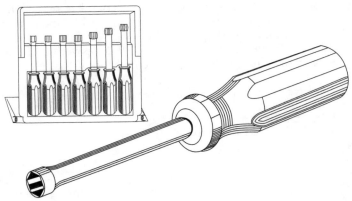

Figure 19-20   Nut-driver set. *(Courtesy Stanley)*

Other Tools

In motor repair it is necessary, in some cases, to remove bearings from the end bell of the motor. A bearing tool (Figure 19-21) makes the task somewhat easier. The illustrated set has nine adapters that easily and quickly remove or insert any sleeve motor bearing or bushing with a ½- to 1-in. inside diameter. It eliminates the chance of broken bearings or end bells.

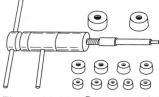

Figure 19-21 Bearing tool with adapters.

Some bearing removals need a different approach. The pulley or gear puller can, in some instances, be used to remove a bearing that has stuck to the armature. These pullers come in a variety of sizes and styles (Figure 19-22). They fit almost any application that the motor repairperson may have.

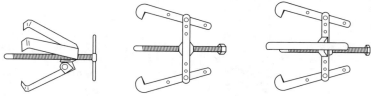

Figure 19-22 Gear pullers.

Bushing Tools

Bushing tools have been designed for removing or inserting bushings in motors. They are handy time savers. A complete set usually consists of 20 pieces: the box, three drivers, and 16 adapters that cover a ⅜- to 1¼-in. range (Figure 19-23).

The solderless connector crimper (Figure 19-24) is very useful in motor repair work. It is used tighten connections that do not

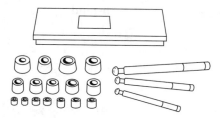

Figure 19-23 Bushing tool set.

require soldering, and a good connection is necessary to withstand the vibration of a motor. A number of connectors have been designed for electrical work. The tool and a kit of connectors and lugs of various sizes are available at most electrical supply houses.

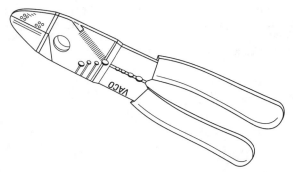

Figure 19-24   Solderless connector crimper.

The wire nut is another type of solderless connector that does not need to be applied by a special tool. It makes a connection by virtue of a piece of copper coiled inside an insulated cover. By twisting the wire nut after the wires are inserted, it is possible to make a good electrical connection (Figure 19-25).

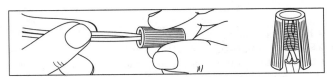

Figure 19-25   Wire-nut solderless connector.

Soldering Iron
The soldering iron (Figure 19-26) comes in handy for making a solder connection that will take vibration and withstand corrosion. Soldering irons are available from about 15 W to over 600 W. The best all-purpose size for use in the shop is about 100 W. This will do the job in almost all cases where larger wires are concerned. The small 15-W irons are very useful for electronics work.

Figure 19-26   Soldering iron.

The soldering gun (Figure 19-27) is very handy for making quick disconnects of soldered joints. It can also be used for many cold-solder joints when the person using it heats up the tip, places solder on the tip, and lets it cool on the joint. The wires being soldered or the metal surface and the wire being connected to it must be of sufficient temperature to melt the solder. This means that the gun must be left in one spot long enough to cause the joint to be heated to the temperature needed to melt the solder. The secret is to heat the material and not the solder.

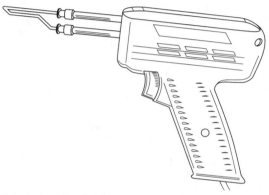

Figure 19-27   Soldering gun.

In some isolated cases it may be necessary to melt silver solder or to braze a joint. A miniature torch (Figure 19-28) welds, brazes, solders, and cuts metal with pinpoint accuracy. It uses butane gas and oxygen to produce a very hot flame in a very restricted area.

Wire Gauges

Wire gauges are needed to measure the wire used in the repair business. There are numbers on the gauge (Figure 19-29) that tell you the size of the wire. Keep in mind, however, that the insulation on the wire—in the case of FORMVAR insulation—will read one size larger on the gauge than the wire. Also keep in mind that the wire is moved through the slot in the gauge. The hole is there

Figure 19-28   Gas welding torch.

to pass the wire through. The slot does the measuring. Pull the wire free of the slot and through the hole. Decimal equivalents are usually stamped on the metal disc on the opposite side of the gauge numbers. No repair shop should be without a wire gauge.

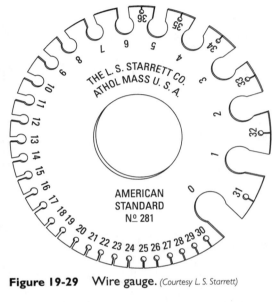

Figure 19-29 Wire gauge. *(Courtesy L. S. Starrett)*

Stator-Holding Stand
The stator-holding stand is a necessary item if you are going to rewind motors. Figure 19-30 is one example of such a stand. It is self-explanatory. In most cases the stand is bolted to a table or mounted in a vise to secure it while the work is being done.

Armature-Holding Stand
To work freely on an armature, you must have an armature-holding stand. Figure 19-31 shows how the stands are used in the process of removing the wooden or plastic wedges of the armature.

Reel Rack
The reel rack (Figure 19-32) holds your reel of wire so that it can be unreeled easily without getting tangled or kinked. The term used here is *de-reeled*. The tension device keeps the wire from falling on the floor or becoming tangled. A number of different types are available. Figure 19-32 illustrates only one type.

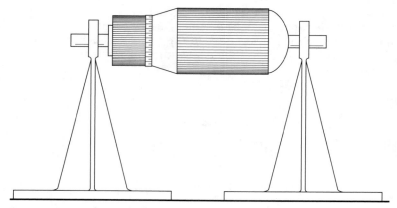

Figure 19-30   Stator-holding stand.

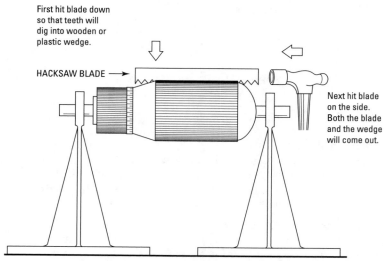

First hit blade down
so that teeth will
dig into wooden or
plastic wedge.

HACKSAW BLADE →

Next hit blade
on the side.
Both the blade
and the wedge
will come out.

Figure 19-31   Using a hacksaw to remove wedges on older motors.

Lathe

A lathe, such as the one shown in Figure 19-33, comes in handy for rewinding or turning down commutators. It can be used for any number of machining operations that may be needed to repair some heavily damaged motors. The lathe shown in Figure 19-33

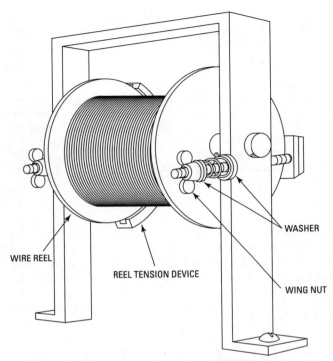

Figure 19-32   Bench setup of a reel rack.

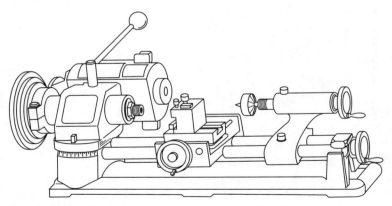

Figure 19-33   Lathe.

can be used for turning, facing, grinding, buffing, sawing, sanding, threading, milling, drilling, and coil winding. The size of the lathe will depend on your budget and the applications for which it is needed.

Vise

No shop would be complete without a vise. It can be used to hold almost everything during the rewiring operation. It can be used to hold armatures, if you line the jaws of the vise with copper or some soft metal so that the clamped piece is not damaged when the jaws are made tight. Vises come in almost any size you can imagine. The size you will need is determined by the size of the motors you handle most often.

Storage Cabinet

Because motors have many parts and pieces, it is best to have some method of storing small parts. The drawers shown in Figure 19-34 make an ideal storage arrangement for nuts, bolts, screws, and other small parts. Each drawer should be labeled and loose parts should be catalogued by placing them into the proper drawer. Then, when you need a part, you will not spend time looking through everything else to find it. An organized shop can make the difference between a profitable business and a losing proposition.

Figure 19-34 Small-item storage cabinet.

Meters and Test Devices

A number of meters are available for use in testing and working with electric motors. The VOM, or volt-ohm-milliammeter (Figure 19-35), is very useful in checking for shorts and opens.

Figure 19-35   Digital
volt-ohm-ammeter.
(Courtesy Weston)

Figure 19-36
Clamp-on
ammeter.
(Courtesy Amprobe)

The best way to check a meter is by its ohms-per-volt (ohms/volt) rating. The higher the ohms/volt rating, the better the meter. It should have at least 20,000 ohms/volt on DC and at least 5000 ohms/volt on AC. This should be sufficient for checking most motor troubles. A shunt ohmmeter that will measure from 0 to 200 ohms is also useful, because most motor windings and armatures are less than 300 ohms. In some cases, it is possible to obtain a VOM with both a series ohmmeter for measuring high resistances and a shunt ohmmeter for measuring low resistances.

The clamp-on type of ammeter (Figure 19-36) can be used to measure current. It will clamp around one wire of an AC line to a motor and read the current being drawn by the motor. The nameplate usually tells how much current the motor will draw during normal operation.

Voltage Tester

The voltage tester is a handy device that checks 10 AC/DC voltage levels. It fits in a shirt pocket. The lighted windows indicate the voltage level, which makes it easy to read in dimly lit areas. The coiled lead cord extends to 50 in. A test button distinguishes normal readings from those due to distributed capacitance or high-resistance leakage. It is also helpful in checking out 115-V AC grounded convenience outlets, and it will operate on 15 to 800 Hz. See Figure 19-37.

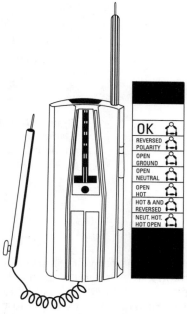

Figure 19-37   Voltage probe.

Insulation Tester

One of the ways to measure insulation resistance is to use a megger or insulation tester (Figure 19-38). A megger quickly locates intermittent shorts, bad electrical connections, insulation breakdowns, and conductor failures due to wear, moisture, and corrosion. Insulation resistance is calibrated directly in megohms (one million ohms). The voltage output of the megger is from 500 volts DC to 2000 volts DC, and can produce a painful shock if not handled properly. However, it is well worth its price since it can locate problems that would otherwise be most difficult or impossible to find.

Tachometer

The tachometer is a very useful device for measuring the speed of a motor. It can help locate possible troubles and can indicate if the motor is operating as it should after it has been repaired.

A hand-held tachometer is shown in Figure 19-39. Its shaft speed will measure from 50 to 4000 rpm. It can be placed on the open end of a motor shaft, or it can be used on motors, saws, compressors, fans, pumps, grinders, and other electric motor-powered tools or equipment. A cone-shaped tip is used for shafts with center holes; a cup-shaped tip is used for flat-end shafts. Other types of tachometers are available that use a strobe light to detect the number of revolutions per minute (see Figure 19-40); however, these are somewhat expensive.

Phase Sequence Adapter

The phase sequence adapter is used in conjunction with any volt/ammeter of appropriate AC range. It lets you determine the phase sequence of any electrical equipment using three-phase lines up to 550 V, 25 to 60 Hz (see Figure 19-41). Because so many motors are three-phase, it is important to be able to check the power source when troubleshooting motors.

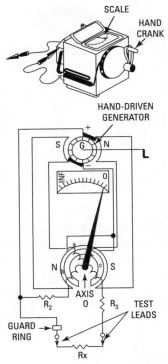

Figure 19-38   Megger.

Oscilloscope

The oscilloscope is also a voltage indicator (see Figure 19-42). It shows the shape of the power being used. It can aid in tracing pulses on the line that may cause timing problems with programmable

Figure 19-39   Tachometer.

(Courtesy Stewart-Warner)

Figure 19-40 Testing a motor with a tachometer.

controllers or other computer-operated machines. In conjunction with a function reference signal generator it is possible to properly adjust and tune the stability circuit of a high-gain motor drive and regulators. The appropriate stability circuits can be optimized by placing a step function into the regulator and observing the feedback loop output with an oscilloscope or chart recorder.

Micrometer

One other device should be considered as necessary for any well-equipped motor repair facility—the micrometer (Figure 19-43). The micrometer is a very accurate instrument that is used to check the thickness of insulation, the diameter of wire, and other such applications. Instructions for using a micrometer are furnished with the instrument or can be found in almost any textbook on metal-working.

Figure 19-41 Phase sequence adapter.

Figure 19-42 Oscilloscope.

Electronic Digital Micrometer

The electronic micrometer is much easier to read and manipulate than the regular micrometer. The electronic micrometer shown in Figure 19-44 has controls for measuring, analyzing, and recording information. It has a smooth friction thimble for uniform pressure and graceful movements with great accuracy. One battery lasts for over a year of normal

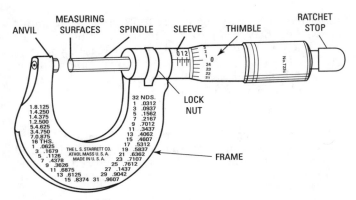

Figure 19-43 Micrometer.

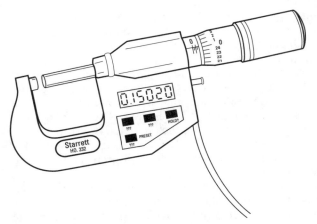

Figure 19-44 Digital micrometer. *(Courtesy of Starrett)*

usage and it automatically turns itself off after 30 minutes of nonuse. The keyboard control has a shift button. It has instant inch/millimeter conversion. The ME or millimeter models will turn on the millimeter mode after installation of a new battery. There is a measurement hold button and it has the ability to zero at any position as well as retain and return to true zero readings of the micrometer. The output jack allows data transmission to the peripherals that analyze, collect data, and make hard copy documentation.

This chapter has included some of the tools used in the motor repair business. You may want to buy them all at one time or acquire them slowly as they are needed. In some cases, you may want to develop your own designs to fit your own habits. In any case, this is not an exhaustive listing but an attempt to show some of the tools needed to do a good job in motor repair.

Summary

Whenever you need to repair an electric motor, you will need standard tools, as well as some tools that are not ordinarily found in a home workshop. In some instances, you may need to locate or borrow or rent special tools, such as holders, pullers, or winders.

Flat-blade is the most common type of screwdriver, with a tip that fits slots of $\frac{1}{8}$ to $\frac{1}{4}$ in. They usually have plastic handles with a shaft of 4 to 8 in. The Phillips-head screwdriver is another common tool. It comes in point sizes from No. 0 to No. 4. The cross-like tip makes it easier to use this type of screwdriver to power drive screws and on mass production lines where the screwdriver slipping out of the slot can be time consuming.

Pliers come in a number of sizes and shapes and each type has its own function. There are diagonal pliers, needle-nose pliers, long-nose pliers, bent-nose pliers, and side-cutter pliers. Some are chrome plated and others have serrated upper and lower jaws. The four-position, 10-in. utility pliers with forged rib and lock design has serrated jaws.

Of the many types of hammers available, three are best suited for motor repair work: the claw hammer, the ball-peen hammer, and the mallet type hammer.

The hacksaw can be very helpful when working with metal. Being able to install a blade and cutting in the forward direction is important. The number of teeth in the blade should be appropriate for the job being done.

Wrenches are used to tighten and loosen nuts and bolts. There are two general types of wrenches: adjustable and nonadjustable. They come in a number of sizes and must be matched to the particular job being done at the time. Allen wrenches are designed to be used with headless screws, which are used in many devices as set screws, especially in motor installations with pulleys and other devices attached to the drive shaft of a motor.

Socket wrenches may be used in some locations that are not easily accessible by a box-end wrench or an open-end wrench. Torque wrenches are made so that you can apply the proper torque to various bolts and nuts, and are made to fit various socket drives. They are made to measure in inch-pounds and foot-pounds.

Nut drivers are nothing more than a socket attached to a screwdriver handle. Sometimes they are color-coded according to size.

In some cases, where you have to remove a bearing or a pulley, a pulley or gear puller may be necessary. These pullers come in a variety of sizes and styles. Bushing tools have been designed for removing or inserting bushings in motors.

The solderless connector crimper is very useful in motor repair work. The soldering iron comes in handy for making a solder connection that will take vibration and withstand corrosion. For making quick connections, the soldering gun is handy. In some instances, it may be necessary to melt silver solder or to braze a joint. A miniature torch welds, brazes, solders, and cuts metal with pinpoint accuracy. It uses butane and oxygen to produce a very hot flame in a very restricted area.

Wire gauges are needed to measure the wire used in the repair business. A number of gauges are available. Decimal equivalents are usually stamped on the metal disc on the opposite side of the gauge numbers.

A stator-holding stand is a necessary item if you are going to rewind motors. An armature-holding stand is also very important for working on armatures.

The reel rack is necessary to hold the reel of wire used in rewinding motor stators and armatures. A lathe is a handy tool to have around when turning commutators and when rewinding various motors. A vise is a necessary tool for any repair facility. It can be used to hold almost everything during the rewiring operation.

Storage cabinets with small-item storage capacity, or many small drawers, is an absolutely essential device. A number of meters are available for testing motors and their windings. The electronic or digital volt-ohm-ammeter is the ideal one to use because it can measure in small parts of the ohm for wire resistances. A clamp-on ammeter is essential when checking the working conditions and performance of a motor.

One of the ways to measure insulation resistance is to use a megger or insulation tester. A megger quickly locates intermittent shorts, bad electrical connections, insulation breakdowns, and conductor failures due to wear, moisture, and corrosion.

The tachometer is very useful in measuring the speed of a motor. It can help locate possible troubles and can indicate if the motor is operating as it should after it has been repaired.

The micrometer is a very necessary device even for a newly opened repair facility. It is a very accurate instrument that is used to check the thickness of insulation, the diameter of wire, and other such applications.

Review Questions

1. Why are tools needed to repair an electric motor?
2. What are the minimum tools needed to start a repair facility?
3. List some types of pliers needed in a repair shop.
4. What are three types of useful hammers in a repair shop?
5. What is the purpose of Allen wrenches?
6. What is the purpose of socket wrenches? Where are they used in working with motors?
7. What are nut drivers?
8. What does a miniature butane torch do in the field of motor repair?
9. How can wire gauges be used to improve your repair of a motor?
10. What is a reel rack? Why is it useful in a repair shop?
11. How is the lathe utilized in the shop?
12. Why would you need storage cabinets in a motor repair facility?
13. How is the megger used in motor repair?
14. How is the micrometer used in a motor repair shop?
15. Why do you need to test insulation?

Chapter 20

Motor Repair Supplies

A good repair shop has enough supplies on hand to avoid repeated trips to the supplier. It takes time for a new repair service to build up a complete supply of parts. However, there are commonly used parts that should always be in stock. This chapter deals with the motor repair supplies most often needed by motor repair facilities.

Setscrews

Setscrews are used to fasten the pulley to the motor shaft in electric motors. An Allen wrench will remove or loosen the setscrew sufficiently to remove the pulley. Care should always be exercised in loosening a setscrew because if it is loosened too much, it may drop out and be lost as the pulley is removed from the shaft of the motor.

Setscrews should be a standard item in a shop's supplies. They can be purchased in 80-piece sets. The sets come in a compartmentalized box in various lengths of ¼–20, ⁵⁄₁₆–18, and ⅜–16 thread. Each size of screw should be kept in its proper section of the storage box. Several forms of setscrews are shown in Figure 20-1.

Carbon Brushes

Carbon brushes connect the revolving armature to the external circuit of a motor. The brushes can be square, rectangular, or round. They are usually sold as a part of a kit. See Figures 20-2 and 20-3 for brush definitions, connections, hammers and lifting clips, and terminals, as well as their actual size and shape.

Electric motors that use portable tool-type carbon brushes include drills, buffers, saws, sanders, and grinders. Differently shaped brushes are needed for other kinds of electric appliance motors. These include motors for fans, mixers, and AC/DC motors. Along with a plentiful supply of brush kits, a large number of springs should also be available. There are always many uses for a variety of different size springs (Figure 20-4).

Bushings and Bearings

Bronze bushings make up bearings for some motors. They are easily damaged by over-tightening a belt. This puts undue stress on one

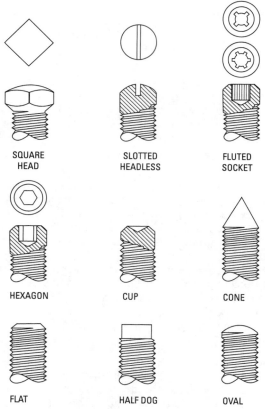

SQUARE
HEAD

SLOTTED
HEADLESS

FLUTED
SOCKET

HEXAGON

CUP

CONE

FLAT

HALF DOG

OVAL

Figure 20-I   Types of setscrews.

side of a bearing surface. The bushing or bearing can wear and
cause the armature to strike the field. This, in turn, will damage the
armature and the field pole.

Bronze bushings, which are used as bearings, can be purchased
in 56-piece kits (see Figure 20-5). The kits are made of styrene and
have sections that keep the various sizes separate. There are nine
different sizes of bushings in a kit. They range from $\frac{3}{8}$-in. ID $\times$ $\frac{1}{2}$-
in. OD to $\frac{3}{4}$-in. ID $\times$ 1-in. OD.

The lack of lubrication is a major cause of wear in sleeve bearings.
Lubricants used with sleeve bearings serve a different purpose from
those used for ball bearings. The lubricant must actually provide an

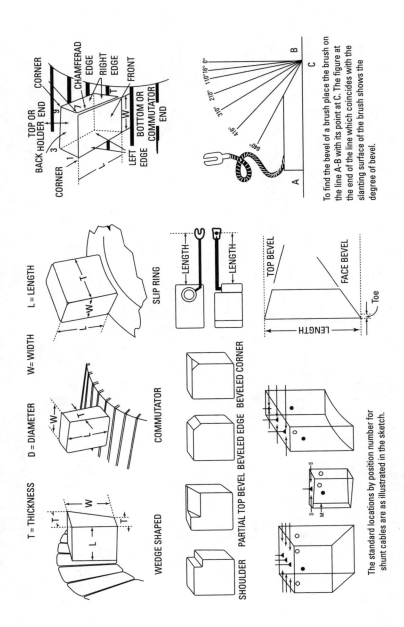

T = THICKNESS   D = DIAMETER   W = WIDTH   L = LENGTH

WEDGE SHAPED

SLIP RING

COMMUTATOR

SHOULDER   PARTIAL TOP BEVEL   BEVELED EDGE   BEVELED CORNER

TOP BEVEL

FACE BEVEL

Toe

LENGTH

The standard locations by position number for shunt cables are as illustrated in the sketch.

TOP OR BACK HOLDER END

CORNER

CHAMFERED EDGE

RIGHT EDGE

FRONT

LEFT EDGE

BOTTOM OR COMMUTATOR END

To find the bevel of a brush place the brush on the line A-B with its point at C. The figure at the end of the line which coincides with the slanting surface of the brush shows the degree of bevel.

Figure 20-2 Brush definitions.

525

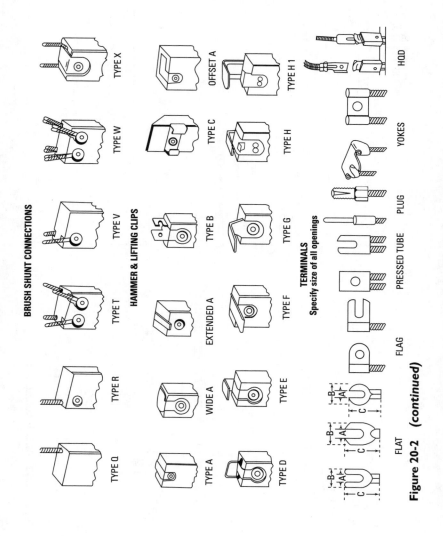

BRUSH SHUNT CONNECTIONS

TYPE Q   TYPE R   TYPE T   TYPE V   TYPE W   TYPE X

HAMMER & LIFTING CLIPS

TYPE A   WIDE A   EXTENDED A   TYPE B   OFFSET A

TYPE D   TYPE E   TYPE F   TYPE G   TYPE C

TYPE H   TYPE H 1

TERMINALS
Specify size of all openings

FLAT   FLAG   PRESSED TUBE   PLUG   YOKES   HOD

Figure 20-2 (continued)

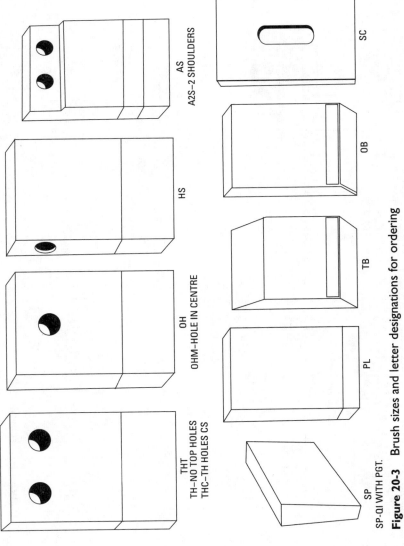

THT
TH–NO TOP HOLES
THC–TH HOLES CS

OH
OHM–HOLE IN CENTRE

HS

AS
A2S–2 SHOULDERS

SP
SP–QI WITH PGT.

PL

TB

OB

SC

Figure 20-3 Brush sizes and letter designations for ordering replacements.

527

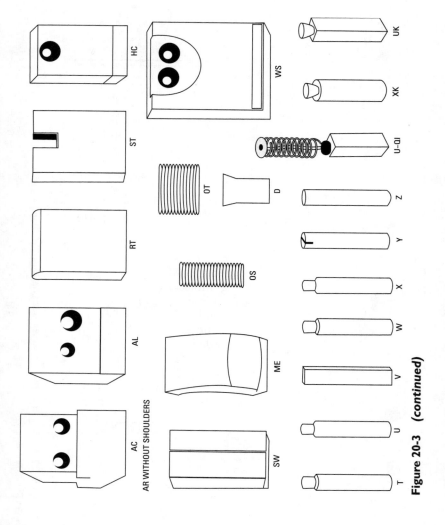

Figure 20-3 *(continued)*

528

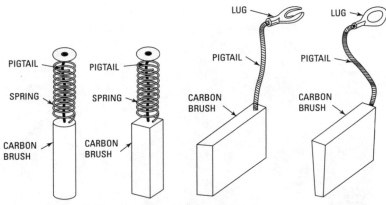

Figure 20-4 Carbon brushes for electric motors.

oil film that completely separates the bearing surface from the rotating shaft member, thus eliminating metal-to-metal contact. The oil film is automatically formed when the shaft begins to turn, and is sustained by the motion of the shaft. The rotational force of motion sets up pressure in the oil film wedge that supports the load. The wedge-shaped film of oil is absolutely essential for effective sleeve-bearing lubrication as it prevents the destruction of the bearing. If the highest-quality oil is used, there will be longer wear and less trouble from the motor. On the other hand, over-lubrication should be avoided since the oil may enter the centrifugal switch and cause shorting. Also, excess oil always attracts dust and dirt.

It should be noted that the bearing also serves as a guide for shaft alignment (Figure 20-6). Consequently, a replacement bearing must be an exact replacement, not just an approximate size.

Sleeve bearings are less sensitive to a limited amount of abrasive or for-

Figure 20-5 Assortment of various-size springs.
(Courtesy General Electric)

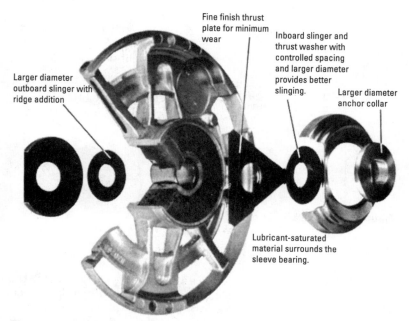

Fine finish thrust plate for minimum wear

Inboard slinger and thrust washer with controlled spacing and larger diameter provides better slinging.

Larger diameter anchor collar

Larger diameter outboard slinger with ridge addition

Lubricant-saturated material surrounds the sleeve bearing.

Figure 20-6 Method of lubricating a bronze-bushing sleeve bearing.

eign material than are ball bearings. They have the ability to absorb small, hard particles into the soft undersurface of the bearing. However, good maintenance practice demands that the bearings be kept clean. In small motors, dirty oil or insufficient lubrication can add enough friction to cause the bearings to seize or freeze in place.

A good practice is to replace the oil every six months. If the motor is operated under very dirty conditions, it should be cleaned more often. Keep in mind that sleeve-bearing motors tend to lose the oil film when stored for one year or more.

Shaft Adapters

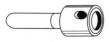

Figure 20-7
Shaft adapter.

Shaft adapters are used to convert the shaft of a motor to a more usable size. The shaft adapter shown in Figure 20-7 has an Allen-head setscrew to secure the adapter to a threaded motor shaft. The illustrated adapter has a ¼-in. diameter and is 1 in. long.

Rubber Mounts

Rubber mounts (Figure 20-8) fit between the base of the motor and mounting rails, and reduce the noise level of the motor. Two motor-mounting rails are shown in Figure 20-9. Many sizes of rubber mounts are available to fit a number of frames. Some types of frame mounts are used to make a motor fit other forms of installation. Rubber mounts should be bought as needed.

Figure 20-8 Rubber motor mounts.

Fiber Washers

Fiber washers are used to remove the end play from motors. They take up the space that exists due to manufacturing tolerances. In the process of disassembling a motor it is easy to

Figure 20-9 Motor-mounting rails.

misplace a fiber washer, so it is wise to have a ready supply in stock. They are sold to repair shops in kits containing 48 different sizes.

Rubber Grommets

Rubber grommets prevent the electrical cord from becoming shorted by contacting the frame of the motor. They fit into the metal case and insulate the hole by providing a rubber covering. A $\frac{3}{8}$-in. and a $\frac{1}{2}$-in. bore are usually sufficient for most uses. The $\frac{3}{8}$-in. bore needs a $\frac{1}{2}$-in. hole and the $\frac{1}{2}$-in. bore grommet requires a $\frac{3}{4}$-in. hole. This is necessary because the outside diameter is $\frac{7}{8}$ in. A good supply of rubber grommets should be kept on hand.

Replacement Switches

In split-phase, capacitor-start, and permanent-split capacitor motors, centrifugal switches are needed to remove the start winding from the circuit after the motor has reached run speed. These switches can fail for a number of reasons. Wear, too much oil, dirt, pitted contacts, and overload on the motor can all cause failure.

Centrifugal switches have various shapes, as shown in Figure 20-10. The ten switches in the figure are made for Westinghouse motors. One style of centrifugal switch with two operative parts is shown in close-up in Figure 20-11. In Figure 20-12 the switch is shown in both the start and running positions. The contacts are closed when the motor is not operating. When the motor is started, the

Factory parts for split phase and capacitor motors, Nema 48 and 56 frames

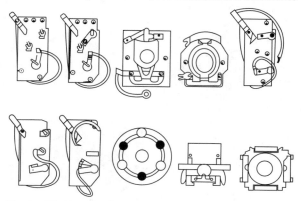

Figure 20-10 Replacement switches for Westinghouse motors.

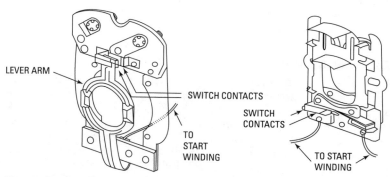

LEVER ARM

SWITCH CONTACTS

TO
START
WINDING

SWITCH
CONTACTS

TO START
WINDING

Figure 20-11 One type of centrifugal switch (both sides).

movement of the rotor (centrifugal force) causes the round slider to move inward on the shaft. As the slider moves inward, it causes the switch contacts to open and the start-winding circuit is broken. If the motor is turned off, the slider will move out again. It is forced by spring action to close the switch contacts of the start winding. The motor is then ready to start again with the start winding in the circuit.

If the centrifugal switch does not close when the motor is stopped, the rotor will not move when the power is turned on. The motor will hum but will not actually start. However, if you give the motor shaft a start by hand—move it either clockwise or counterclockwise—the motor will start.

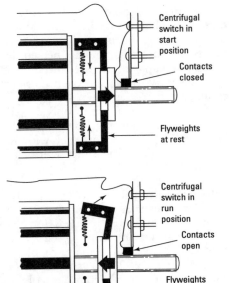

Figure 20-12 Operation of the centrifugal switch.

Centrifugal switch in start position

Contacts closed

Flyweights at rest

Centrifugal switch in run position

Contacts open

Flyweights have over-come spring tension.

Electrolytic Capacitors

The capacitor-start motor is basically a split-phase motor with two separate windings in parallel connection: the main winding and the start winding. In the capacitor-start motor, an electrolytic capacitor is inserted in series, with the start winding in the start position, to increase the starting torque and reduce the starting current. When replacing the capacitor in the motor, use only the electrolytic capacitor marked for motor use.

There are two types of electrolytic capacitors: polarized and nonpolarized. The polarized type is used in electronics. It has a positive (+) symbol on one lead and negative (−) symbol on the other. It may also have a red dot or some other type of code for the positive lead. The polarized type will operate on DC only, and should not be used with AC at all.

A nonpolarized electrolytic capacitor can be made by placing two polarized types in series. However, they must be set up in series opposition; in other words, connect them so that positive (+) goes to positive (+) and the two negative (−) terminals go the AC circuit.

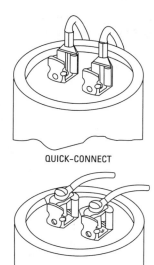

QUICK-CONNECT

SCREW

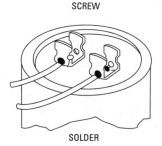

SOLDER

Figure 20-13 Three types of connections to an electrolytic capacitor used in a motor circuit.

One of the problems with a *substitute* electrolytic capacitor is making sure that the capacitance is correct. If the two capacitors are connected in series, the capacitance is reduced. For instance, if two 100 μF capacitors were connected in series, they would produce a total capacitance of only 50 μF. However, the working voltage DC is increased. Two capacitors each with 50 WVDC and placed in series will combine to make 100 WVDC capability.

Capacitor-start (CS) motors use only one electrolytic to start. The capacitor is usually round in shape and has the size indicated in μF. The voltage is marked in white letters and numbers if the case is made of black Bakelite. Figure 20-13 shows three types of connectors for an electrolytic capacitor used on electric motors.

The permanent-split capacitor (PSC) motor may use one oil-type capacitor (usually an oblong type) for both starting and running (Figure 20-14). Some motors require two capacitors to start and one capacitor to run. These motors use one oil-type and one dry electrolytic-type capacitor in the start position. Only the oil type is needed while the motor is running.

There are two types of electrolytics: dry and wet. The dry type uses a paste material and is not truly dry. The wet type uses oil as the electrolyte.

The capacitor is usually located on top of the motor and has a metal cover plate or enclosure. For refrigerators, air conditioners (large units), and washing machines, the capacitor may be mounted somewhere other than on the motor itself (Figure 20-15).

Insulating Enamel

Insulating enamel can be used as a protective coating for frames and end shields and as a gasket cement and sealant for oil, water,

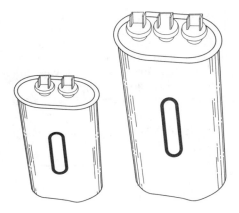

Figure 20-14   Oil-filled capacitor for use with motors.

and connections. The enamel can also be used on windings, coils, commutator ends, and bus bars. It produces a tough, flexible, oil-proof and waterproof film. Insulating enamel comes in spray cans and is quick drying.

Figure 20-15   Capacitor located on top of a motor. *(Courtesy Leeson)*

Cord

A supply of heavy-duty cord should be on hand to replace any cords that may be dried out or cracked. Three-conductor cord will have white, black, and green wire. The green wire is the safety wire and should be connected to the frame of the motor. The rating of the cord needs to be checked. It is also important to ensure that the current rating of the motor matches the current rating of the cord.

Every cord has a maximum allowable current-carrying capacity (Table 20-1). Cords for heavy motors and high-wattage appliances must be heavy enough to carry the needed current. Overloaded

Table 20-1   Cord Sizes and Current-Carrying Capabilities

Cord Sizes and Uses

	Type	Wire Size	Use
Ordinary lamp cord	POSJ SPT	No. 16 or 18	In residences for lamps or small appliances
Heavy-duty— with thicker covering	S or SJ	No. 10, 12, 14, or 16	In shops, and outdoors for larger motors, lawn mo- wers, outdoor lighting, etc.

Ability of Cord to Carry Current (2- or 3-wire cord)

Wire Size	Type	Normal Load	Capacity Load
No. 18	S, SJ, or POSJ	5.0 amp (600 W)	7 amp (840 W)
No. 16	S, SJ, or POSJ	8.3 amp (1000 W)	10 amp (1200 W)
No. 14	S	120.5 amp (1500 W)	15 amp (1800 W)
No. 12	S	16.6 amp (1900 W)	20 amp (2400 W)

Selecting the Length of Wire

Light Load (to 7 amp)	Medium Load (7–10 amp)	Heavy Load (10–15 amp)
To 15 ft— Use No. 18	To 15 ft— Use No. 16	To 15 ft— Use No. 14
To 25 ft— Use No. 16	To 25 ft— Use No. 14	To 25 ft— Use No. 12
To 35 ft— Use No. 14		To 45 ft— Use No. 10

Note: As a safety precaution, be sure to use only cords that are listed by Underwriters Laboratories. Look for the UL seal.

cords over-heat, wasting power and often causing the motor to run at lower than normal speeds. Of course, there is always the possibility of fire from an overheated cord.

Motor Wiring for Installation

All wiring and electrical connections should comply with the National Electrical Code (NEC) and with local codes and sound practices.

Use of undersized wire between the motor and the power source will adversely limit the starting and load-carrying abilities of a motor. Recommended minimum wire sizes for motor branch circuits are given in Tables 20-2 and 20-3.

Table 20-2   Individual Branch Circuit Wiring for Single-Phase Induction Motors

Motor Data		Copper Wire Size (Minimum AWG No.)				
		Branch Circuit Length				
H.P.	Volts	0–25 ft	50 ft	100 ft	150 ft	200 ft
1/6	115/230	14/14	14/14	14/14	12/14	10/14
1/4	115/230	14/14	14/14	12/14	10/14	8/14
1/3	115/230	14/14	12/14	10/14	8/14	6/12
1/2	115/230	14/14	12/14	10/14	8/14	6/12
1/4	115/230	12/14	10/14	8/14	6/12	4/10
1 1/2	115/230	12/14	10/14	8/14	6/14	4/10
1	115/230	10/14	10/14	6/12	4/10	4/8
2	115/230	10/14	8/12	6/12	4/10	8
3	115/230	6/10	6/10	4/10	8	8
5	230	8	8	8	6	4

Table 20-3   Individual Branch Circuit Wiring for Three-Phase, Squirrel-Cage Induction Motors

Motor Data		Copper Wire Size (Minimum AWG No.)				
		Branch Circuit Length				
H.P.	Volts	0–25 ft	50 ft	100 ft	150 ft	200 ft
1/2	230/460	14/14	14/14	14/14	14/14	14/14
3/4	230/460	14/14	14/14	14/14	14/14	14/14
1	230/460	14/14	14/14	14/14	14/14	14/14

Table 20-3 *(continued)*

Motor Data		Copper Wire Size (Minimum AWG No.)				
		Branch Circuit Length				
H.P.	Volts	0–25 ft	50 ft	100 ft	150 ft	200 ft
1½	230/460	14/14	14/14	14/14	14/14	12/14
2	230/460	14/14	14/14	14/14	12/14	12/14
3	230/460	14/14	14/14	12/14	10/14	10/14
5	230/460	12/14	12/14	10/14	10/14	8/14
7½	230/460	10/14	10/14	10/14	8/14	6/12
10	230/460	8/12	8/12	8/12	6/12	4/12
15	230/460	6/10	6/10	6/10	4/10	4/10
20	230/460	4/8	4/8	4/8	4/8	3/8
25	230/460	4/8	4/8	4/8	3/8	2/8
30	230/460	3/6	3/6	3/6	2/6	1/6
40	230/460	1/6	1/6	1/6	1/6	0/6
50	230/460	00/4	00/4	00/4	00/4	00/4

Fuses

One time, one element fuses. If a current of more than rated load is continued long enough, the fuse link becomes overheated. This causes the center portion to melt. The melted portion drops away. However, due to the short gap, the circuit is not immediately broken. An arc continues and burns the metal at each end until the arc is stopped because of the very high increase in resistance. The material surrounding the link tends to break the arc mechanically. The center portion melts first, because it is farthest from the terminals that have the highest heat conductivity.

Overloads

Fuses will carry a 10 percent overload indefinitely under laboratory-controlled conditions. However, they will blow promptly if materially overloaded. They will stand 150 percent of the rated amperes for the following time periods.

1 minute (fuse is 30 A or less)

2 minutes (fuse of 31 to 60 A)

4 minutes (fuse of 61 to 100 A)

Time-Delay, Two-Element Fuses

Two-element fuses use the burnout link described previously. They also use a low-temperature soldered connection that will open under overload. This soldered joint has mass, so it does not heat quickly enough to melt if a heavy load is imposed for only a short time. However, a small but continuous overload will soften the solder so that the electrical contact can be broken.

With this type of protection against light overloads, the fusible link can be made heavier, yet blow quickly to protect against heavy overloads. This results in fewer nuisance burnouts and equipment shutdowns. Types of dual-element fuses are shown in Figure 20-16.

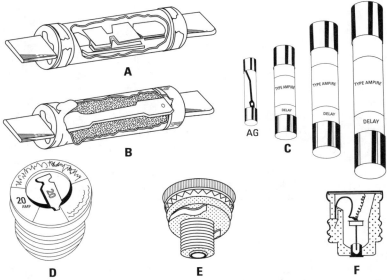

Figure 20-16   Fuse Types. (A) Two-element renewable (cutaway view). The end caps screw off for replacement of the fusible element. (B) Ordinary one-time fuse (cutaway view). This type cannot be opened for replacement of the fusible element. (C) Cartridge-type fuses. AG stands for automotive glass. This type is used in electronics equipment and instruments. (D) Plug-type fuse used for a screw-in type receptacle. This is an old type used in home wiring circuits. (E) Fustat®. Special threads make this fit only a particular type of fuse holder. (F) Fusetron® is a dual-element, plug-type fuse. The spring increases the possibility that it will trip at the correct rating.

Types of Fuses

In addition to those fuses just described, there are three general categories based on shape and size.

The AG (*automobile glass*) fuse consists of a glass cylinder with metallic end caps between which is connected a slender metal element that melts on current overload. This fuse has a length of $1^{15}/_{16}$ in. and a diameter of $1/4$ in. It is available for only a few amperages. While used in special appliances, it is not used to protect permanently installed wiring.

Cartridge fuses are similar to AG fuses, but they are larger. The cylindrical tube is fiber, rather than glass. The metallic end pieces may be formed as lugs, blades, or cylinders to meet a variety of fuse-box socket requirements. The internal metal fusible link may be enclosed in sand or powder to quench the burnout arc.

Cartridge fuses are made in a variety of dimensions, based on amperage and voltage. Blade type terminals are common above 60 A. Fuses used to break 600 V arcs are longer than those available in many capacities other than the listed standard capacities, particularly in the two-element, time-delay variety. Often, they are dimensioned in such a way that they are not interchangeable with fuses of other capacities.

Plug fuses are limited in maximum capacity to 30 A. They are designed for use in circuits of not more than 150 V above ground. Two-element time-delay types are available to fit standard screw lamp sockets. They are also available with nonstandard threads made especially for various amperage ratings. The best way to keep up with the latest fuse technology is to keep yourself listed with the local supplier and get his updated catalogs as they come out.

Summary

Repair shops should have enough supplies on hand to avoid repeated trips to the supplier. It takes time for a new repair service to build up a complete stock of parts. However, there are commonly used parts that should always be in stock.

Setscrews should be a standardized item in the shop. They are used to fasten the pulley to the motor shaft in almost all electric motors. They are available in a number of heads from slotted-headless to fluted socket. They may be cup-shaped, cone-shaped, flat, or half-dog, as well as oval- or square-head.

Carbon brushes connect the revolving armature to the external circuit of a motor. The brushes can be square, rectangular, or round. They are usually sold as part of a kit, and are used in electric motors that are portable and drive tool-type machines. These brushes, in such tools as drills, buffers, saws, sanders, and grinders, are often short-lived due to overloads. Different shaped brushes are needed for other kinds of electric appliance motors. Fans, mixers, and AC/DC motors all require brushes. There should also be an ample supply of springs to hold the brushes in place. Bushings and bearings are also important for the proper repair and maintenance of electric motors. Bronze bushings, which are used as bearings, can be purchased in 56-piece kits. They are organized according to size in small plastic boxes with partitions. Sleeve bearings wear out quickly if not properly lubricated. Because they serve an important role in keeping the shaft properly aligned, they should be replaced when visible wobble or vibration occurs. If stored for over a year they tend to lose their oil film.

Shaft adapters are used to convert the shaft of a motor to a more usable size. They are usually secured to the shaft by Allen-head setscrews. Rubber mounts fit between the base of the motor and the mounting rails, and reduce the noise level of the motor. Many sizes of rubber mounts are available. Fiber washers are used to remove the end play from motors. They take up the space that exists due to manufacturing tolerances. Rubber grommets prevent the electrical cord from becoming shorted by contacting the frame of the motor. The centrifugal switch used in capacitor-start, split-phase, and permanent-split capacitor motors is subject to pitting and becoming troublesome. A supply of them should be kept handy for repair purposes. Westinghouse motors have at least 10 types used in the NEMA 48 and 56 frames.

Electrolytic capacitors are needed in capacitor-start motors and can easily become overheated and explode when AC is applied for longer than their rated time. Electrolytics can also dry out if not used for long periods of time. They can also change value if left idle for a long amount of time. There are at least three ways to connect the leads of an electrolytic capacitor that is used in motor circuits. Some motors use two capacitors, one for starting and one for running. Keep in mind that there are two types of capacitors: dry and wet. The wet type uses oil as an electrolytic. The dry type is not really dry, but has a paste material for an electrolyte.

Insulating enamel can be a protective coating for frames and end shields and a gasket cement and sealant for oil, water, dirt, and dust. Insulating enamel comes in spray cans and is quick drying. Of course, heavy-duty cord should be on hand to replace any that may be dried out or cracked. Three-conductor cord will have white, black, and green wires. The green wire is used as a safety feature and is connected to an earth-ground and the motor frame. Keep in mind that every cord has a maximum current carrying capacity. Cords for heavy motors and high-wattage appliances must be heavy enough to carry the needed current. Overloaded cords overheat, wasting power and often causing the motor to run at lower than normal speeds. There is always the possibility of a fire being created by an overheated cord. Also make sure that all wiring and electrical connections comply with the National Electrical Code.

Review Questions

1. Why should a motor repair facility have enough supplies on hand to avoid repeated trips to the supplier?

2. Why are set screws important in a motor repair facility?

3. What are the shapes of carbon brushes for small powered hand tools?

4. What type of devices use carbon brushes?

5. What is a limitation of motors with bronze bushings or bearings?

6. What happens to motors with bronze bearings when not used in one year or more?

7. What is the value of shaft adapters?

8. How are rubber mounts used to reduce motor noises?

9. What is the function of fiber washers in a motor?

10. How do rubber grommets protect a power cord?

11. What is a centrifugal switch used for?

12. What are the two types of electrolytic capacitors?

13. What is insulating enamel used for, other than to insulate?

14. What are the colors of the three wires in a motor's power cord?

15. What can happen when a motor is overloaded and the power cord is too small?

Chapter 21

General Maintenance

Proper maintenance procedures mean the difference between a short and a long motor life. Trouble-free operation is the main objective of the manufacturer and owner of a motor. If the motor does not operate properly, it means a loss of money or effort. Manufacturers specify procedures that should be followed for proper operation of electric motors. The procedures are, in most instances, very simple and easy to follow. Whenever these precautions are neglected, the obvious problems result. This chapter deals with these problems and how to remedy them once they have been identified.

One of the most often encountered troubles in electric motors is low voltage. Low voltage can be caused by the wrong wire size in the circuit. Table 21-1 shows the circuit wire sizes for individual single-phase motors. Note how the length of the run from the distribution panel to the motor affects the size of wire used. Selecting the wrong size wire can cause a voltage drop along the wire that will result in the motor receiving lower voltage. As a result of low voltage, the motor will attempt to draw more current. As the current increases, the heat generated increases by the square of the current. This further increases the voltage drop along the wire and, in turn, again reduces the line voltage at the motor.

Table 21-1   Circuit Wire Sizes for Individual Single-Phase Motors

Horse-power of Motor	Volts	Approxi-mate Starting Current Amperes	Approxi-mate Full-Load Current Amperes	Length of Run in Feet (from Main Switch to Motor)								
				Feet	25	50	75	100	150	200	300	400
1/4	120	20	5	Wire Size	14	14	14	12	10	10	8	6
1/3	120	20	5.5	Wire Size	14	14	14	12	10	8	6	6
1/2	120	22	7	Wire Size	14	14	12	12	10	8	6	6
3/4	120	28	21.5	Wire Size	14	12	12	10	8	6	4	4
1/4	240	10	2.5	Wire Size	14	14	14	14	14	14	12	12
1/3	240	10	3	Wire Size	14	14	14	14	14	14	12	10
1/2	240	11	3.5	Wire Size	14	14	14	14	14	12	12	10
3/4	240	14	4.7	Wire Size	14	14	14	14	14	12	10	10
1	240	16	5.5	Wire Size	14	14	14	14	12	10	8	8
1 1/2	240	22	7.6	Wire Size	14	14	14	14	12	10	8	8

(continued)

Table 21-1   *(continued)*

Horse-power of Motor	Volts	Approxi-mate Starting Current Amperes	Approxi-mate Full-Load Current Amperes	Feet	Length of Run in Feet (from Main Switch to Motor)							
					25	50	75	100	150	200	300	400
2	240	30	10	Wire Size	14	14	14	12	10	10	8	6
3	240	42	14	Wire Size	14	12	12	12	10	8	6	6
5	240	69	23	Wire Size	10	10	10	8	8	6	4	4
7½	240	100	34	Wire Size	8	8	8	8	6	4	2	2
10	240	130	43	Wire Size	6	6	6	6	4	4	2	1

In general, motors should be checked for a number of wear indicators at least twice a year. In the case of heavy usage, the motor should be checked more often. A maintenance schedule can be set up according to previous experience with the various types of motors. Manufacturers' recommendations should be followed closely for the best operation with the least amount of trouble.

Each type of small motor is covered in this chapter for its maintenance and troubleshooting procedures. The split-phase, the capacitor-start, the permanent-split capacitor, the shaded-pole, the three-phase, and the brush-type (series, permanent-magnet, shunt, and compound) motors are covered in detail to aid you in locating the possible source of trouble and in maintaining the motors so that fewer troubles will appear during normal operation.

General Troubleshooting Procedures

Before servicing or working on equipment, disconnect the power source. This applies especially when servicing equipment with thermally protected, automatic-restart devices instead of manual-restart devices and when examining or replacing brushes on brush-type motors.

Clean the motor environment regularly to prevent dirt and dust from interfering with the ventilation or clogging the moving parts.

Capacitor-Equipped Motors

The capacitors used for capacitor-start and capacitor-start, capacitor-run motors are dangerous when charged. Before servicing motors with capacitors, always discharge the capacitor by placing a conductor (a screwdriver, usually) across its terminals before you touch the terminals with any part of your body.

In many cases, easy-to-detect symptoms will indicate exactly what is wrong with your fractional-horsepower motor. However, since general types of motor trouble have similar symptoms, it is necessary to check each possible cause separately. The tables provided in this chapter list some of the more common ailments of small motors, with suggestions as to probable causes.

Most common motor troubles can be checked by some test or inspection. While the order of these tests rests with the troubleshooter, it is advisable to make the easy ones first. In diagnosing troubles, a combination of symptoms will often give a definite clue to the source of the trouble. For example, if a motor will start but heating occurs, there is a good likelihood that a short or ground exists in one of the windings.

Centrifugal starting switches, found on many types of fractional-horsepower motors, are occasionally the source of motor trouble. Such switches have a finite life and they wear in many ways, depending on the design and usage. If the switch sticks in the open position, the motor will not start. When stuck in the closed position, the motor will normally operate at a slightly reduced speed and the start winding will quickly overheat. The motor may also fail to start if the contact points of the switch are out of adjustment or coated with oxide. It is important to remember, however, that any adjustment of the switch or contacts should be made only at the factory or by an authorized service center.

Brush-Type Motors

Because of wear, brushes and commutators on commutated motors require more maintenance than nonbrush types. The wear rate of brushes depends on many parameters (armature speed, amperage conducted, duty cycle, and humidity, to name a few). For optimum performance, brush-type motors need periodic user-maintenance. The maintenance interval is best determined by the user. Inspect the brushes regularly for wear (replace in same axial position). Replace brushes when their length is less than 1/4 in. (7 mm). Periodically remove carbon dust from the commutator and from inside the motor. This can be accomplished by occasionally wiping them with a clean, dry, nonlinting cloth. Do not use lubricants or solvents on the commutator. If necessary, use No. 0000 or finer sandpaper to dress the commutator. However, if pitted spots still appear, the commutator should be reground by an experienced electric-motor service center.

Maintenance of the Split-Phase Motor

The split-phase motor will perform well with little attention as long as it is not overloaded. For a motor that performs efficiently and with little trouble, these are a few maintenance suggestions:

1. Inspect the motor at least twice a year. Check for obvious wear. With a belt-driven arrangement, the belts may be too tight and may cause uneven wearing of the two bearings by placing undue pressure on them. Furthermore, as the bearings wear down, the rotor can strike the stator poles and permanently damage the rotor. This can distort the stator windings and cause them to become damaged severely enough to require rewinding (Figure 21-1).

RESET CIRCUIT BREAKER

Figure 21-1   Stator windings and circuit breaker of a motor. This stator should be rewound.

2. Check the centrifugal switch (Figure 21-2). It should shut off before the motor has come to final running speed, at 75 percent of the rated speed. This check can be made with some degree of precision by using a tachometer to test the operating speed and the click-out speed of the switch. On the 1800-rpm motor, the switch should click out at approximately 1350 rpm. On the 3600-rpm motor, the start switch should click out at 2700 rpm. The nameplate speeds will be 1725 rpm for

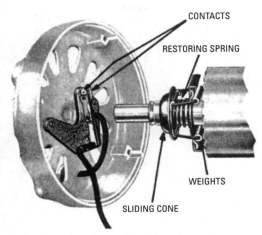

CONTACTS

RESTORING SPRING

WEIGHTS

SLIDING CONE

Figure 21-2 **Centrifugal switch.** *(Courtesy Bodine)*

the 1800-rpm motor and 3450 rpm for the 3600-rpm motor. The switch should click in just before the motor coasts to a dead stop after it has been disconnected from the power line.

Do not overlook the screws that hold the centrifugal switch in place when the motor is disassembled for inspection or maintenance. Make sure the switch screws have not become loose. Because the switch is used each time the motor is turned on or off, check the mounting screws when cleaning the points of the switch and tighten the screws. See Figure 21-3 for a typical centrifugal switch.

3. Check for overloads. Sometimes a load will build up because of increased friction and wear throughout the driven system. Check the motor temperature for any indication that it is running hot. The motor should feel warm when you place your hand on it, but it should not feel hot after a few seconds. Make sure you check the load again before continuing to use the motor under these conditions.

Overload fuses, or cut-outs, should be part of the motor design. If they are not, make sure a fuse or cut-out is inserted in series with the line that is supplying power to the motor. If the motor's reset button keeps cutting out and turning the motor off, make sure you find the cause of the overload and correct it before the motor becomes shorted and the windings are burned. Some motors have a relay to open the auxiliary

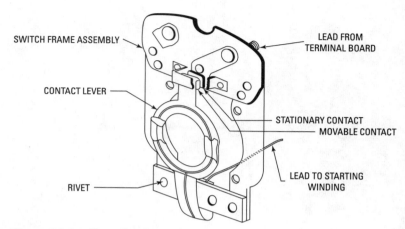

SWITCH FRAME ASSEMBLY

CONTACT LEVER

RIVET

LEAD FROM TERMINAL BOARD

STATIONARY CONTACT
MOVABLE CONTACT

LEAD TO STARTING WINDING

Figure 21-3 Centrifugal switch.

coil circuit (Figure 21-4). When power is applied to the line, current flows through the heavy main winding. This closes the contacts connecting the start winding to the supply line voltage (Figure 21-5). With the pickup in speed, the current is reduced. This reduction in current causes the current-sensitive relay solenoid to de-energize (drop out), which causes the contacts to open, removing the auxiliary (start) winding from the circuit (see schematic in Figure 21-6).

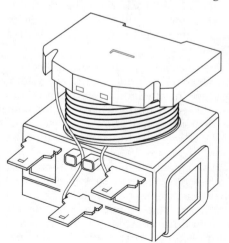

Figure 21-4 Current-sensitive relay used to take the start winding out of the circuit once the motor has started and come up to 75 percent of synchronous speed.

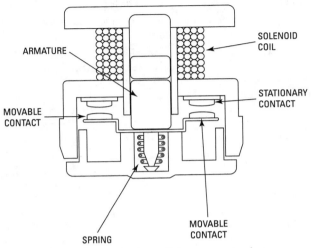

ARMATURE

SOLENOID
COIL

MOVABLE
CONTACT

STATIONARY
CONTACT

MOVABLE
CONTACT

SPRING

Figure 21-5   Cutaway view of the current relay.

4. Lubricate the motor, especially sleeve-type motor bearings. A
motor that is run continuously requires more attention than
one that is run occasionally. However, it is easy to forget to
lubricate a motor. A typical example would be the split-phase
fan motor that is mounted inside the furnace cabinet in hot-air
furnaces. As the saying goes, "Out of sight, out of mind."
Lubrication for this motor is usually forgotten.

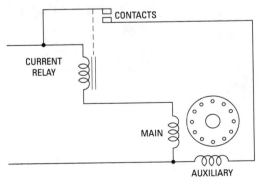

CONTACTS

CURRENT
RELAY

MAIN

AUXILIARY

Figure 21-6   Schematic of the current relay in the motor circuit.

The motor should be lubricated to the manufacturer's specifications. Provide enough oil, but do not overdo it. Too much oil collects dust and creates problems of another sort.

5. Check the centrifugal switch. The sliding member of the switch assembly should move freely on the motor shaft. If the contact points are pitted or burned, use a fine-grade sandpaper to polish them so that they sit against one another properly. Do not use metallic abrasive papers; they will leave particles of metal that can cause excessive wear if they get into the bearings, or even shorts if they stick to parts of the switch.

6. Check the alignment and pulley mechanism. Watch for an out-of-alignment condition where the shaft is not matched with the load. Sometimes the motor is damaged by an increased load that is caused by misalignment of the drive shaft and the driven load. Check to make sure the pulley on the motor and the pulley on the load are aligned and that the belt is not twisted in any way (Figure 21-7).

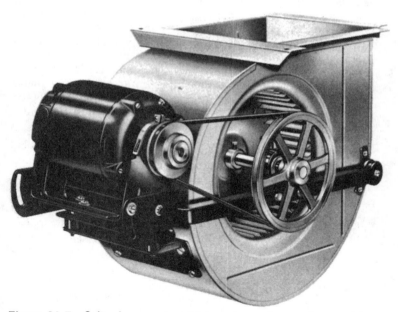

Figure 21-7 Split-phase motor driving a home hot-air furnace blower. Note the belt drive. Also note the adjustments that were made to the base so that the motor can be aligned properly to prevent excessive belt wear or motor bearing damage. *(Courtesy Lennox)*

7. Check the wiring when installing a motor for the first time or when moving a motor to a different load condition. Make sure the rating of the motor and the wires match. If the motor draws more current than the wires can handle safely, there may be a drop in the voltage reaching the motor. This means that the motor is operating under reduced voltage conditions and there is an increase in heat generated by the motor.

A common problem with saw motors and portable motor-driven tools is that the extension line is too small to handle the current drawn by the motor when started. Too much voltage is dropped along the line and the available voltage for the motor is severely limited. This often happens in on-the-job building situations where carpenters use extension cords from one newly built house to the house currently under construction when the power is not yet available to the new house under construction.

Identification of Split-Phase Motor
In a classroom situation you may be given a motor that has been pretty well stripped down and is hard to identify. Just remember that the split-phase motor will have a start winding made of small-diameter wire and a run winding made of a larger wire. The capacitor-start motor has windings with wires of the same diameter.

Troubleshooting
Table 21-2 is a quick reference for troubleshooting a split-phase motor.

Table 21-2   Troubleshooting the Split-Phase Motor

Trouble	Probable Cause
Will not start at all.	Open circuit in the connection to the line. Open circuit in the motor winding. Contacts of the centrifugal switch are not closed. Start winding open.
Will not always start, even with no load, but will run in either direction when started manually.	Contacts of centrifugal switch are not closed. Start winding open. Centrifugal switch is not opening. Winding short-circuited or grounded.
Starts, but heats rapidly.	
Starts, but runs too hot.	Winding short-circuited or grounded.

(continued)

Table 21-2   *(continued)*

Trouble	Probable Cause
Will not start, but will run in either direction when started manually; overheats.	Contacts of centrifugal switch are not closed. Start winding open. Winding short-circuited or grounded.
Motor gets too hot; reduction in power evident.	Winding short-circuited or grounded. Sticky or tight bearings. Interference between stationary and rotating members.
Motor blows fuse, or will not stop when it is turned to the OFF position.	Winding short-circuited or grounded. Grounded near the switch end of the winding.

Maintenance of Capacitor Motors

Capacitor motors (Table 21-3), like the split-phase types, will operate for years with little or no maintenance, but can develop some troubles with the capacitors and the start switches. Inspect the motor twice a year. Check for wear that may be obvious to the eye. Belts may be too tight and cause wear on the bearings if they are the sleeve type. The bearings may be damaged by prolonged use with a belt drive that is too tight.

Check the centrifugal switch. It should click out before the motor has reached full speed. The ideal place for this to occur is at 75 percent of rated speed. This can be checked with some precision

Table 21-3   Troubleshooting the Capacitor-Start Motor

Trouble	Probable Cause
Will not start.	Open circuit in connection to the line. Open circuit in motor winding. Contacts of centrifugal switch are not closed. Defective capacitor. Start winding open.
Will not always start, even with no load, but will run in either direction when started manually.	Contacts of centrifugal switch are not closed. Defective capacitor. Start winding open.
Starts, but heats rapidly.	Centrifugal switch is not opening. Winding is short-circuited or grounded.
Starts, but runs too hot.	Winding is short-circuited or grounded.

Table 21-3   (continued)

Trouble	Probable Cause
Will not start, but will run in either direction when started manually; overheats.	Contacts of the centrifugal switch are not closed. Defective capacitor. Start winding open. Winding short-circuited or grounded.
Reduction in power; motor runs too hot.	Winding short-circuited or grounded. Sticky or tight bearings. Interference between the stationary and rotating members.
Motor blows fuse or will not stop when the switch is turned to OFF position.	Winding short-circuited or grounded. Grounded near the switch end of the winding.

if a tachometer is used to obtain the running speed. For example, the start switch on a 1800-rpm motor should click out at 1350 rpm. The nameplate speeds will be 1725 rpm for 1800-rpm motors and 3450 rpm for 3600-rpm motors. The switch should click in just before the motor coasts to a dead stop after it has been disconnected from the power line.

Check for overloads. Sometimes a load will build up due to increased friction and wear throughout the driven system. Watch the motor temperature for any indication that it is running hot. In most cases, if you place your hand on the motor and have to pull it back quickly because the motor is too hot for your hand, the motor is too hot in general.

Make sure the overload protection is working properly. A fuse or circuit breaker of the proper size should be in the circuit to protect it in case of an overload.

Check the centrifugal switch to see if the sliding part is moving freely. If the contact points are pitted, clean them with a piece of sandpaper. Do not separate the points too much or they may not serve their purpose properly.

Do not overlook the screws that hold the centrifugal switch in place when you have the motor disassembled for service or to look for damage. In some instances, be sure that the switch screws have not worked loose. Since the switch is used each time the motor is turned on or off, it is best to check it when you clean the points of the switch. Just tighten the screws to be safe.

If pulleys are used, check their alignment to be sure that the belt is not out of line. This causes undue wear on the bearings of the motor and the driven device.

Check the wiring when you install a motor for the first time. It is a good idea to check the wire size if you are changing the location of a motor (Figure 21-8). If the motor draws too much current, the wire will cause a voltage drop, which means that there is less voltage than the motor specifications call for.

Figure 21-8 Capacitor-start motor used on a table saw. Note the size of the flexible cord used as a power cord.

Keep in mind that the starting current of the capacitor motor is more than the running current. Make sure the protective device (fuse or circuit breaker) is capable of handling the start current for the short start-up period without failing (Figure 21-9).

Electrolytic Capacitors

The device that gives the capacitor-start motor its ability to start under load is the electrolytic capacitor. The electrolytic capacitor comes in a black case in most instances. This Bakelite case has two snap connectors that connect to the leads from the motor start windings. Either lead can go to either terminal on the capacitor since it is an AC electrolytic. No polarity is needed. However, it is necessary to obtain the right size of electrolytic for the motor. The

Figure 21-9   Magnetic start switch for a 1-horsepower motor, such as shown in Figure 21-8.

correct microfarads and the correct working voltage should be obtained for best utilization of the motor's design qualities.

Table 5-1, in Chapter 5, will help you to match the ability of the electrolytic to the motor. There are three voltage ratings, so the correct voltage will have to be selected to make sure the capacitor is not damaged when connected into the start circuit.

It is a good rule to make sure that you use the identical size capacitor as came out of the motor. If you cannot find the markings on the capacitor, look at Table 5-1, in Chapter 5, and select one that matches the characteristics of the motor as shown on its name-plate.

Tests for Capacitors

Defective capacitors are very often the cause of trouble in capacitor-start and capacitor-run motors. Shorts, grounds, and insufficient capacity in microfarads are conditions for which capacitors should be tested.

To determine a grounded capacitor, set the instrument on the proper voltage range and connect the instrument and capacitor to the line, as shown in Figure 21-10. A full line voltage indication on the meter signifies that the capacitor is grounded to the can. A high resistance ground will be evident by a voltage reading that is somewhat below line voltage. A negligible reading or a reading of no voltage will indicate that the capacitor is not grounded.

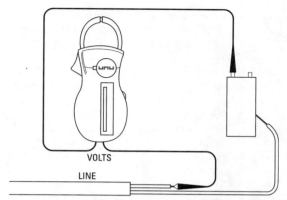

Figure 21-10   Test for finding a grounded capacitor.

To measure the capacity of the capacitor, set the test unit's switch to the proper voltage range and read the line voltage indication. Then set the appropriate current range and read the capacitor current indication. During the test, keep the capacitor on the line for a very short period of time, because motor starting electrolytic capacitors are rated for intermittent duty (see Figure 21-11). The capacity in microfarads is then computed by substituting the voltage and current readings in the following formula, assuming that a full 60-Hz line was used:

$$\text{Microfarads} = \frac{2650 \times \text{amperes}}{\text{volts}}$$

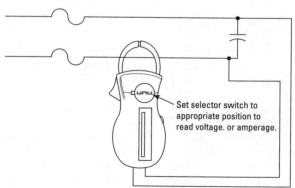

Figure 21-11   Measuring the capacity of a capacitor.

An open capacitor will be evident if there is no current indication in the test. A shorted capacitor is easily detected. It will blow the fuse when the line switch is turned on to measure the line voltage.

The Permanent-Split Capacitor Motor

This type of motor has a capacitor in the circuit even during the run period of the motor: a capacitor inserted in series with one of the two motor windings. The main winding is in parallel with the series coil-capacitor winding arrangement. The capacitor, which is somewhat expensive and bulky, is not taken out of the circuit when the motor has started. This is not a practical motor for heavy-duty starting. The troubles encountered with this type of motor are similar to those of the capacitor-start motor, with a few exceptions. Table 21-4 indicates some of the more common problems and their probable causes.

Table 21-4   Troubleshooting the Permanent-Split Capacitor Motor

Trouble	Probable Cause
Will not start at all.	Open circuit in connection to line. Blown fuses; overload protector tripped or faulty. Open circuit in motor winding. Defective capacitor. Start winding open. Overloaded motor. Winding short-circuited or grounded. One or more windings open. Tight or seized bearings. Interference between stationary and rotating member. Wrong connection to motor. Improper or low voltage.
Will not always start, even with no load, but will run in either direction when started manually.	Defective capacitor. Start winding open. One or more windings open. Wrong connection to motor.
Starts, but heats rapidly.	Defective capacitor. Overloaded motor. Winding short-circuited or grounded. Tight or seized bearings. Interference between stationary and rotating member. Wrong connection to motor.
Runs too hot after extended operation.	Overloaded motor. Tight or seized bearings. Failure of ventilation (blocked or obstructed ventilation openings). Improper or low voltage from line. Worn bearings. High ambient temperature.

(continued)

Table 21-4 *(continued)*

Trouble	Probable Cause
Excessive noise (mechanical).	Interference between stationary and rotating member. Worn bearings. Unbalanced rotor or armature (vibration). Poor alignment between motor and load; loose motor mounting. Amplified motor noises.
Reduction in power; motor gets too hot.	Defective capacitor. Winding short-circuited or grounded. Tight or seized bearings. Interference between rotating and stationary member. Wrong connection of motor. Improper or low line voltage.

The Shaded-Pole Motor

The shaded-pole motor has some very interesting characteristics, one of which is the ability to start and run without two coils. The simple shaded-pole motor used for clocks and small fans is a one-coil device. It has a winding that is connected across the line at all times. It relies upon its core windings and the iron in the laminations to prevent over-heating. Other types of shaded-pole motors may have four or six windings. These are closer in appearance to the regular split-phase-type motor. However, shaded-pole motors do not have start windings as such. They do have a large piece of metal (usually coat-hanger-size wire) that shorts a portion of the pole of each coil. This will identify the motor for troubleshooting purposes. Table 21-5 indicates the probable causes of shaded-pole motor troubles.

Table 21-5   Troubleshooting the Shaded-Pole Motor

Trouble	Probable Cause
Motor will not start at all.	Open connection to the line. Open circuit in motor winding. Overloaded motor. Winding short-circuited or grounded. Tight or seized bearings. Interference between stationary and rotating member. Wrong connection to motor. Improper line voltage. Low line voltage.
Starts, but heats rapidly.	Overloaded motor. Winding short-circuited or grounded. Tight or seized bearings. Interference between stationary and rotating member. Wrong connection to motor.

Table 21-5 *(continued)*

Trouble	Probable Cause
Runs too hot after extended operation.	Overloaded motor. Tight or seized bearings. Failure of ventilation. Improper or low line voltage. Worn bearings. High ambient temperatures.
Excessive noise (mechanical).	Interference between stationary and rotating member. Worn bearings. Unbalanced rotor (vibration). Poor alignment between motor and load or loose motor mounting. Amplified motor noises.
Reduction in power; motor gets too hot.	Winding short-circuited or grounded. Tight or seized bearings. Interference between stationary and rotating member. Wrong connection to motor. Improper or low line voltage.

To repair, just remedy the cause. Remove grounds or shorts, replace the sticky bearings, or make the proper clearance between rotating members and the stationary members of the motor. A drop of oil is all that is needed in some cases to ease the sticky or tight bearing.

Maintenance of the Three-Phase Motor

In order to completely check the three-phase motor, it is necessary to take it apart and look at each component part. Both the inside and the outside should be checked. In Figure 21-12 you see a rather old, three-phase motor, still operational after about 40 years. Its basic design is the same as those made today. If we look at it closely, we will see it is ½ horsepower and was made by Peerless Electric Products.

There are four bolts holding the motor together. Remove these to take the end bells off. However, before going farther, look at the type of terminals used to connect power into the motor. The screw-on type, where the wire is pushed through a hole in the terminal and the cap is screwed down to hold it in place, is no longer used. Another thing to look at while it is still completely assembled is the way it is lubricated. This is a sleeve-bearing-type motor, which means that it has two places to put oil.

Once you remove the end bell, you will be able to see the rotor with its fan (Figure 21-13). Do not lose any of the thin washers that may be on the shaft. These washers determine end

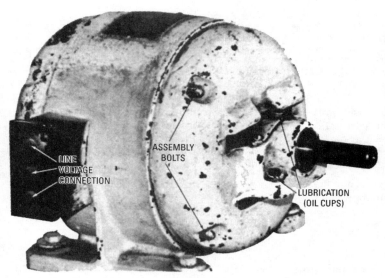

Figure 21-12 A ½-horsepower, three-phase motor.

Figure 21-13 Motor with one end bell removed, showing the fan that is part of the rotor assembly.

play and can make a difference in the fan blades hitting the windings or not.

In Figure 21-14, you can see the windings. They should be checked for damage that is the result of the fan scraping them. In this view you can see that the insulation paper has deteriorated somewhat, but performance is not affected. Check the poles inside the motor to see if the rotor may have been touching them at any point. There could be a shiny spot, which would indicate that the bearings need replacing.

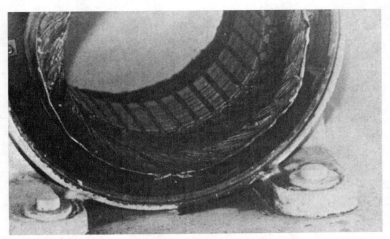

Figure 21-14   Stator windings on a three-phase motor.

Figure 21-15 shows what the motor looks like with both end bells removed. It is now possible to inspect both ends of the motor for possible motor winding damage. Figure 21-15 shows some possible insulation damage where it extended past the windings. This may have been done when the motor was disassembled previously and the fan pushed back a little too far. It does not cause any problem with operation of the motor, however.

Figure 21-16 shows a close-up view of the inside of an end bell. Note that the bearing is sleeve-type. Check for indications of wear and lack of lubrication.

In Figure 21-17 you can get a better idea of what the rotor of a three-phase motor looks like. Note that it does not have a centrifugal switch; the switch is not needed in a three-phase motor. However, be sure not to lose any of the thin washers on the shaft. Also make a note of the location and the number of washers on

Figure 21-15 Three-phase motor with both end bells and rotor removed.

Figure 21-16 Inside of the end bell.

Figure 21-17   Three-phase motor rotor.

Figure 21-18   Balancing weights on the rotor fan.

each end. Figure 21-18 shows the balancing weights that are placed in the fan to make the motor run smoothly.

Note the two oil cups in the end bell (Figure 21-12). One is for the top of the bearing and the other is for making sure the lower half of the bearing is properly lubricated. In most cases you will find only one oil cup or one grease fitting per bearing. Better methods of distributing oil have been devised.

General Maintenance

Maintenance of the three-phase motor is rather simple. Oil it if it has sleeve bearings. Make sure it is not overloaded if it has ball bearings. It is almost trouble-free if properly loaded and operated on the correct line voltages. If it has flow-through ventilation, make sure the intakes are kept clean of dust, dirt, and the collection of fuzz.

Repair

It will become necessary to replace the bearings when they are worn—whether they are sleeve or ball types. It is a simple matter to replace either, so no special instructions are needed; just make sure they are replaced with the proper size. In some cases where the windings are burned or the insulation has been removed by over-heating, it will be necessary to rewind the windings.

As you can see from Figure 21-18, the rotor does not need any rewinding, because it has no windings. Just make sure the spacer washers are the correct thickness and *number* when you reassemble the motor. The proper number refers to the number of washers needed to make sure there isn't too much end play in the rotor. Table 21-6 shows how to troubleshoot the three-phase motor.

Maintenance of Brush-Type Motors

This heading refers to motors that can be used on AC or DC and are most commonly referred to as the universal type. This also includes permanent-magnet, shunt, and compound DC motors. Keep in mind that a reduction in speed of a motor will also increase its life expectancy. A rough rule of thumb is that when the speed is reduced by 50 percent, the brush life is tripled. Table 21-7 shows the symptoms, probable causes, and possible remedies for the small brush-type motor. In the previous tables the remedy was rather obvious. In this type of motor the remedy may not be so obvious, so the *Remedy* column has been added for your convenience.

To remedy, take the obvious steps. Lubricate if the bearings are sticking. Check the clearance to make sure the rotating and stationary members are not touching by checking bearings and proper armature for this motor. Check for shorts and remove any if possible; if not, rewind the motor. If a line is open, complete the circuit in a manner approved by the National Electrical Code recommendations for that motor.

Table 21-6   Troubleshooting Three-Phase Motors

Trouble	Probable Cause
Will not start.	Open circuit in connection line. Open circuit in motor winding. One or more windings open.
Will not always start, even with no load, but will run in either direction when started manually.	One or more windings open.
Starts, but heats rapidly.	Winding short-circuited or grounded.
Starts, but runs too hot.	Winding short-circuited or grounded.
Will not start, but will run in either direction when started manually; overheats.	Winding short-circuited or grounded. One or more windings open.
Reduction in power; motor gets too hot.	Winding short-circuited or grounded. Sticky or tight bearings. Interference between stationary and rotating members.
Motor blows fuse or will not stop when switch is turned to OFF position.	Winding short-circuited or grounded. Grounded near switch end of winding.

Table 21-7   Troubleshooting the Permanent-Magnet Motor

Symptom	Probable Cause	Remedy
Blows circuit breaker and shaft turns very hard.	Bad bearing.	Replace the bearing. Also check the armature.
Blows circuit breaker and armature heats up.	Check for machine overload.	Replace armature and possibly rear-end bell assembly.
Does not run.	Worn brushes or brush hang-up.	Replace or release brushes and check the commutator.
Does not run or runs too fast.	Open armature connection.	Replace the armature.
Motor runs in the wrong direction.	Reversed motor connections.	Reverse the polarity of the power supply.

Troubleshooting universal, series, shunt, and compound DC motors is slightly different, because the permanent-magnet type does not have a field winding. This is covered in Table 21-8.

Table 21-8   Troubleshooting Universal, Series, Shunt, or Compound DC Motors

Trouble	Probable Cause
Will not start.	Open circuit in connection to the line. Open circuit in motor winding. Worn brushes and/or annealed brush springs. Open circuit or short circuit in the armature winding.
Starts, but heats rapidly.	Winding short-circuited or grounded.
Starts, but runs too hot.	Winding short-circuited or grounded.
Sluggish; sparks severely at the brushes.	High mica between commutator bars. Dirty commutator or commutator is out of round. Worn brushes or annealed brush springs. Open circuit or short circuit in the armature winding. Oil-soaked brushes.
Abnormally high speed; sparks severely at the brushes.	Open circuit in the shunt winding.
Reduction in power; motor gets too hot.	Open circuit or short circuit in the armature windings. Sticky or tight bearings. Interference between the stationary and rotating members.
Motor blows fuse or does not stop when switch is turned to OFF position.	Grounded near switch end of winding. Shorted or grounded armature winding.
Jerky operation, severe vibration.	High mica between commutator bars. Dirty commutator or commutator is out of round. Worn brushes and/or annealed brush springs. Open circuit or short circuit in the armature winding. Shorted or grounded armature winding.

Summary

Proper maintenance procedures will make the difference between a short motor life and a long motor life. One of the troubles encountered most often in electric motors is low voltage. Low voltage can be caused by the wrong size wire in the circuit.

In general, motors should be checked for a number of wear indicators at least twice a year. In the case of heavy usage, the motor should be checked more often. Manufacturers' recommendations should be followed.

Before servicing or working on equipment, disconnect the power source. Clean the motor environment regularly to prevent dirt and dust from interfering with the ventilation or clogging the moving parts.

Capacitors used for capacitor-start and capacitor start, capacitor-run motors are dangerous when charged. Before servicing motors with capacitors, always discharge the capacitor by placing a conductor (screwdriver blade, usually) across its terminals before you touch the terminals with any part of your body.

Most common motor troubles can be checked by some test or inspection. Centrifugal starting switches, found on many types of fractional-horsepower motors, are occasionally the source of motor trouble.

Because of wear, brushes and commutators on commutated motors require more maintenance than non-brush types. The wear rate of brushes depends on many parameters. Do not use oil or solvents on the commutator.

The split-phase motor will perform well with little attention as long as it is not overloaded. However, it should be inspected and checked at least once a year. The centrifugal switch should also be checked. The switch should click in just before the motor coasts to a dead stop after it has been disconnected from the power line.

Check for overloads and be sure to lubricate sleeve-type motors. Do not overoil. Also check the alignment of the pulley mechanism. On portable motors, check the size of the power cord in reference to its distance from the power source.

The start winding of a split-phase motor has a smaller diameter than the run winding. However, in the capacitor-start motor, both windings are of the same size wire.

Bearings on the capacitor-start motor, especially sleeve bearings, may be worn down by a belt that has been adjusted too tightly. Check the centrifugal switch periodically for proper operation. Keep in mind that the starting current of the capacitor motor is more than the running current. Make sure the protective is capable of handling the start current for the short start-up period without failing. The electrolytic capacitors used for motors are AC electrolytics; they have no polarity, unlike most capacitors over 1 microfarad. They are usually constructed by placing two electrolytics back-to-back, or in other words, by connecting positive to positive and negative to negative, instead of negative to positive, as is the usual electrolytic usage. If you don't have the right size and try to use two DC electrolytics on the AC line, you should remember that the capacity of the combination will be reduced because they are placed in series; in other words, two 100 MFD capacitors in series equals only 50 MFD total capacitance. Some capacitors are labeled MFD for microfarad and others are more updated to μF.

The split-capacitor has a capacitor in the circuit when running. The troubles with this type motor are similar to those of the capacitor-start type.

Troubleshooting the shaded-pole motor is covered in Table 21-5. One of the most often encountered troubles is the lack of lubrication for the bronze bearings. This is especially true if the motor is used where there is a constant source of heat, such as an oven timer.

The three-phase motor is less prone to problems because it has no start windings and no centrifugal switch. It is plagued most of the time with opens that occur in one or the other of its three power lines. General maintenance includes keeping the environment clean and dust-free if it is not a sealed motor. If noise is a concern, sleeve bearings should be oiled. Correct line voltage is also important for proper operation.

Brush motors present most of the maintenance and repair problems encountered by a motor repair shop. Brushes are subject to wear and commutators are subject to scoring and damage from various contaminants.

Troubleshooting the series, universal, shunt and compound motors is somewhat different than working with the shaded-pole and the split-phase types. Table 21-7 can be very helpful for troubleshooting these motors.

Review Questions

1. What can increase the life of an electric motor?
2. How often should motors be checked?
3. Why should a motor be cleaned and checked for lubrication?
4. Why are charged capacitors dangerous?
5. How are capacitors selected for motors? How are charged capacitors discharged?
6. What is a centrifugal switch? Why do they fail to operate properly?
7. What type of motors requires the most maintenance?

8. What is the worst problem for split-phase motors, in terms of maintenance?

9. How do you check a centrifugal switch for proper operation?

10. Which type of motor, sleeve or ball-bearing, requires the most attention?

11. How is an AC electrolytic different from a DC electrolytic?

12. What happens to electrolytics when they are connected back to back and in series?

13. How much total capacitance is there in a two-100 μF capacitor series hook-up?

14. What is the most often encountered problem with shaded-pole motors?

15. What does the three-phase motor require in the way of maintenance? What is its most often encountered problem?

Chapter 22

Shafts and Bearings

All motors have shafts and bearings. Shafts can be long, short, or any size in between. The shaft length, finish, diameter, and design are all subject to the demands of the buyer. Some general types are available for fans, pulleys, and direct drives that are lubricated by bearings. The bearings may be either of the sleeve or ball design; each has its advantages and disadvantages. Lubrication becomes a very important part of maintaining a motor. The type of lubricant and the frequency of application determines, in most instances, where the motor will be useful. Therefore, it is very important for the buyer to select the proper bearings and shaft configuration for the job to be done.

Motor Shafts

The unit heater motor shown in Figure 22-1 has a shaft with a flat spot. The flat can be used to make sure the setscrew in the pulley will not slip and eat into the shaft. Some pulleys have a flat spot to fit directly over this type of shaft.

Some motors have one end of the shaft that extends past the body of the motor. Others have an extended shaft on both ends (Figure 22-2). This allows for attaching two separate devices to the motor. It also increases the load on the bearings. Figure 22-3 shows how the shaft extends through the rotor. It is usually ground to a precision fit and is either press-fit onto the rotor or welded in place. Note how the shaft has been machined to fit the bearing surfaces and the pulley it will eventually drive.

Gearmotors have short shafts because the end of the shaft has been grooved to match the gears in the gearbox. See Figure 22-4 for the short shaft that drives a gearbox for the reduction of the motor speed. The gearbox is bolted onto the motor and the gears mesh to provide a greater torque at a slower speed for a special application.

A typical single-reduction, right-angle gearmotor is shown in Figure 22-5. Here the shaft is a little longer and a gear is mounted on its end to mesh with the gearbox right-angle shaft. Note how the bearings are placed and lubricated.

Rotor and Shaft

To present a typical specialty-motor design characteristic, let's take a look at Figure 22-6. This is the rotor and shaft from a Westinghouse

Figure 22-1 Enclosed, capacitor-start unit heater motor with sleeve bearings and resilient mounting. *(Courtesy Westinghouse)*

Figure 22-2 A fan-coil motor with rubber end pieces to absorb vibration when mounted. Hexagonal holes fit over the end of the motor. The motor is thermally protected. *(Courtesy Westinghouse)*

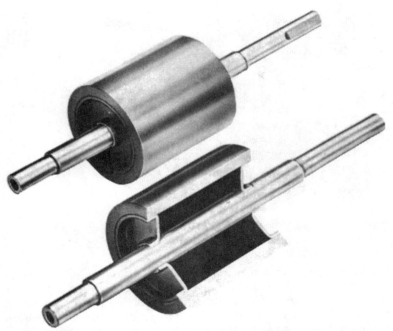

Figure 22-3   The shaft is machined to fit the bearings and to be press-fit through the rotor. *(Courtesy Bodine)*

Figure 22-4   Gearbox with a straight-through shaft. *(Courtesy Bodine)*

Figure 22-5 A gearbox with a 90° drive shaft. *(Courtesy Bodine)*

Frame 42 specialty motor. Precision die-casting equipment forces molten primary aluminum through the rotor core to form the rotor conductor bar cage. The preheated rotor is accurately located on the finished shaft and allowed to cool to a tight shrink-fit. Precision rotor machining and uniform aluminum die-casting ensure excellent rotor balance and concentricity.

Figure 22-6 A shaft with a flat side for setscrews. *(Courtesy Bodine)*

A number of rotor skew options are available to reduce noise and vibration. Shafts are precision-finished to exacting tolerances to prevent or reduce bearing wear. Phosphate is used to treat each carbon-steel shaft for protection against rust and corrosion. A clean corrosion-free shaft ensures easy removal of shaft-mounted devices.

Shaft Adapters

The motor you have may not have the correct size shaft or slot mechanism. This can be remedied by obtaining a transition shaft adapter (Figure 22-7). Two sizes of shaft adapters with keys are available for instances where the shaft diameter of a new motor is smaller than the shaft of the motor being replaced. One adapter will increase the shaft diameter from $\frac{1}{2}$ to $\frac{5}{8}$ in. The other adapter will change the shaft diameter from $\frac{5}{8}$ to $\frac{3}{4}$ in.

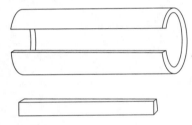

Figure 22-7 Transition shaft adapter. (*Courtesy General Electric*)

Selection of Bearings

It is extremely important to select a motor or gearmotor with the correct bearings. Severe conditions of operation dictate the type of bearing to be used. Since metal-to-metal contact during rotation causes friction and heating, care in the selection of a drive unit with the appropriate bearings for the intended application is an essential factor in the life and effectiveness of any driven machine. Among the many considerations that affect the selection of bearings are speed requirements, temperature limits, lubrication, load capacity, noise and vibration, tolerance, space and weight limitations, end thrust, corrosion resistance, infiltration of dirt or dust, and, of course, cost.

There are two principal types of bearings to fit the needs of the various motors under varying conditions. Fractional-horsepower motors use sleeve (journal) and ball bearings.

In gearheads for fractional-horsepower motors you will find sleeve, ball, tapered-roller, needle-thrust, and draw-cup, full-complement needle bearings.

Characteristics of ball and journal (sleeve) bearings are given in Table 22-1. This data and the application data will aid in the selection of the proper motor.

Table 22-1 Comparison of Ball- and Sleeve-Bearing Characteristics

Characteristics	Sleeve Bearing	Ball Bearing
Load		
Unidirectional	Good	Excellent
Cyclic	Good	Excellent
Starting and Stopping	Poor	Excellent
Unbalanced	Good	Excellent
Shock	Fair	Excellent
Thrust	Fair	Excellent
Overhung	Fair	Excellent
Speed Limited by	Turbulence of oil. Usual limit (5000 rpm max.)	20,000 rpm max.
Misalignment Tolerance	Poor (unless of the self-alignment type)	Fair
Starting Friction	High	Low
Space Requirement		
Radial	Small	Large
Axial	Large	Small
Damping of Vibration	Good	Poor
Type of Lubrication	Oil	Oil or grease
Lubricant, Amount Required	Large	Small
Noise	Quiet	Depends upon quality of bearing and resonance of mounting
Low Temperature Starting	Poor	Good
High Temperature Operation	Limited by lubricant	Limited by lubricant
Maintenance	Relubrication required periodically	Relubrication required only occasionally

Table 22-1 *(continued)*

Characteristics	Sleeve Bearing	Ball Bearing
		Grease-lubricated bearings often last for the life of the application without attention

(Courtesy Bodine)

Sleeve Bearings

Sleeve, or journal, bearings are the simplest in construction. They are the most widely used when low initial cost is a factor. They are quiet in operation and have a good radial load capacity. They can be used over a fairly wide temperature range. Sleeve bearings also have virtually unlimited storage life if the motor is to remain unused for extended periods of time. They show good resistance to humidity, to mild dirt infiltration, and to corrosion if made of bronze. Under light loads, the static friction of sleeve bearings is very nearly as low as grease-packed ball bearings (Figure 22-8).

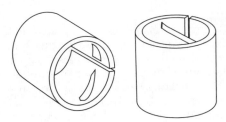

Figure 22-8 Large, babbitt-lined sleeve bearings are precision-machined to extremely close tolerances for accurate alignment.

(Courtesy General Electric)

The main disadvantage of the sleeve bearing is the need for relubrication. They are longer than ball bearings and may be somewhat at a disadvantage if space or size is a factor. An oil reservoir, felt, or similar oil-retaining material must also be incorporated into the end shield, and the lubricating oil must be replenished periodically (Figure 22-9).

Caution
Sleeve bearings cannot be allowed to run dry!

Graphitized, Self-Lubricating Sleeve Bearings
Graphitized, self-lubricating sleeve bearings are made of solid bronze. They have graphite as an inner recess filler. The recesses are often in

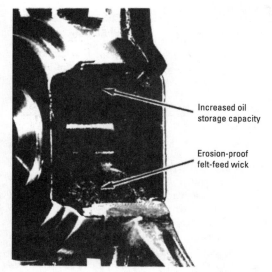

Increased oil
storage capacity

Erosion-proof
felt-feed wick

Figure 22-9   Oil reservoir for a sleeve-bearing motor. *(Courtesy General Electric)*

the shape of two figure-eights. They may also employ graphite-filled holes to conduct oil between the reservoir and the inner bearing surface. The bronze bearing or body of such bearings provides strength and resistance to shock or vibration, while the presence of graphite helps form a lubricating film on the bearing surface. It also prevents metal-to-metal contact when the motor is stopped. The graphite will also act as an emergency lubricant if the oil level is allowed to run low. Graphited bearings will also usually withstand higher operating temperatures than ordinary sleeve bearings.

Porous Bronze Sleeve Bearings

Another type of self-lubricating bearing (sleeve type) is made from porous bronze. The porous bronze sleeve bearing is oil-impregnated and can be used with a felt washer around its periphery to hold additional oil in suspension, making frequent relubrication unnecessary (Figure 22-10).

Porous bronze bearings are more compact and offer more freedom from attention than solid bronze bearings. They are often constructed to be self-aligning, to reduce friction, and to be shaft-binding. The porous bearing is generally more economical than the graphited or solid bronze types.

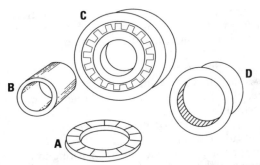

Figure 22-10 Comparison of bearing types: (A) Needle-thrust bearing; (B) sleeve bearing; (C) ball bearing; (D) full-complement, drawn-cup needle bearing. *(Courtesy Bodine)*

Sleeve-Bearing Lubrication

Lubricants for sleeve bearings must provide an oil film that completely separates the bearing surface from the rotating shaft member and ideally eliminates metal-to-metal contact. Oil, because of its adhesion properties and its viscosity (or resistance to flow), is dragged along by the rotating shaft of the motor and forms a wedge-shaped film between the shaft and the bearing (Figure 22-11). The oil film forms automatically when the shaft

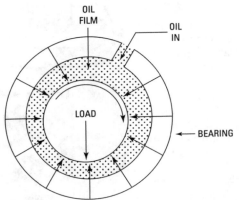

Figure 22-11 The oil film in a hydrodynamic bearing. A wedge-shaped gap is formed between the shaft and bearing. Higher pressures are developed where the gap narrows, which lifts the shaft and its load.

(Courtesy Bodine)

begins to turn and is maintained by the motion. The rotational motion sets up a pressure in the oil film wedge, which, in turn, supports the load. That means that the oil selected to carry this load is very important.

High temperatures and high motor operating temperatures will have a destructive effect on sleeve bearings lubricated with standard-temperature-range oils. Special oils are available for motor applications at high temperature, and also for applications at lower-than-normal temperatures. Motor performance and bearing life are directly related to lubrication.

Sleeve bearings are not as sensitive to abrasive or foreign matter as are ball bearings. They are able to absorb some of the small particles in the soft undersurface of the bearings. However, good maintenance practice dictates that you keep the lubrication oil as clean as possible.

A conservative lubrication and maintenance program should call for periodic inspection of the oil level and cleaning and refilling with new oil every six months.

Note
Sleeve-bearing motors may tend to lose their oil film when stored for extended periods (one year or more).

Ball Bearings
Virtually all sizes and types of electric motors use ball bearings. They have low friction loss when lubricated with oil, are suited for high-speed operation, and can be used in relatively wide ranges of temperature.

Ball bearings can accommodate thrust loads and permit end play to be conveniently minimized. Compared to sleeve bearings, ball bearings require significantly less maintenance, especially if grease-packed (Figure 22-12).

Ball bearings are slightly more expensive than sleeve bearings. They are also noisier. This is due to the nature of their rolling action. Recent developments have noticeably reduced the noise levels in ball-bearing motors (Figure 22-13).

Figure 22-12 A double-shielded ball bearing. *(Courtesy SKF Industries)*

Figure 22-13   Cutaway view showing shielded ball bearing installed in a general-purpose cage motor. *(Courtesy Allis-Chalmers)*

This type of bearing is more susceptible to rust since it is made of steel. If it is kept in storage for some time, it may show a tightening of the shaft due to the lubricant hardening. This can also be caused by low temperatures. Giving the motor some warm-up time will, in some instances, rejuvenate the ball-bearing grease to a suitable condition. Recently, long-life grease has been developed.

Figure 22-14 shows how the ball bearing is mounted in the motor. Compare this with Figure 22-15, where the sleeve bearing is used. Note the differences in lubrication methods and design of the bearings.

Ball-Bearing Lubrication
The main purpose of the ball-bearing lubricant is to keep out dirt or moisture. It also helps to dissipate the heat that builds up in the bearing, but it does not provide an oil film to reduce bearing friction.

Overlubrication of ball-bearing motors is more of a problem than insufficient lubrication or motor overload. Both sleeve- and ball-bearing motors can suffer from overlubrication. Too much lubricant will eventually find its way into the stator windings and break down the windings' insulation.

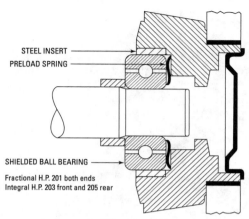

STEEL INSERT

PRELOAD SPRING

SHIELDED BALL BEARING

Fractional H.P. 201 both ends
Integral H.P. 203 front and 205 rear

Figure 22-14 Ball bearing installed in a motor with a preload spring.
(Courtesy Westinghouse)

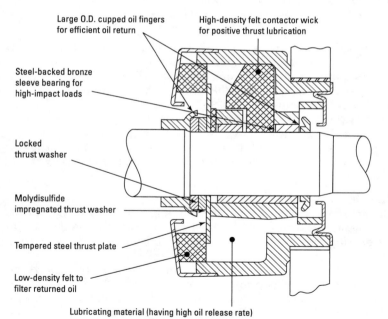

Large O.D. cupped oil fingers
for efficient oil return

High-density felt contactor wick
for positive thrust lubrication

Steel-backed bronze
sleeve bearing for
high-impact loads

Locked
thrust washer

Molydisulfide
impregnated thrust washer

Tempered steel thrust plate

Low-density felt to
filter returned oil

Lubricating material (having high oil release rate)

Figure 22-15 A sleeve-bearing installation in a motor. *(Courtesy Westinghouse)*

Either oil or grease may be used in lubricating ball bearings. Check the manufacturer's specifications. The life of the bearings will depend on the operating conditions. Generally, relubrication of the bearings during the lifetime of the motor is not necessary.

Ball bearings can be mounted in a motor in almost any position. If the bearing has seals on both sides, it is not designed to be relubricated. Instead, if the bearing is worn or needs attention, it should be replaced.

Oil is the most efficient lubricant. Oiling, however, requires some rather elaborate oiling devices. Seals and closures are used to prevent leakage of the oil. Lubricant levels must be maintained and only good-quality oils should be used.

Sleeve bearings are less sensitive to a limited amount of abrasive or foreign materials than are ball bearings. Oil and bearings should be kept clean. In larger, industrial-type motors, the oil should be changed at regular intervals. The split-phase motor usually requires that the wick be oiled and that a few drops be placed into the cup once every six months (Figure 22-16).

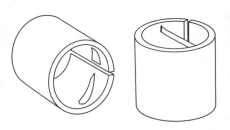

Figure 22-16 Sleeve bearing that shows oil grooves.

Caution
Never overlubricate bearings because the oil will seep into the insulation and settle on the windings of the motor.

Bearing Failure
The most common type of motor problem is bearing failure. This can lead to shaft failure, as well, if not caught in time. Sleeve-bearing replacement consists of pressing out the old bearing and pressing in a new one. This can be done with a drill press fitted with a dowel rod that is the same diameter as the sleeve bearing. The dowel rod may have to be turned slightly to fit into the drill press chuck. However, a downward pressure from the drill press arm will cause the bearing to slip out of the bell housing of the motor. If this doesn't work, use the bearing mandrels made for the removal of bearings. A slight tap of the ball-peen hammer on the mandrel will remove the bearing.

The new (or replacement) bearing must be an exact factory replacement, or it must be installed and line-reamed for an exact fit. If not, repeated failure is certain.

Ball bearing replacement is easier than sleeve bearing replacement. A bearing puller and inserter is almost a necessity. Take a look at Chapter 19 to see what tools are needed. See especially Figures 19-21 and 19-22.

Shaft Repair

Shafts may be damaged if the bearings are worn too much. If the shaft is damaged for this or any other reason, it is possible to regrind it so that the bearings will fit properly. This usually takes some special skills with metalworking tools and machines. It is not advisable to undertake the task if you are inexperienced. In some instances, however, the shaft can be damaged by a pulley that does not fit properly, which may require the removal of some parts by a hand file or light grinding. You may need to replace the pulley if it no longer fits the refinished surface of the shaft, depending on the amount of metal removal that is involved. Not all shafts are repairable or replaceable. When some go bad, the whole armature or rotor must be replaced.

Bearings can be worn by an unbalanced armature or rotor. Figure 22-17 shows how the armature of a hand drill is balanced by the drilling of holes to remove material in the laminations. Three-phase motors use the addition of small weights in the fan blades to

Figure 22-17   Holes drilled in the armature laminations aid in the balance of the rotating member.

help balance the rotating member of the motor. Modern motor manufacturing techniques use computers to aid in balancing the rotor and the armature. The whole job is usually done by removing metal from the laminations as shown in Figure 22-17. This is the easier and faster method for production purposes. The balanced rotor or armature adds considerably to the quietness of the motor. Armatures of series (or universal) motors can rotate at speeds up to 20,000 rpm. This is sufficient to cause all kinds of trouble with vibration noise if the armature is not balanced.

Motor Problems Caused by Bearing Wear or Damage

A number of problems can be observed as being caused by bearing wear. Bearings can be worn excessively if the belt to the pulley on the shaft is too tight. This places undue pressure on the bearing surface. A severe blow to the shaft or a pulley on the shaft can cause damage to the bearings. Once the bearings are damaged, the rotating members of the motor will start to strike parts of the stationary members within a few revolutions of the rotor or armature. This can cause a number of symptoms and problems.

Worn bearings can cause:

- All motors to run too hot after extended operation
- All motors to have excessive mechanical noise

Unbalanced rotor or armature vibration can cause:

- All motors to have excessive mechanical noise

Interference between stationary and rotating member can cause:

- All motors not to start
- All motors to start, but heat up rapidly
- All motors to have excessive mechanical noise
- All motors to have a reduction in power and to get too hot

Summary

All motors have shafts and bearings. The shaft length, finish, diameter, and design are all subject to the demands of the buyer. Some general types are available for fans, pulleys, and direct drives that are driven by lubricated bearings. The bearings may be either sleeve or ball in design. There are advantages and disadvantages to each. Lubrication becomes a very important part of keeping a motor running.

Motor shafts can extend from one end of the motor or from both ends. The shaft may be round or it may have a flat spot where a pulley or other device is attached. Gearmotors have short shafts because the end of the shaft is grooved to match the gears in the gearbox. Gearbox motors are available with a straight-through shaft or with a 90° shaft. Phosphate is used to treat each carbon-steel shaft for protection against rust and corrosion.

Two sizes of shaft adapters with keys are available for use in instances where the shaft diameter of a new motor is smaller than the shaft of the motor being replaced. The shaft can be increased from ½ to ⅝ in. Another type can be used to change the shaft diameter from ⅝ to ¾ in.

Sleeve, or journal, bearings are the simplest in construction. They are the most widely used when low initial cost is a factor. They are quiet in operation and have a good radial load capacity. They also have an unlimited storage life if the motor isn't used for a long time. Sleeve bearings cannot be allowed to run dry.

Graphitized, self-lubricating sleeve bearings are made of solid bronze. They have graphite as an inner recess filler. They usually withstand higher operating temperatures than ordinary sleeve bearings. Porous bronze sleeve bearings are more compact and offer more freedom from attention than solid bronze bearings. They are often constructed so as to be self-aligning, which reduces friction and shaft-binding. They are generally more economical than the graphited or solid bronze types. High temperatures and high motor operating temperatures will have a destructive effect on sleeve bearings lubricated with standard-temperature-range oils. Special oils are available for motor applications at high temperature, and also for applications at lower-than-normal temperatures. Motor performance and bearing life are directly related to lubrication. Sleeve bearings are not as sensitive to abrasive or foreign matter as are ball bearings. They are able to absorb some of the small particles in the soft undersurface of the bearings.

Almost all sizes and types of electric motors use ball bearings. They have low friction loss when lubricated with oil, are suited for high-speed operation, and can be used in a relatively wide range of temperatures. They are slightly more expensive than sleeve bearings. They are also noisier. Grease for ball bearings has been improved for longer life and better lubrication quality.

The main purpose of the lubricant in a ball bearing is to keep out dirt and moisture. It also helps to dissipate the heat that builds up in the bearing. It does not provide an oil film to reduce bearing friction. Over-lubrication of ball-bearing motors is more of a problem than insufficient lubrication or overloading. Both sleeve- and ball-bearing motors can suffer when over-lubricated. Either oil or grease may be used for lubricating ball bearings. Generally, relubrication of the ball bearings during the lifetime of the motor is not necessary. Keep in mind that over-lubrication of bearings causes the oil to seep into the insulation and settle on the windings of the motor.

The most common type of motor problem is bearing failure. This can lead to shaft failure, as well, if it is not caught in time. Sleeve bearing replacement consists of pressing out the old bearing and pressing in a new one. This can be done with a drill press fitted with a dowel rod that is the same diameter as the sleeve bearing. Ball bearing replacement is easier than sleeve bearing replacement. A bearing puller and inserter is almost a necessity. The new bearing must be an exact factory replacement or it must be installed and line-reamed for exact fit. If it is not, repeated failure is certain. Shafts may be damaged if the bearings are worn too much. Bearings can be worn by an unbalanced armature or rotor. Three-phase motors use additional small weights in the fan blades to help balance the rotating member of the motor. Modern motor manufacturing techniques use computers to aid in balancing the rotor and the armature, which is usually done by removing metal from the laminations. A balanced rotor or armature adds considerably to the quietness of the motor.

Worn bearings can cause all motors to run too hot after extended operation. Worn bearings can also cause all motors to have excessive mechanical noise. Unbalanced rotor or armature vibration can cause all motors to have excessive mechanical noise. Interference between stationary and rotating members can cause all motors not to start. It can cause all motors to start, but heat up rapidly. It can also cause all motors to have excessive mechanical noise and a reduction in power and to become too hot to touch.

Review Questions

1. What are the two types of bearings used in motors?
2. What is the most important part of the proper operation of an electric motor?
3. What is a gearbox? Why is it necessary?
4. What is phosphate treatment used for in electric motors?

5. What are the two most commonly used sizes of shaft adapters?

6. What is a journal? How is the word used in motor terminology?

7. What is the purpose of graphite in motor operation?

8. Which electric motors use ball bearings?

9. Where would you want to specify a sleeve-bearing motor?

10. What is the main purpose of the lubricant in a ball bearing?

11. What is the most common type of motor problem?

12. How are sleeve bearings removed from the motor?

13. When do you need a bearing puller?

14. How are bearings worn excessively?

15. How are three-phase motors balanced during manufacture?

Part VII

Reference

Appendix A

Miscellaneous Information

This appendix contains useful reference information, including the following:

- Fractional-horsepower motors and their applications
- NEMA motor-frame dimension standards
- Motor schematics
- Formulas for motor applications
- Horsepower to kilowatt equivalents
- Horsepower/watts versus torque conversion chart
- Temperature conversion table
- Electrical formulas
- SI (metric) conversion table
- Metric conversion table
- Mechanical and electrical characteristics
- NEMA frame dimensions and specifications
- Electronic symbols for devices used with electric motors
- Symbols used in motor schematics

Fractional-Horsepower Motors and Their Applications

Applications	Motor Diameter, in.							
	← 2½ × 3 →		← 3.4 →		← 3.8 →		← 4.9 →	← 5.6 →
	33&44 Frame	51 Frame	59 Frame	11 Frame	19 Frame	29 Frame	39 Frame	Form G2(GE)
Refrigerators & Freezers								
Commercial refrigeration Vending machines, display cases, unit coolers, water coolers, condensing packages, carbonator pumps								
Room air conditioners Portables, window and casement units, through-the-wall units								

(continued)

**Central heat
& air
conditioning**
Furnaces,
outdoor
condensers,
heat pumps,
blower
packages

Fans & vents
Window and
floor fans,
attic and
room vents,
kitchen and
bathroom
vents, range
hoods

**Pumps &
blowers**
Sump pumps,
swimming
pool pumps,
commercial
and industrial
blowers

(continued)

Applications	Motor Diameter, in.							
	← 2½ × 3 ←	← 3.4 →			← 3.8 →	← 4.9 →		← 5.6 →
	33&44 Frame	51 Frame	59 Frame	11 Frame	19 Frame	29 Frame	39 Frame	Form G2(GE)
Other home Shoe polisher, juicer, compact washer, garage-door openers, health equipment, microwave ovens								
Other specialty markets Room heaters, hot-water circulators, oil burners, fan coil units, business machines, hand dryers, tools								

(Courtesy General Electric)

NEMA Motor-Frame Dimension Standards

Standardized motor dimensions as established by the National Electrical Manufacturers Association (NEMA) are tabulated below and apply to all base-mounted motors listed herein that carry a NEMA frame designation.

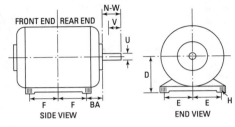

(Courtesy Bodine)

NEMA Frame	D*	2E	2F	BA	H	N-W	U	V§ Min.	Key Wide	Thick	Long	NEMA Frame
42	2⅝	3½	1¹¹⁄₁₆	2¹⁄₁₆	⁹⁄₃₂ slot	1⅛	⅜	—	—	³⁄₆₄ flat	—	42
48	3	4¼	2¾	2½	¹¹⁄₃₂ slot	1½	½	—	—	³⁄₆₄ flat	—	48
56	3½	4⅞	3	2¾	¹¹⁄₃₂ slot	1⅞†	⅝†	—	³⁄₁₆†	³⁄₁₆†	1⅜†	56
56H			3&5‡									56H
56HZ	3½	**	**	**	**	2¼	⅞	2	³⁄₁₆	³⁄₁₆	1⅜	56Hz
66	4⅛	5⅞	5	3⅛	¹³⁄₃₂ slot	2¼	¾	—	³⁄₁₆	³⁄₁₆	1⅞	66
143T	3½	5½	4	2¼	¹¹⁄₃₂ dia.	2¼	⅞	2	³⁄₁₆	³⁄₁₆	1⅜	143T
145T			5									145T
182			4½			2¼	⅞	2	³⁄₁₆	³⁄₁₆	1⅜	182
184			5½									184
	4½	7½		2¾	¹³⁄₃₂ dia.							
182T			4½			2¾	1⅛	2½	¼	¼	1¾	182T
184T			5½									184T
213			5½			3	1⅛	2¾	¼	¼	2	213
215			7									215
	5¼	8½		3½	¹³⁄₃₂ dia.							
213T			5½			3⅜	1⅜	3⅛	⁵⁄₁₆	⁵⁄₁₆	2⅜	213T
215T			7									215T
254U			8¼			3¾	1⅜	3½	⁵⁄₁₆	⁵⁄₁₆	2¾	254U
256U			10									256U
	6¼	10		4¼	¹⁷⁄₃₂ dia.							
254T			8¼			4	1⅝	3¾	⅜	⅜	2⅞	254T
256T			10									256T

(continued)

NEMA Frame	D*	2E	2F	BA	H	N-W	U	V§ Min.	Wide	Thick	Long	NEMA Frame
			—All Dimensions in Inches—						Key			
284U			9½			4⅞	1⅝	4⅝	⅜	⅜	3¾	284U
286U			11									286U
	7	11		4¾	17/32 dia.							
284T			9½			4⅝	1⅞	4¾	½	½	3¼	284T
286T			11									286T
324U			10½			5⅜	1⅞	5⅝	½	½	4¼	324U
326U			12									326U
	8	12½		5¼	21/32 dia.							
324T			10½			5¼	2⅛	5	½	½	3⅞	324T
326T			12									326T
326TS			12			3¾d	1⅞d	3½d½	½		2d	326TS
364U	9	14	11¼	5⅞	21/32 dia	6⅜	2⅛	6⅛	½	½	5	364U
365U			12¼									365U

* Dimension D will never be greater than the above values on rigid-mount motors, but it may be less so that shims up to ¹⁄₃₂˝ thick (¹⁄₁₆˝ on 364U and 365U frames) may be required for coupled or geared machines.

‡ Dayton motors designated 56H have two sets of 2F mounting holes—3˝ and 5˝.

ᵈ Standard short shaft for direct-drive applications.

** Base of the 56-Hz frame motors has holes and slots to match NEMA 56, 56H, 143T and 145T mounting dimensions.

† Certain NEMA 56Z frame motors have ½˝ dia. ω 1½˝ long shaft with ³⁄₆₄˝ flat. These exceptions are noted in the motor manufacturer's catalog.

§ Dimension V is shaft length available for coupling, pinion, or pulley hub—this is a minimum value.

NEMA Letter Designations Following Frame Number

C—Face mount

H—Has 2F dimension larger than same frame without H suffix

J—Face mount to fit jet pumps

K—Has hub for sump pump mounting

M—Flange-mount for oil burner

N—Flange mount for oil burner

T,U—Integral HP motor dimension standards set by NEMA in 1964 and 1953, respectively

Y—Non-standard mounting; see manufacturer's drawing for mounting dimensions

Z—Non-standard shaft extension (NW,U dimensions)

For their own identification, manufacturers may use a letter before the NEMA frame number. This has no reference to mounting dimensions.

Motor Schematics

AC Single-Phase

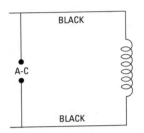

Figure A-1   Shaded-pole:
Nonreversible.

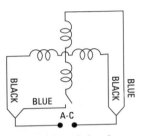

Figure A-2   Split-phase:
To reverse (from rest only),
transpose blue leads or black leads.

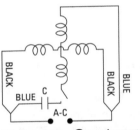

Figure A-3   Capacitor-start:
To reverse (from rest only),
transpose blue leads or
black leads.

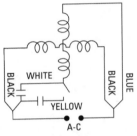

Figure A-4   Two-valve capacitor:
Two-capacitor-start, one-capacitor-
run. To reverse (from rest only),
transpose the black leads.

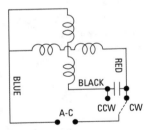

Figure A-5   Permanent-split
capacitor (3-lead): To reverse,
connect either side of capacitor
to line.

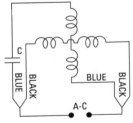

Figure A-6   Permanent-split
capacitor (4-lead): To reverse,
transpose the black leads.

AC Polyphase

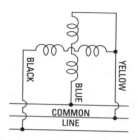

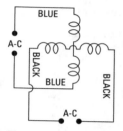

Figure A-7 Two-phase (three-wire): To reverse, transpose blue and black leads.

Figure A-8 Two-phase (four-lead): To reverse, transpose blue or black leads.

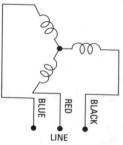

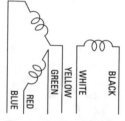

Figure A-9 Three-phase (single-voltage): To reverse, transpose any two leads.

Figure A-10 Three-phase (star or delta): For 440V, connect together white, yellow, and green. Connect to line black, red, and blue. To reverse, transpose any two line leads. For 220V, connect white to blue, black to green, and yellow to red. Then connect each junction point to line. To reverse, transpose any two junction points with line.

DC

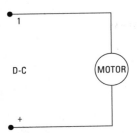

Figure A-11 Permanent-magnet, brushless DC, printed circuit, shell-type armature: To reverse, transpose motor leads.

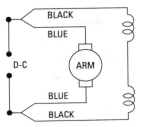

Figure A-12 Shunt-wound: To reverse, transpose blue or black leads.

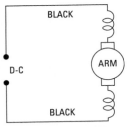

Figure A-13 Series-wound (two-lead): Non-reversible.

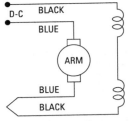

Figure A-14 Series-wound (four-lead): To reverse, transpose blue leads.

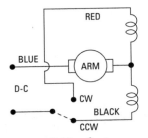

Figure A-15 Series-wound (split-field): To reverse, connect other field to lead line.

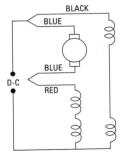

Figure A-16 Compound-wound (five-wire reversible): To reverse, transpose blue leads.

Formulas for Motor Applications

T = torque or twisting moment (force × moment arm length)

π = 3.1416

N = revolutions per minute

HP = horsepower (33,000 ft − lbs per min.) applies to power output

R = radius of pulley, in feet

E = input voltage

I = current in amperes

P = power input in watts

$$HP = \frac{T(\text{lb} - \text{in}) \times N(\text{rpm})}{63,025}$$

$$HP = T\,(\text{oz} - \text{in.}) \times N \times 9.917 \times 10^{-7}$$

$$= \text{approx. } T\,(\text{oz} - \text{in.}) \times N \times 10^{-6}$$

$$P = EI \times \text{power factor} = \frac{HP \times 746}{\text{motor efficiency}}$$

Power to Drive Pumps

$$HP = \frac{\text{gal. per min} \times \text{total head (inc. friction)}}{3,960 \times \text{eff. of pump}}$$

where
Approx. friction head (ft.) =

$$\frac{\text{pipe length (ft.)} \times [\text{velocity of flow (fps)}]^2 \times 0.02}{5.367 \times \text{diameter (in.)}}$$

Eff. = Approx. 0.50 to 0.85

Time to Change Speed of Rotating Mass

$$\text{Time (sec.)} = \frac{WR^2 \times \text{change in rpm}}{308 \times \text{torque (ft} - \text{lb.)}}$$

where

$$WR^2 \text{ (disc)} = \frac{\text{weight (lbs.)} \times [\text{radius (ft.)}]^2}{2}$$

$$WR^2 \text{ (rim)} = \frac{\text{wt. (lbs.)} \times [(\text{outer radius in ft.})^2 + \text{inner radius in ft.})^2]}{2}$$

Power to Drive Fans

$$\text{HP} = \frac{\text{cu. ft. air per min.} \times \text{water gauge pressure (in.)}}{6.350 \times \text{Eff.}}$$

Horsepower to kW Equivalents

Horsepower	Kilowatt (kW)*
$\frac{1}{20}$	0.025
	0.035
	0.05
	0.071
$\frac{1}{8}$	0.1
$\frac{1}{6}$	0.14
$\frac{1}{4}$	0.2
$\frac{1}{3}$	0.28
$\frac{1}{2}$	0.4
1	0.8
$1\frac{1}{2}$	1.1
2	1.6
3	2.5
5	4.0
7.5	5.6
10	8.0

*James W. Polk, A Preview of Metric Motors, Westinghouse Electric Corporation.

Horsepower/Watts vs. Torque Conversion Chart

Hp	Watts	@ 1125 rpm		@ 1200 rpm		@ 1425 rpm	
		Oz.-in.	mN.in.	Oz.-in.	mN.m	Oz.-in.	mN.m
1/2000	0.373	0.4482	3.1649	0.4202	2.9670	0.3538	2.4986
1/1500	0.497	0.5976	4.2198	0.5602	3.9561	0.4718	3.3314
1/1000	0.746	0.8964	6.3297	0.8403	5.9341	0.7077	4.9971
1/750	0.994	1.1951	8.4396	1.1205	7.9121	0.9435	6.6628
1/500	1.49	1.7927	12.6594	1.6807	11.8682	1.4153	9.9943
1/200	3.73	4.4818	31.6485	4.2017	29.6705	3.5383	24.9857
1/150	4.97	5.9757	42.1980	5.6023	39.5606	4.7177	33.3142
1/100	7.46	8.9636	63.2970	8.4034	59.3409	7.0765	49.9713
1/75	9.94	11.9515	84.3960	11.2045	79.1212	9.4354	66.6284
1/70	10.70	12.8052	90.4243	12.0048	84.7727	10.1093	71.3876
1/60	12.40	14.9393	105.4950	14.0056	98.9015	11.7942	83.2855
1/50	14.90	17.9272	126.5940	16.8086	118.6818	14.1531	99.9426
1/40	18.60	22.4090	158.2425	21.0085	148.3523	17.6913	124.9283
1/30	24.90	29.8787	210.9899	28.0113	197.8031	23.5884	166.5710
1/25	29.80	35.8544	253.1879	33.6135	237.3637	28.3061	199.8852
1/20	37.30	44.8180	316.4849	42.0169	296.7046	35.3827	249.8565
1/15	49.70	59.7574	421.9799	56.0225	395.6061	47.1769	333.1420
1/12	62.10	74.6967	527.4748	70.0282	494.5077	58.9711	416.4275
1/10	74.6	89.6361	632.9698	84.0338	593.4092	70.7653	499.7130
1/8	93.2	112.0451	791.2123	105.0423	741.7615	88.4566	624.6413
1/6	124.0	149.3934	1054.9497	140.0563	989.0153	117.9422	832.8550
1/4	186.0	224.0902	1582.4245	210.0845	1483.5230	176.9133	1249.2825
1/3	249.0	298.7869	2109.8994	280.1127	1978.0307	235.8844	1665.7101

Hp	Watts	@ 1500 rpm		@ 1725 rpm		@ 1800 rpm	
		Oz.-in.	mN.m	Oz.-in.	mN.m	Oz.-in.	mN.m
1/2000	0.373	0.3361	2.3736	0.2923	2.0640	0.2801	1.9780
1/1500	0.497	0.4482	3.1648	0.3897	2.7520	0.3735	2.6374
1/1000	0.746	0.6723	4.7473	0.5846	4.1281	0.5602	3.9561
1/750	0.994	0.8964	6.3297	0.7794	5.5041	0.7470	5.2747
1/500	1.490	1.3445	9.4945	1.1692	8.2561	1.1205	7.9121
1/200	3.730	3.3614	23.7364	2.9229	20.6403	2.8011	19.7803
1/150	4.97	4.4818	31.6485	3.8972	27.5204	3.7348	26.3737
1/100	7.46	6.7227	47.4727	5.8458	41.2806	5.6023	39.5606
1/75	9.94	8.9636	63.2970	7.7944	55.0409	7.4697	52.7475
1/70	10.70	9.6039	67.8182	8.3512	58.9723	8.0032	56.5152

Hp	Watts	@ 1500 rpm		@ 1725 rpm		@ 1800 rpm	
		Oz.-in.	mN.m	Oz.-in.	mN.m	Oz.-in.	mN.m
1/60	12.40	11.2045	79.1212	9.7431	68.8011	9.3371	65.9344
1/50	14.90	13.4454	94.9455	11.6917	82.5613	11.2045	79.1212
1/40	18.60	16.8068	118.6818	14.6146	103.2016	14.0056	98.9015
1/30	24.90	22.4090	158.2425	19.4861	137.6021	18.6742	131.8687
1/25	29.80	26.8908	185.8909	23.3833	165.1226	22.4090	158.2425
1/20	37.3	33.6135	237.3637	29.2292	206.4032	28.0113	197.8031
1/15	49.7	44.8180	316.4849	38.9722	275.2043	37.3484	263.7374
1/12	62.1	56.0225	395.6061	48.7153	344.0053	46.6854	329.6718
1/10	74.6	67.2270	474.7274	58.4583	412.8064	56.0225	395.6061
1/8	93.2	84.0338	593.4092	73.0729	516.0080	70.0282	494.5077
1/6	124.0	112.0451	791.2123	97.4305	688.0107	93.3709	659.3436
1/4	186.0	168.0676	1186.8184	146.1458	1032.0160	140.0563	989.0153
1/3	249.0	224.0902	1582.4245	194.8610	1376.0213	186.7418	1318.6871

Hp	Watts	@ 3000 rpm		@3450 rpm		@ 3600 rpm	
		Oz.-in.	mN.m	Oz.-in.	mN.m	Oz.-in.	mN.m
1/2000	0.373	0.1681	1.1868	0.1461	1.0320	0.1401	0.9890
1/1500	0.497	0.2241	1.5824	0.1949	1.3760	0.1867	1.3187
1/1000	0.746	0.3361	2.3736	0.2923	2.0640	0.2801	1.9780
1/750	0.994	0.4482	3.1648	0.3897	2.7520	0.3735	2.6374
1/500	1.490	0.6723	4.7473	0.5846	4.1281	0.5602	3.9561
1/200	3.730	1.6807	11.8682	1.4615	10.3202	1.4006	9.8902
1/150	4.97	2.2409	15.8242	1.9486	13.7602	1.8674	13.1869
1/100	7.46	3.3614	23.7364	2.9229	20.6403	2.8011	19.7803
1/75	9.94	4.4818	31.6485	3.8972	27.5204	3.7348	26.3737
1/70	10.70	4.8019	33.9091	4.1756	29.4862	4.0016	28.2576
1/60	12.40	5.6023	39.5606	4.8715	34.4005	4.6685	32.9672
1/50	14.90	6.7227	47.4727	5.8458	41.2806	5.6023	39.5606
1/40	18.6	8.4034	59.3409	7.3073	51.6008	7.0028	49.4508
1/30	24.9	11.2045	79.1212	9.7431	68.8011	9.3371	65.9344
1/25	29.8	13.4454	94.9455	11.6917	82.5613	11.2045	79.1212
1/20	37.3	16.8068	118.6818	14.6146	103.2016	14.0056	98.9015
1/15	49.7	22.4090	158.2425	19.4861	137.6021	18.6742	131.8687
1/12	62.1	28.0113	197.8031	24.3576	172.0027	23.3427	164.8359
1/10	74.6	33.6135	237.3637	29.2292	206.4032	28.0113	197.8031
1/8	93.2	42.0169	296.7046	36.5364	258.0040	35.0141	247.2538

(continued)

(continued)

Hp	Watts	@ 3000 rpm		@3450 rpm		@ 3600 rpm	
		Oz.-in.	mN.m	Oz.-in.	mN.m	Oz.-in.	mN.m
1/6	124.0	56.0225	395.6061	48.7153	344.0053	46.6854	329.6718
1/4	186.0	84.0338	593.4092	73.0729	516.0080	70.0282	494.5077
1/3	249.0	112.0451	791.2123	97.4305	688.0107	93.3709	659.3436

Hp	Watts	@ 5000 rpm		@7500 rpm		@ 10,000 rpm	
		Oz.-in.	mN.m	Oz.-in.	mN.m	Oz.-in.	mN.m
1/2000	0.373	0.1008	0.7121	0.0672	0.4747	0.0504	0.3560
1/1500	0.497	0.1345	0.9495	0.0896	0.6330	0.0672	0.4747
1/1000	0.746	0.2017	1.4242	0.1345	0.9495	0.1008	0.7121
1/750	0.994	0.2689	1.8989	0.1793	1.2659	0.1345	0.9495
1/500	1.490	0.4034	2.8484	0.2689	1.8989	0.2017	1.4242
1/200	3.730	1.0084	7.1209	0.6723	4.7473	0.5042	3.5605
1/150	4.97	1.3445	9.4945	0.8964	6.3297	0.6723	4.7473
1/100	7.46	2.0168	14.2418	1.3445	9.4945	1.0084	7.1209
1/75	9.94	2.6891	18.9891	1.7927	12.6594	1.3445	9.4945
1/70	10.70	2.8812	20.3455	1.9208	13.5636	1.4406	10.1727
1/60	12.40	3.3614	23.7364	2.2409	15.8242	1.6807	11.8682
1/50	14.90	4.0336	28.4836	2.6891	18.9891	2.0168	14.2418
1/40	18.60	5.0420	35.6046	3.3614	23.7364	2.5210	17.8023
1/30	24.90	6.7227	47.4727	4.4818	31.6485	3.3614	23.7364
1/25	29.80	8.0672	56.9673	5.3782	37.9782	4.0336	28.4836
1/20	37.30	10.0841	71.2091	6.7227	47.4727	5.0420	35.6046
1/15	49.70	13.4454	94.9455	8.9636	63.2970	6.7227	47.4727
1/12	62.10	16.8068	118.6818	11.2045	79.1212	8.4034	59.3409
1/10	74.6	20.1681	142.4182	13.4454	94.9455	10.0841	71.2091
1/8	93.2	25.2101	178.0228	16.8068	118.6818	12.6051	89.0114
1/6	124.0	33.6135	237.3637	22.4090	158.2425	16.8068	118.6818
1/4	186.0	50.4203	356.0455	33.6135	237.3637	25.2101	178.0228
1/3	249.0	67.2270	474.7274	44.8180	316.4849	33.6135	237.3637

(Courtesy Bodine)

Temperature Conversion Table—°F← →°C

The numbers in italics in the center column refer to the temperature, either in Celsius or Fahrenheit, which is to be converted to the other scale. If converting Fahrenheit to Celsius, the equivalent temperature will be found in the left column. If converting Celsius to Fahrenheit, the equivalent temperature will be found in the column on the right.

—100 to 30			31 to 71			72 to 212			213 to 620			621 to 1000		
C		F	C		F	C		F	C		F	C		F
−73	(−)100	−148	−0.6	31	87.8	22.2	72	161.6	104	220	428	332	630	1166
−68	(−)90	−130	0	32	89.6	22.8	73	163.4	110	230	446	338	640	1184
−62	(−)80	−112	0.6	33	91.4	23.3	74	165.2	116	240	464	343	650	1202
−57	(−)70	−94	1.1	34	93.2	23.9	75	167.0	121	250	482	349	660	1220
−51	(−)60	−76	1.7	35	95.0	24.4	76	168.8	127	260	500	354	670	1238
−46	(−)50	−58	2.2	36	96.8	25.0	77	170.6	132	270	518	360	680	1256
−40	(−)40	−40	2.8	37	98.6	25.6	78	172.4	138	280	536	366	690	1274
−34.4	(−)30	−22	3.3	38	100.4	26.1	79	174.2	143	290	554	371	700	1292
−28.9	(−)20	−4	3.9	39	102.2	26.7	80	176.0	149	300	572	377	710	1310
−23.3	(−)10	14	4.4	40	104.0	27.2	81	177.8	154	310	590	382	720	1328
−17.8	0	32	5.0	41	105.8	27.8	82	179.6	160	320	608	388	730	1346
−17.2	1	33.8	5.6	42	107.6	28.3	83	181.4	166	330	626	393	740	1364
−16.7	2	25.6	6.1	43	109.4	28.9	84	183.2	171	340	644	399	750	1382
−16.1	3	37.4	6.7	44	111.2	29.4	85	185.0	177	350	662	404	760	1400
−15.6	4	39.2	7.2	45	113.0	30.0	86	186.8	182	360	680	410	770	1418
−15.0	5	41.0	7.8	46	114.8	30.6	87	188.6	188	370	698	416	780	1436
−14.4	6	42.8	8.3	47	116.6	31.1	88	190.4	193	380	716	421	790	1454
−13.9	7	44.6	8.9	48	118.4	31.7	89	192.2	199	390	734	427	800	1472
−13.3	8	46.4	9.4	49	120.0	32.2	90	194.0	204	400	752	432	810	1490
−12.8	9	48.2	10.0	50	122.0	32.8	91	195.8	210	410	770	438	820	1508
−12.2	10	50.0	10.6	51	123.8	33.3	92	197.6	216	420	788	443	830	1526
−11.7	11	51.8	11.1	52	125.6	33.9	93	199.4	221	430	806	449	840	1544
−11.1	12	53.6	11.7	53	127.4	34.4	94	201.2	227	440	824	454	850	1562
−10.6	13	55.4	12.2	54	129.2	35.0	95	203.0	232	450	842	460	860	1580
−10.0	14	57.2	12.8	55	131.0	35.6	96	204.8	238	460	860	466	870	1598
−9.4	15	59.0	13.3	56	132.8	36.1	97	206.6	243	470	878	471	880	1616
−8.9	16	60.8	13.9	57	134.6	36.7	98	208.4	249	480	896	477	890	1634
−8.3	17	62.6	14.4	58	136.4	37.2	99	210.2	254	490	914	482	900	1652
−7.8	18	64.4	15.0	59	138.2	37.8	100	212.0	260	500	932	488	910	1670
−7.2	19	66.2	15.6	60	140.0	43	110	230	266	510	950	493	920	1688
−6.7	20	68.0	16.1	61	141.8	49	120	248	271	520	968	499	930	1706
−6.1	21	69.8	16.7	62	143.6	54	130	266	277	530	986	504	940	1724
−5.6	22	71.6	17.2	63	145.4	60	140	284	282	540	1004	510	950	1742
−5.0	23	73.4	17.8	64	147.2	66	150	302	288	550	1022	516	960	1760
−4.4	24	75.2	18.3	65	149.0	71	160	320	293	560	1040	521	970	1778
−3.9	25	77.0	18.9	66	150.8	77	170	338	299	570	1058	527	980	1796
−3.3	26	78.8	19.4	67	152.6	82	180	356	304	580	1076	532	990	1814
−2.8	27	80.6	20.0	68	154.4	88	190	374	310	590	1094	538	1000	1832
−2.2	28	82.4	20.6	69	156.2	93	200	392	316	600	1112			
−1.7	29	84.2	21.1	70	158.0	99	210	410	321	610	1130			
−1.1	30	86.0	21.7	71	159.8	100	212	414	327	620	1148			

Electrical Formulas

Ohm's Law

Ohm's Law: $\text{amperes} = \dfrac{\text{volts}}{\text{ohms}}$

Power in DC Circuits

$\text{Watts} = \text{volts} \times \text{amperes}$

$$\text{Horsepower} = \dfrac{\text{volts} \times \text{amperes}}{746}$$

$$\text{Kilowatts} = \dfrac{\text{volts} \times \text{amperes}}{1000}$$

$$\text{Kilowatt-hours (kWh)} = \dfrac{\text{volts} \times \text{amperes} \times \text{hours}}{1000}$$

Power in AC Circuits

$$\text{Apparent power kilovolt-amperes(kVA)} = \dfrac{\text{volts} \times \text{amperes}}{1000}$$

$$\text{Power factor} = \dfrac{\text{kilowatts}}{\text{kilovolt-amperes}}$$

Single–Phase Kilowatts (kW) =

$$\dfrac{\text{volts} \times \text{amperes} \times \text{power factor}}{1000}$$

Two–Phase (kW) =

$$\dfrac{\text{volts} \times \text{amperes} \times \text{power factor} \times 1.4142}{1000}$$

Three–Phase (kW)=

$$\dfrac{\text{volts} \times \text{amperes} \times \text{power factor} \times 1.7321}{1000}$$

SI (Metric) Conversion Table

	SI Unit	Imperial/Metric to SI	SI to Imperial/Metric
Length	meter (m)	1 inch=2.54 × 10⁻²ᵐ 1 foot = 0.305 m 1 yard = .914 m	1 m = 39.37 inches = 3.281 feet = 1.094 yards
Mass	kilogram (kg)	1 ounce (mass) = 28.35 × 10⁻³kg 1 pound (mass) = 0.454 kg 1 slug = 14.59 kg	1 kg = 35.27 ounces = 2.205 pounds = 68.521 × 10⁻³ slug
Area	square meter (m²)	1 sq. in. = 6.45 × 10⁻⁴m² 1 sq. ft = 0.93 × 10⁻¹m² 1 sq. yd = 0.836 m²	1 m² = 1550 sq. in. = 10.76 sq. ft = 1.196 sq. yd
Volume	cubic meter (m³)	1 cu. in. = 16.3 × 10⁻⁶m³ 1 cu. ft = 0.028 m³	1 m³ = 6.102 × 10⁴ cu. in. = 35.3 cu. ft
Time	second (s)	same as Imperial/Metric	same as Imperial/Metric
Electric Current	ampere (A)	same as Imperial/Metric	same as Imperial/Metric
Plane Angle	radian (rad)	1 angular deg. = 1.745 × 10⁻² rad 1 Revolution = 6.283 rad	1 r = 57.296 rad
Frequency	hertz (Hz)	1 cycle/sec = 1 Hz	1 Hz = 1 cps
Force	newton (N)	1 oz. (f) = 0.278 N 1 lb (f) = 4.448 N 1 kilopond = 9.807 N 1 kgf = 9.807 N	1 N = 3.597 oz. (f) = 0.225 lb (f) = 0.102 kp = 0.102 kgf
Energy (Work)	joule (J)	1 Btu = 1055.06 J 1 kWh = 3.6 × 10⁶J 1 Ws = 1 J 1 kcal = 4188.8	1 J = 9.478 × 10⁻⁴ Btu = 2.778 × 10⁻⁷kWh =1 Ws = 2.389 × 10⁻⁴ kcal
Power	watt (W)	1 hp (electric) = 746 W	1 W = 1.341 × 10⁻³hp (electric)
Quantity of Electricity	coulomb (C)	same as Imperial/Metric	same as Imperial/Metric
EMF	volt (V)	same as Imperial/Metric	same as Imperial/Metric
Resistance	ohm (Ω)	same as Imperial/Metric	same as Imperial/Metric
Electric Capacitance	farad (F)	same as Imperial/Metric	same as Imperial/Metric

(continued)

	SI Unit	Imperial/Metric to SI	SI to Imperial/Metric
Electric Induction	henry (H)	same as Imperial/Metric	same as Imperial/Metric
Magnetic Flux	weber (Wb)	1 line = 10^{-8} Wb 1 Mx = 10^{-8} Wb 1 Vs = 1 Wb	1 Wb = 10^8 lines = 10^8 Mx = 1 Vs
Magnetic Flux Density	tesla (T)	1 line/in.^{2} = 1.55×10^{-5}T 1 gauss = 10^{-4}T	1 t = 6.452×10^4 lines/in.^{2} = 10^4 gauss
Linear Velocity	meter/sec (m/s)	1 inch/sec = 2.54×10^{-2}m/s 1 mph = 1.609 km/s	1 m/s = 39.37 in./sec = 3.281 ft/sec
Linear Accel.	meter/sec^2 (m/s^2)	1 in./sec^2 = 2.54×10^{-4}m/s^2	1 m/s^2 = 39.37 in./sec^2
Torque	newtonmeter (Nm)	1 lb. ft = 1.356 Nm 1 oz.-in. = 7.062×10^{-3}Nm 1 kilopondmeter = 9.807 Nm 1 lb-in. = 0.113 Nm	1 Nm = 0.738 lb-ft = 8.851 lb-in. = 0.102 kpm = 141.61 oz.-in.
Temperature	degree Celsius (°C)	F = (C × ⁹⁄₅) + 32	C = (F − 32) × ⁵⁄₉

Note: There is at press-time no international equivalent for revolutions per minute (rpm). Commonly used expressions are rpm = r/min = t/min = U/min = Rev/min = min⁻¹ (f) = force

Metric Conversion Table

Fractions	Inches	mm	Fractions	Inches	mm
1/64	.0156	.3969	33/64	.5156	13.097
1/32	.0312	.7937	17/32	.5312	13.494
3/64	.0468	1.191	35/64	.5468	13.891
1/16	.0625	1.588	9/16	.5625	14.288
5/64	.0781	1.984	37/64	.5781	14.684
3/32	.0937	2.381	19/32	.5937	15.081
7/64	.1093	2.778	39/64	.6093	15.478
1/8	.125	3.175	5/8	.625	15.875
9/64	.1406	3.572	41/64	.6406	16.272
5/32	.1562	3.969	21/32	.6562	16.669
11/64	.1718	4.366	43/64	.6718	17.066
3/16	.1875	4.763	11/16	.6875	17.463
13/64	.2031	5.159	45/64	.7031	17.859
7/32	.2187	5.556	23/32	.7187	18.256
15/64	.2343	5.953	47/64	.7343	18.653
1/4	.25	6.350	3/4	.75	19.050
17/64	.2656	6.747	49/64	.7656	19.447
9/32	.2812	7.144	25/32	.7812	19.844
19/64	.2968	7.541	51/64	.7969	20.241
5/16	.3125	7.938	13/16	.8125	20.638
21/64	.3281	8.334	53/64	.8281	21.034
11/32	.3437	8.731	27/32	.8437	21.431
23/64	.3593	9.128	55/64	.8593	21.828
3/8	.375	9.525	7/8	.875	22.225
25/64	.3906	9.922	57/64	.8906	22.622
13/32	.4062	10.319	29/32	.9062	23.019
27/64	.4219	10.716	59/64	.9218	23.416
7/16	.4375	11.113	15/16	.9375	23.813
29/64	.4531	11.509	61/64	.9531	24.209
15/32	.4687	11.906	31/32	.9687	24.606
31/64	.4843	12.303	63/64	.9843	25.003
1/2	.5	12.700			

1 mm = 03937 inches
1 inch = 25.4 mm
1 meter = 3.2809 feet

Mechanical and Electrical Characteristics

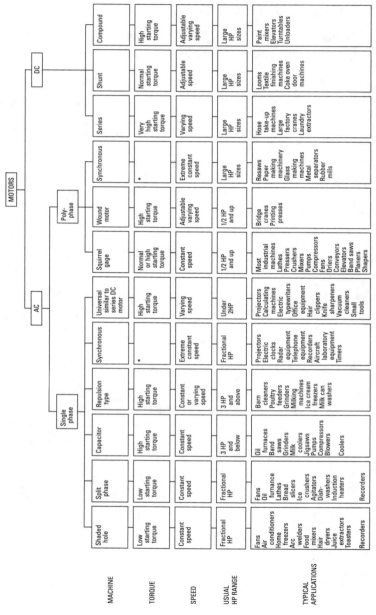

MACHINE	Shaded pole	Split phase	Capacitor	Repulsion type	Synchronous	Universal similar to series DC motor	Squirrel cage	Wound motor	Synchronous	Series	Shunt	Compound
							Poly-phase			DC		
	Single phase											
	AC											
								MOTORS				
TORQUE	Low starting torque	Low starting torque	High starting torque	High starting torque	*	High starting torque	Normal or high starting torque	High starting torque	*	Very high starting torque	Normal torque	High starting torque
SPEED	Constant speed	Constant speed	Constant speed	Constant or varying speed	Extreme constant speed	Varying speed	Constant speed	Adjustable varying speed	Extreme constant speed	Varying speed	Adjustable speed	Adjustable varying speed
USUAL HP RANGE	Fractional HP	Fractional HP	3 HP and below	3 HP and above	Fractional HP	Under 2HP	1/2 HP and up	1/2 HP and up	Large HP sizes	Large HP sizes	Large HP sizes	Large HP sizes
TYPICAL APPLICATIONS	Fans Air conditioners Home freezers Arc welders Food mixers Hair dryers Juice extractors Toasters Recorders	Fans Oil furnace Lathes Bread slicers Ice crushers Agitators Dish-washers Induction heaters Recorders	Oil furnaces Band saws Grinders Milk coolers Jigsaws Pumps Compressors Blowers Coolers	Barn cleaners Poultry feeders Grinders Milking machines Ice cream freezers Milk can washers	Projectors Electric clocks Radar equipment Telephone equipment Recorders Aircraft laboratory equipment Timers	Projectors Calculating machines Electric typewriters Office equipment Hair clippers Knife sharpeners Vacuum cleaners Small tools	Most industrial machines Lathes Pressers Crushers Mixers Pumps Compressors Fans Driers Conveyors Elevators Band saws Planers Shapers	Bridge cranes Printing presses	Resaws Paper making machinery Glass making machines Metal separators Rubber mills	Hose take-up machines Large factory cranes Laundry extractors	Looms Textile finishing machines Coke oven door machines	Paint mixers Elevators Turntables Unloaders

NEMA Frame Dimensions and Specifications

Motor Frame Dimensions

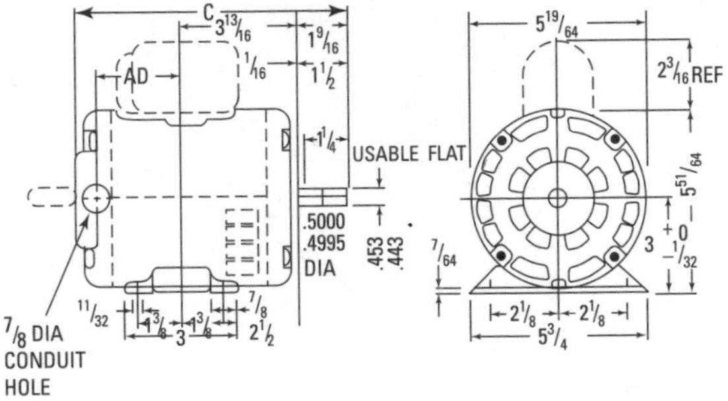

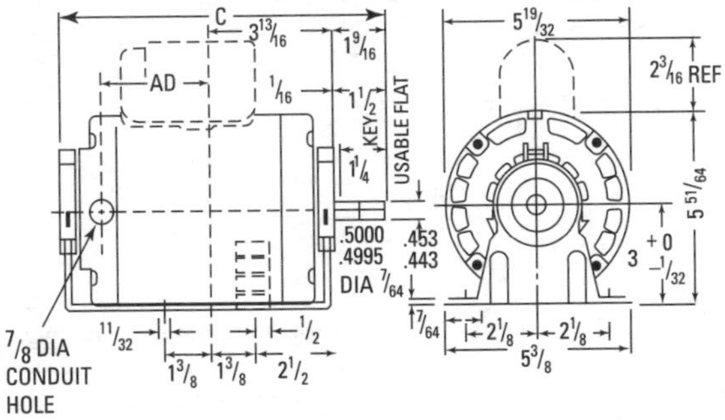

NEMA Frame Specifications

NEMA 48 Frame Rigid Base

OPEN DRIP-PROOF

FRAME	G48	J48	L48	N48	R48
C	8 1/2	9	9 1/2	10	10 1/2
AD	2 3/8	2 7/8	3 3/8	3 7/8	4 3/8

T.E.N.V.

FRAME	J48	L48	N48	R48	U48
C	9 7/2	10 3/8	10 7/8	11 3/8	11 7/8
AD	3 1/4	3 3/4	4 1/4	4 3/4	5 1/4

NEMA 48 Frame Resilient Base

OPEN DRIP-PROOF

FRAME	G48	J48	L48	R48
C	9 3/8	9 7/8	10 3/8	11 3/8
AD	2 3/4	3 1/4	3 3/4	4 3/4

For totally enclosed nonyentilated designs use open drip-proof dimensions.

(continued)

Motor Frame Dimensions

NEMA Frame Specifications

NEMA 56 Frame Rigid Base

OPEN DRIP-PROOF

FRAME	JS56	LS56	NS56	RS56	C56	D56
C	$9\frac{1}{2}$	10	$10\frac{1}{2}$	11	$9\frac{7}{8}$	$10\frac{3}{8}$
AD	$2\frac{7}{16}$	$2\frac{15}{16}$	$3\frac{7}{16}$	$3\frac{15}{16}$	$2\frac{7}{8}$	$3\frac{3}{8}$
P	$5\frac{19}{32}$	$5\frac{19}{32}$	$5\frac{19}{32}$	$5\frac{19}{32}$	$6\frac{19}{32}$	$6\frac{19}{32}$

FRAME	E56	F56	G56	H56	J56	K56
C	$10\frac{7}{8}$	$11\frac{3}{8}$	$11\frac{7}{8}$	$12\frac{3}{8}$	$12\frac{7}{8}$	$13\frac{3}{8}$
AD	$3\frac{7}{8}$	$4\frac{3}{8}$	$4\frac{7}{8}$	$5\frac{3}{8}$	$5\frac{7}{8}$	$6\frac{3}{8}$
P	$6\frac{19}{32}$	$6\frac{19}{32}$	$6\frac{19}{32}$	$6\frac{19}{32}$	$6\frac{19}{32}$	$6\frac{19}{32}$

T.E.F.C.

FRAME	C56	D56	E56	F56	G56	H56	J56	K56
C	$10\frac{13}{16}$	$11\frac{5}{16}$	$11\frac{13}{16}$	$12\frac{5}{16}$	$12\frac{13}{16}$	$13\frac{5}{16}$	$13\frac{13}{16}$	$14\frac{5}{16}$
AD	$-\frac{9}{16}$	$-\frac{1}{16}$	$\frac{7}{16}$	$\frac{15}{16}$	$1\frac{7}{16}$	$1\frac{15}{16}$	$2\frac{7}{16}$	$2\frac{15}{16}$

For totally enclosed nonventilated designs use open drip-proof dimensions. For SF6 totally enclosed nonventilated designs only, add 1" to "c" dimension.

NEMA 56 Frame "C" Face

OPEN DRIP-PROOF

FRAME	GS56	JS56	LS56	NS56	RS56	C56	D56
C	9	9½	10	10½	11	9⅞	10⅜
AD	6	6½	7	7½	8	6 15/16	7 7/16
P	5 19/32	5 19/32	5 19/32	5 19/32	5 19/32	6 19/32	6 19/32

FRAME	E56	F56	G56	H56	J56	K56
C	10⅞	11⅜	11⅞	12⅜	12⅞	13⅜
AD	7 15/16	8 7/16	8 15/16	9 7/16	9 15/16	10 7/16
P	6 19/32	6 19/32	6 19/32	6 19/32	6 19/32	6 19/32

For totally enclosed nonventilated designs use open drip-proof dimensions. For S56 totally enclosed nonventilated designs only, add ½″ to "C" dimension.

T.E.F.C.

FRAME	C56	D56	E56	F56	G56	H56	J56	K56
C	10 13/16	11 5/16	11 13/16	12 5/16	12 13/16	13 5/16	13 13/16	14 5/16
AD	3½	4	4½	5	5½	6	6½	7

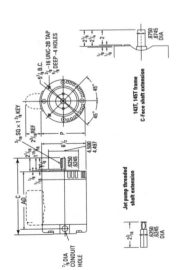

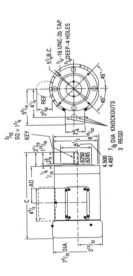

(continued)

(continued)

Motor Frame Dimensions

NEMA Frame Specifications

NEMA 56 Frame Resilient Base
OPEN DRIP-PROOF

FRAME	GS56	JS56	LS56	NS56	RS56	C56
C	$9\frac{13}{16}$	$10\frac{5}{16}$	$10\frac{13}{16}$	$11\frac{5}{16}$	$11\frac{13}{16}$	$10\frac{13}{16}$
AD	$2\frac{7}{16}$	$2\frac{15}{16}$	$3\frac{7}{16}$	$3\frac{15}{16}$	$4\frac{7}{16}$	$3\frac{15}{32}$
P	$5\frac{19}{32}$	$5\frac{19}{32}$	$5\frac{19}{32}$	$5\frac{19}{32}$	$5\frac{19}{32}$	$6\frac{19}{32}$

FRAME	D56	E56	F56	G56	J56
C	$11\frac{5}{16}$	$11\frac{13}{16}$	$12\frac{5}{16}$	$12\frac{13}{16}$	$13\frac{13}{16}$
AD	$3\frac{31}{32}$	$4\frac{15}{32}$	$4\frac{31}{32}$	$5\frac{15}{32}$	$6\frac{15}{32}$
P	$6\frac{19}{32}$	$6\frac{19}{32}$	$6\frac{19}{32}$	$6\frac{19}{32}$	$6\frac{19}{32}$

For totally enclosed nonventilated designs use open drip-proof dimensions.

NEMA Combination Base 56HZ, 143T, 145T
OPEN DRIP-PROOF

FRAME	E143T	F143T	F145T	G145T	H145T	J145T	K145T	L145T
C	$11\frac{1}{16}$	$11\frac{9}{16}$	$11\frac{9}{16}$	$12\frac{1}{16}$	$12\frac{9}{16}$	$13\frac{1}{16}$	$13\frac{9}{16}$	$14\frac{1}{16}$
AD	$2\frac{3}{16}$	$2\frac{11}{16}$	$2\frac{9}{16}$	$3\frac{1}{16}$	$3\frac{9}{16}$	$4\frac{1}{16}$	$4\frac{9}{16}$	$5\frac{1}{16}$

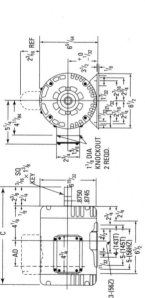

T.E.F.C.

FRAME	E143T	F143T	G143T	E145T	F145T	G145T	H145T	J145T	K 145T
C	$12^{3}/_{16}$	$12^{11}/_{16}$	$13^{3}/_{16}$	$12^{3}/_{16}$	$12^{11}/_{16}$	$13^{3}/_{16}$	$13^{11}/_{16}$	$14^{3}/_{16}$	$14^{11}/_{16}$
AD	$2^{3}/_{16}$	$2^{11}/_{16}$	$3^{3}/_{16}$	$2^{3}/_{16}$	$2^{11}/_{16}$	$3^{3}/_{16}$	$3^{11}/_{16}$	$4^{3}/_{16}$	$4^{11}/_{16}$

NEMA Combination Resilient Base 56HZ, 143T, 145T
OPEN DRIP-PROOF

FRAME	E143T	F143T	F145T	G145T
C	$12^{3}/_{16}$	$12^{11}/_{16}$	$12^{11}/_{16}$	$13^{3}/_{16}$
AD	$2^{3}/_{4}$	$3^{1}/_{4}$	$3^{1}/_{4}$	$3^{3}/_{4}$

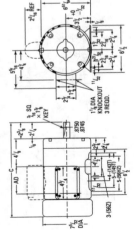

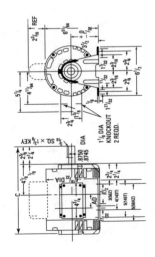

(continued)

(continued)

NEMA Frame Specifications

NEMA 56 Frame Rigid Base Brakemotor
OPEN DRIP-PROOF

FRAME	C56	D56	E56	F56	G56
C	13⁹⁄₁₆	14¹⁄₁₆	14⁹⁄₁₆	15¹⁄₁₆	15⁹⁄₁₆
AD	−⁹⁄₁₆	−¹⁄₁₆	⁷⁄₁₆	¹⁵⁄₁₆	1⁷⁄₁₆

For totally enclosed nonventilated designs use open drip-proof dimensions.

NEMA 56 Frame "C" Face Brakemotor
OPEN DRIP-PROOF

FRAME	C56	D56	E56	F56	G56
C	13⁹⁄₁₆	14¹⁄₁₆	14⁹⁄₁₆	15¹⁄₁₆	15⁹⁄₁₆
AD	−⁹⁄₁₆	−¹⁄₁₆	⁷⁄₁₆	¹⁵⁄₁₆	1⁷⁄₁₆

For totally enclosed nonventilated designs use open drip-proof dimensions.

Motor Frame Dimensions

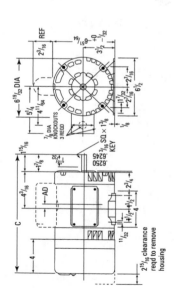

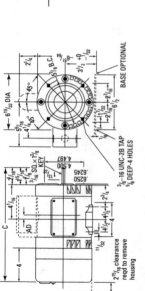

NEMA 56 Frame Swim Pool Motor
OPEN DRIP-PROOF

FRAME	C56	D56	E56	F56	G56	H56	K56
C	$10^{3}/_{16}$	$10^{11}/_{16}$	$11^{3}/_{16}$	$11^{11}/_{16}$	$12^{3}/_{16}$	$12^{11}/_{16}$	$3^{11}/_{16}$
AD	$7^{3}/_{16}$	$7^{11}/_{16}$	$8^{3}/_{16}$	$8^{11}/_{16}$	$9^{3}/_{16}$	$9^{11}/_{16}$	$10^{11}/_{16}$

Electronic Symbols for Devices Used with Electric Motors

Name	Symbol
DIODE	ANODE ▷ CATHODE
ZENER DIODE	
TRANSISTOR	C / B / E
SILICON CONTROLLED RECTIFIER (SCR)	ANODE / GATE / CATHODE
BATTERY	—⊣∣∣⊢— OR —⊣∣∣⊢—
CAPACITOR FIXED	NEW / OLD
CIRCUIT BREAKERS Air Circuit Breaker	
Three-Pole Power Circuit Breaker (Single-Throw) (with Terminals)	

(continued)

Name	Symbol
Thermal Trip Air Circuit Breaker	

COILS

Nonmagnetic Core-Fixed

Magnetic Core-Fixed

Magnetic Core-Adjustable
  Tap or Slide Wire

Operating Coil

Blowout Coil

Blowout Coil with Terminals

Series Field

Shunt Field

Commutating Field

CONNECTIONS (MECHANICAL)

Mechanical Connection of Shield

Mechanical Interlock

Direct Connected Units

CONNECTIONS (WIRING)

Electric Conductor—Control

Electric Conductor—Power

Junction of Connectors

Wiring Terminal

Ground

(continued)

(continued)

Name	Symbol
Crossing of Conductors—Not Connected	
Crossing of Connected Conductors	
Joining of Conductors—Not Crossing	

CONTACTS (ELECTRICAL)
Normally Closed Contact (NC)

Normally Open Contact (NO)

NO Contact with Time
Closing Feature (TC)

TC

NC Contact with Time
Opening Feature (TO)

TO

Note: NO (Normally Open) and
NC (Normally Closed) designates
the position of the contacts when
the main device is in de-energized
or nonoperated position.

**CONTACTOR, SINGLE-POLE
ELECTRICALLY OPERATED,
WITH BLOWOUT COIL**

(continued)

Name	Symbol
Note: Fundamental Symbols for contact coils, mechanical connections, and so on, are the basis of contactor symbols.	

FUSE

INDICATING LIGHTS
Indicating Lamp with Leads

Indicating Lamp with Terminals

INSTRUMENTS
Ammeter, with Terminals

OR

Voltmeter, with Terminals

OR

Wattmeter, with Terminals

OR

MACHINES (ROTATING)
Machine or Rotating Armature

Squirrel-Cage Induction Motor

Wound-Rotor Induction Motor or Generator

Synchronous Motor, Generator, or Condenser

(continued)

(continued)

Name	Symbol

DC Compound Motor or
  Generator
Note: Commutating, series, and
shunt fields may be indicated
by 1, 2, and 3 zigzags,
respectively. Series and shunt
coils may be indicted by heavy
and light lines or 1 and 2
zigzags, respectively.

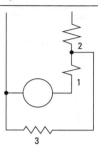

WINDING SYMBOLS

Three-Phase Wye (Ungrounded)

Three-Phase Wye (Grounded)

Three-Phase Delta

Note: Winding symbols may be
shown in circles for all motor
and generator symbols.

RECTIFIER DRY OR ELECTROLYTIC, FULL WAVE

FULL WAVE

RELAYS

Overcurrent or Overvoltage
  Relay with One NO contact

OR

Thermal Overload Relays with
  Two Series Heating Elements
  and One NC Contact

OR

(continued)

Name	Symbol

RESISTORS (OLD SYMBOLS)

Resistor, Fixed, with Leads

Resistor, Fixed, with Terminals

Resistor, Adjustable Tap or
Slide Wire

Resistor, Adjustable by
Fixed Leads

Resistor, Adjustable by
Fixed Terminals

Instrument or Relay Shunt

RESISTORS (NEW SYMBOLS)

Resistor (Fixed)

Resistor (Variable)

SWITCHES

Knife Switch, Single-Pole (SP)

Knife Switch, Double-Pole,
Single-Throw (DPST)

Knife Switch, Triple-Pole,
Single-Throw (TPST)

Knife Switch, Single-Pole
Double-Throw (SPDT)

Knife Switch, Double-Pole,
Double-Throw (DPDT)

Knife Switch, Triple-Pole,
Double-Throw (TPDT)

(continued)

(continued)

Name	Symbol
Field-Discharge Switch with Resistor	
Push Button Normally Open (NO)	
Push Button Normally Closed (NC)	
Push Button Open and Closed (Spring-Return)	
Normally Closed Limit Switch Contact	LS
Normally Open Limit Switch Contact	LS
Thermal Element (Fuse)	

TRANSFORMERS

Single-Phase, Two Winding Transformer   OLD   NEW

Autotransformer Single-Phase   OLD   NEW

Symbols Used in Motor Schematics

RESISTOR

RESISTOR

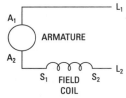

SERIES MOTOR

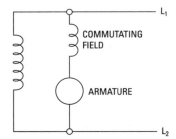

SHUNT WOUND DC
MOTOR WITH INTERPOLES

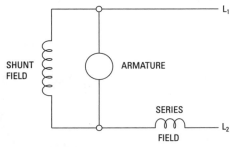

CUMULATIVELY COMPOUND
WOUND DC MOTOR

(continued)

(continued)

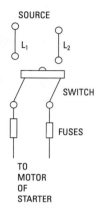

LOCATION OF
SWITCH IN LINE

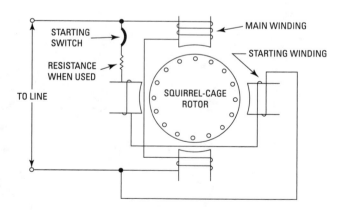

SPLIT-PHASE INDUCTION MOTOR

(continued)

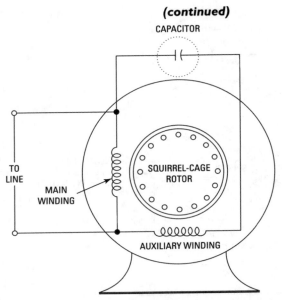

PERMANENT-CAPACITOR SPLIT-PHASE MOTOR

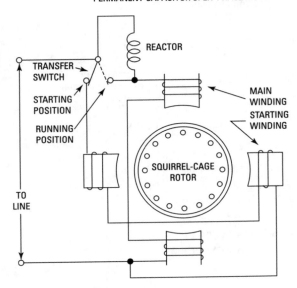

REACTOR-START MOTOR

(continued)

(continued)

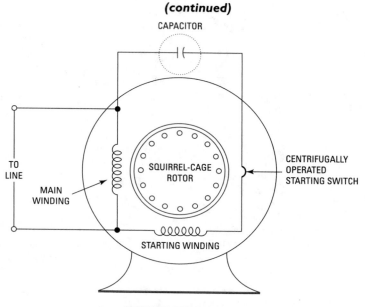

CAPACITOR-START MOTOR

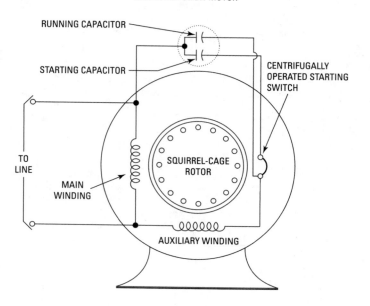

(continued)

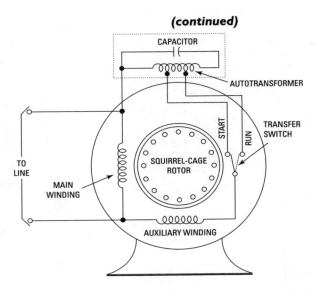

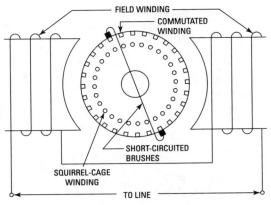

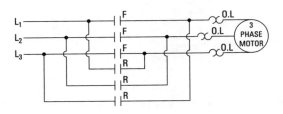

(continued)

(continued)

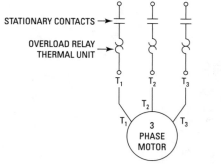

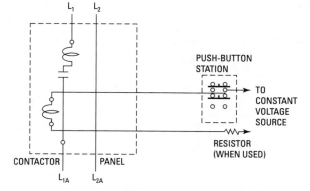

Single-pole shunt-type circuit breaker controlled from a push-button station

(continued)

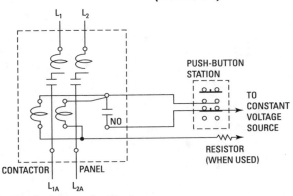

Double-pole shunt-type circuit breaker
controlled from a push-button station

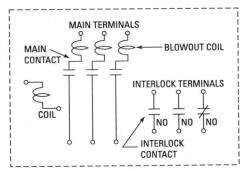

Three-pole magnetic contactor with two normally open (NO) and one normally
closed (NC) electrical interlocks contactor shown in its de-energized position.

Appendix B

Wiring Diagrams

The purpose of any wiring diagram is to indicate electric circuits by means of conventional symbols, thus making it possible to show the connections between various apparatuses and instruments in a diagrammatic form from which actual connections in the field may be made. In addition, wiring diagrams serve to make a permanent record of the wiring and apparatus installed, thus facilitating changes and replacement. Diagrams also assist in the location of any trouble that may develop in the circuit after installation.

Diagram Design

The general procedure in laying out a circuit or diagram, especially one that is complicated, usually consists of first working out an elementary or schematic diagram, and then making the working or general diagram from this layout. In an elementary diagram, no attention need be paid to the physical location of contactors or coils, but the strictest attention must be paid toward the electrical sequence of operation and to see that the equipment is correctly connected to operate properly. After a careful check of the elementary diagram has been made, the circuit is usually transferred to a working diagram where the relays, contactors, and other apparatuses are laid out in a rear-view order, although the actual electrical connections conform in every detail with that of the elementary diagram. Thus, the purpose of the elementary diagram is only to simplify the circuit for construction and checking.

In addition to elementary diagrams, there are several other types, each designed for a particular purpose or installation. The National Electrical Manufacturers Association (NEMA) and the Institute of Electrical and Electronics Engineers (IEEE) have given the following definitions:

> **Controller Wiring Diagram.** A controller wiring diagram shows the electric connections among the parts comprising the controller and indicates the external connections.

> **External Controller Wiring Diagram.** An external controller wiring diagram shows the electric connections between the controller terminals and outside points, such as the connections from the line to the motor and to auxiliary devices.

Controller Construction Diagram. A controller construction diagram indicates the physical arrangement of parts, such as wiring, busses, resistor units, etc.

Elementary Controller Wiring Diagram. An elementary controller wiring diagram uses symbols and a plan of connections to illustrate, in simple form, the scheme of control.

Control Sequence Table. A control sequence table is a tabulation of the connections that are made for each successive position of the controller.

Reading Diagrams

Because wiring diagrams consist of symbols, it is most important that these be thoroughly understood. These basic symbols, which have been generally adopted by electric equipment manufacturers, are shown in Appendix A. A study and understanding of these symbols by engineers, wire workers, maintenance workers, operators, and others who work with wiring diagrams will result in saved time and energy when determining how control equipment should be installed and how it is intended to operate. Knowledge of the symbols will make it easy to identify the devices and apparatuses represented on the print, even though there may be little resemblance between the symbol used and the actual appearance of the device. It is general practice for a diagram to show devices in their de-energized position. Power and main motor circuits are usually indicated by heavy lines, while control circuits are indicated by light lines.

Figure B-1 illustrates the connections of a typical AC reversing magnetic switch, pushbutton station, and induction motor. The magnetic switch contains two three-pole contactors and a temperature overload relay. Each contactor has three normally open power contacts, a normally open electric interlock, and an operating coil. The relay includes two heaters connected in the power circuit, and a normally closed control-circuit contact. The complete symbol for the switch shows its electric parts in the same general relationships as they appear in the actual device.

When the FORWARD pushbutton is depressed (Figure B-1), contactor F closes and applies power to the motor. A holding circuit for the coil of F is established around the FORWARD pushbutton by auxiliary interlock F. The motor continues to run until shut down either by an operator depressing the STOP button, by the tripping of the overload, or by a power failure. Following an overload condition, which causes relay OL to trip, it is necessary

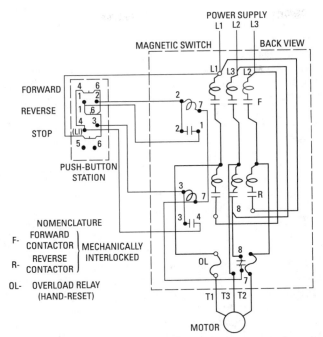

Figure B-1   Wiring diagram of a reversing magnetic switch.

to reset the relay contact by hand before the motor can be restarted. Operation of the motor in the reverse direction is obtained by means of the REVERSE button. The back, or normally closed, contacts of the directional pushbutton units are used for electric interlocking and to prevent the coils of contactors *F* and *R* from being energized at the same time. With this arrangement, it is also possible to reverse the motor directly from the FORWARD and REVERSE buttons without first operating the STOP button.

To understand the operation of control equipment, it is necessary to have a complete idea of the circuits involved. As control systems and circuits become more complicated and require more devices, it is increasingly difficult to check the circuits and understand the operation from the conventional panel type of wiring diagram. For this reason, elementary or schematic diagrams are used. Since the main reason for this type of diagram is simplicity, the various electric elements of the devices are separated. They are shown in their respective functional position in the circuit without regard to their actual relationship in the device.

It is also customary to separate the power circuits from the control circuits. Elementary diagrams, which are often referred to as *one-line diagrams,* make it easy to visualize and understand the operation. They are particularly beneficial when troubleshooting or when changes become necessary in the operation. For example, in some cases when large equipments are involved, the elementary circuits are shown on one print, and the detailed panel wiring and interconnections on separate prints. In other cases, the circuits may be shown in both elementary and detailed forms on a single print. An elementary diagram of the magnetic switch in Figure B-1 is illustrated in Figure B-2.

In controllers of the magnetic-switch type just described, the motor power circuits are handled by magnetically operated contactors under the control of a pushbutton station or other auxiliary device. In contrast to this type, there are different kinds of controllers that are manually operated. In these, the contacts that carry the main motor circuits are closed and opened by hand. Figure B-3 illustrates the connections of a manual drum switch of the cam type for use with a series-wound, direct-current motor. In this diagram, the letter X indicates that the contacts close when the handle moves to that particular position. Auxiliary control contacts UV, LSF, and LSR are closed when the switch handle is at the OFF position.

When the switch handle is moved to its first point forward, contacts $1F$, $2F$, and M close and apply power to the motor. When the handle is moved to the second position, contacts $1F$, $2F$, and M remain closed and, in addition, contact $1A$ closes. In the third position, contact $2A$ also closes, and in the fourth position $3A$ closes to short-circuit the starting resistor completely. Returning the handle to its OFF position causes the previously mentioned contacts to open and disconnect the motor from the line. When the handle is moved to the REVERSE position, contacts $1R$, $2R$, and M close in the first position to energize the motor in the reverse direction.

Two control-type limit switches are shown in Figure B-3. The contacts of these switches are normally closed but will assume an open position when the switches are tripped at their respective ends of travel.

The connections in Figure B-3 show not only the circuits of a manual cam-type drum switch, but also how undervoltage, overload, and

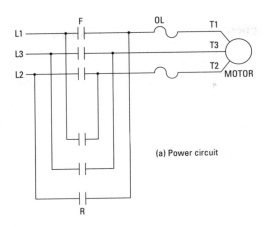

(a) Power circuit

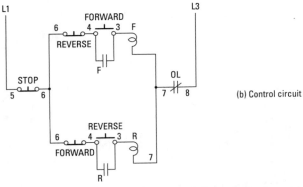

(b) Control circuit

Figure B-2 Elementary diagram of the switch shown in Figure B-1.

directional over-travel protection can be obtained by the use of a protective panel, a pushbutton station, and control-type limit switches.

Abbreviations

Symbols for devices such as relays, contactors, switches, etc., are often marked to indicate the function or use of the particular device. Table B-1 shows abbreviations that are used for device markings and designated as standards by the National Electrical Manufacturers Association (NEMA).

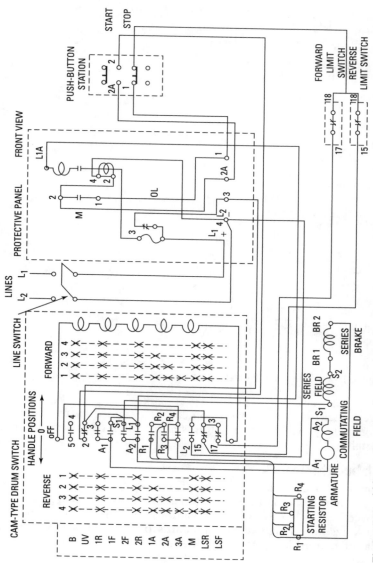

Figure B-3 Wiring diagram of a cam-type drum switch with protective panel and over-travel limit switches.

Table B-1 Abbreviations

Term	Abbreviation
Armature accelerator	A
Armature shunt	AS
Auxiliary switch (breaker), normally open	"a"
Auxiliary switch (breaker), normally closed	"b"
Balanced voltage	BV
Brake	BR
Compensator—running	MR
Compensator—starting	MS
Control	CR
Door switch	DS
Down	D
Dynamic braking	DB
Field accelerator	FA
Field decelerator	FD
Field discharge	FD
Field dynamic brake	DF
Field failure (loss of field)	FL
Field forcing (decreasing on variable voltage)	DF
Field forcing (increasing on variable voltage)	CF
Field protective (field weakened at standstill)	FP
Field reversing	FR
Field weakening	FW
Final limit—forward	FLF
Final limit—reverse	FLR
Final limit—hoist	FLH
Final limit—lower	FLL
Final limit—up	FLU
Final limit—down	FLD
Forward	F
Full field	FF
Generator field	GF
High speed	HS
Hoist	H
Jam	J
Kick off	KO
Landing	LD

(continued)

Table B-1 *(continued)*

Term	Abbreviation
Limit switch	LS
Lowering	LT
Low speed	LS
Low torque	LT
Low voltage	LV
Main or line	M
Master switch	MS
Maximum torque	MT
Middle landing	MLD
Motor field	MF
Overload	OL
Pilot motor	PM
Plug	P
Reverse	R
Series relay	SR
Slow down	SD
Thermostat	TS
Time	T
Up	U
Undervoltage	UV
Voltage relay	VR

Terminal Markings

The purpose of applying markings to the terminals of electric power apparatuses, according to a standard, is to aid in making connections to other parts of the electric power system and to avoid improper connections that may result in unsatisfactory operation or damage. The markings are placed on, or directly adjacent to, terminals to which connection must be made from outside circuits or from auxiliary devices that must be disconnected for shipment. They are not intended to be used for internal machine connections.

Although the system of terminal markings (with letters and subscript numbers) gives information and facilitates the connecting of electrical machinery, you may find terminals that are not marked with a system or are marked according to some system other than standard (especially on old machinery or machinery of foreign manufacture). You may also find that internal connections have been changed or that errors were made in markings. It is therefore advisable before connecting any apparatus to a power source to check for phase rotation, phase relation, polarity, and equality of potential.

The markings consist of a capital letter of the alphabet followed by a number subscript. The letter indicates the character or function of the winding that is brought to the terminal. A terminal letter followed by the subscript number 0 designates a neutral connection. Thus, T_0 would be applied to the terminal connected to the neutral point of a stator winding.

AC Machines

Subscript numbers 1, 2, 3, and so forth, on AC machine terminals indicate the order of the phase succession for standard direction of rotation. *The standard direction of rotation for AC generators and synchronous motors is clockwise when facing the end of the machine opposite the drive end.* It is customary to connect the coil windings and place collector rings on this end.

Figures B-4, B-5, and B-6 show the terminal markings and connections for various types of AC motors and/or generators. The terminal letters assigned to the different windings are listed in the tables in the illustrations.

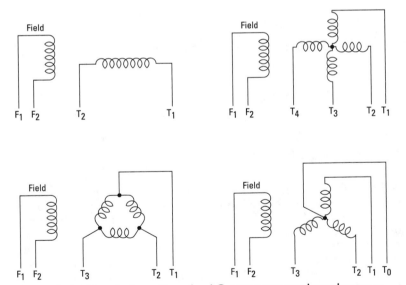

Figure B-4   Terminal markings for AC generators and synchronous motors.

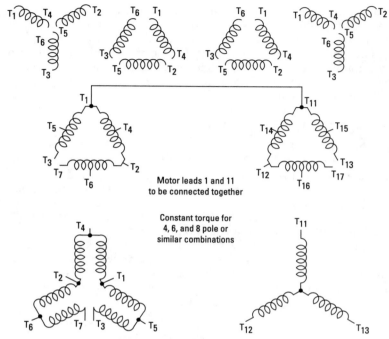

Motor leads 1 and 11
to be connected together

Constant torque for
4, 6, and 8 pole or
similar combinations

Figure B-5 Induction-motor stator connections.

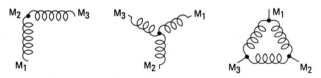

Figure B-6 Induction-motor rotor connections.

For induction motors, subscripts 1, 2, 3, and so on, indicate the order of phase succession. Thus, if induction-motor terminals T_1, T_2, and T_3 are respectively connected to AC generator terminals T_1, T_2, and T_3, the generator will cause the induction motor to turn in the same direction. In a synchronous converter, the sequence of the subscripts 1, 2, and 3 applied to the collector-ring leads M_1, M_2, and M_3 indicates that, when the collector leads are connected to the correspondingly numbered terminals of a three-phase generator, the standard rotation of the generator (clockwise, facing the end opposite the drive) will cause a clockwise rotation when viewing the direct-current or commutator end.

DC Machines

As applied to the terminals of the DC windings of generators, motors, and synchronous converters, the subscript numbers indicate the direction of current in the windings. Thus, with a standard direction of rotation and polarity, the current in all windings will be flowing from 1 to 2, or from a lower to a higher number, for example 3 to 4. Figures B-7, B-8, B-9, and B-10 show the terminal markings for various types of DC motors and generators.

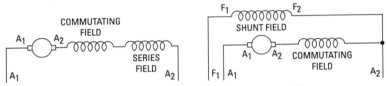

Figure B-7   Terminal markings on DC generators without commutating poles.

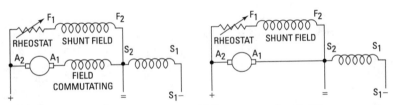

Figure B-8   Terminal markings on DC compound generators.

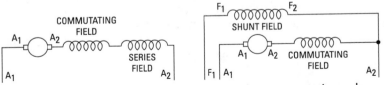

Figure B-9   Terminal markings on nonreversing, commutating-pole types of DC motors.

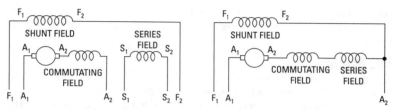

Figure B-10 Terminal markings on compound, commutating-pole types of DC motors.

The standard direction of rotation for DC generators is clockwise when facing the end of the machine opposite the drive (usually the commutator end of the machine). The standard rotation for DC motors is counterclockwise when facing the end opposite the drive (usually the commutator end).

Note
Any DC machine can be used either as a generator or as a motor. For the desired direction of rotation, connection changes may be necessary. The conventions for current flow in combination with the standardization of opposite directions of rotation of DC generators and DC motors is such that any DC machine can be called "generator" or "motor" without changing the terminal markings.

Note
A DC motor and a DC generator, by direct coupling, constitute a "motor-generator." With such coupling, the direction of rotation of the motor and generator is necessarily reversed when each is viewed from the "end opposite the drive." The standardized clockwise rotation for DC generators and counter-clockwise for DC motors meets such coupling requirements without changing the standard connections or rotation for either machine. Similarly, a DC motor may be mechanically coupled to an AC generator, without changing from the standard, from either individual machine. However, the coupling of an AC motor to a DC generator cannot be made without rotation other than standard for one of the two machines. Since the rotation of the AC machine is usually easier to change, it is general practice to operate a motor-generator with a clockwise rotation, viewed from the generator end.

Transformers
On single-phase transformers, the subscript indicates the polarity relation between the terminals on the primary and secondary

windings. Thus, during that part of the AC cycle when high-tension terminal H_1 is positive ($+$) with respect to H_2, the low-tension terminal X_1 is positive with respect to X_2. The idea is further carried out on single-phase transformers that have tapped windings by applying the subscripts 1, 2, 3, 4, 5, and so on, to the taps so that the potential gradient follows the sequence of the subscript numbers. Figures B-11, B-12, and B-13 show the standard terminal markings of various transformers.

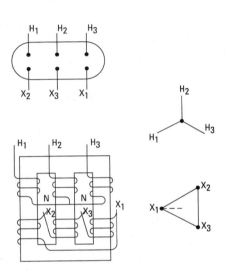

Figure B-11 Terminal markings for a star-delta connected, three-phase transformer that has additive polarity, 30° angular displacement, and standard phase rotation.

Note
When the primary of a transformer receives energy through the connecting leads, the secondary delivers energy to its connected circuit. However, the direction of current in the winding is reversed with respect to the polarity of the voltage at the terminals. It is important to take note of the difference between the practice of applying subscripts to DC generator and motors, where the subscripts are assigned according to the direction of current, and to single-phase transformers, where subscripts are assigned according to terminal voltage.

In the case of polyphase transformers, the terminal subscripts are applied so that if the phase sequence of voltage on the high-voltage side is in the time order H_1, H_2, H_3, and so on, it is in the time order X_1, X_2, X_3, and so forth, on the low-voltage side, and also in the time order Y_1, Y_2, Y_3, and so on, if there is a tertiary winding.

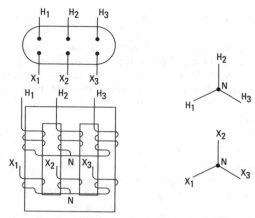

Figure B-12 Terminal markings for a star-star connected, three-phase transformer that has subtractive polarity, 0° phase displacement, and standard phase rotation.

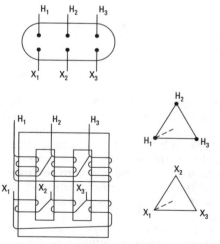

Figure B-13 Terminal markings for a delta-delta connected, three-phase transformer that has subtractive polarity, 0° phase displacement, and standard phase rotation.

Note
The terminal markings of polyphase transformers afford information on how phase rotation is carried through the transformer, but do not disclose completely the phase relations between correspondingly numbered primary and secondary terminals. Consequently, additional information on internal connections is required before polyphase transformers can be safely paralleled.

AC generators driven counterclockwise (clockwise is standard) will generate without change in connections, but the phases will follow the sequence of 3, 2, 1 instead of sequence 1, 2, 3.

Synchronous motors, synchronous condensers, induction motors, and synchronous converters may be operated with reversed rotations by transposing the connections so that the phase sequence of the polyphase supply is applied to the terminals in reversed order, that is, 3, 2, 1.

DC generators with connections properly made for standard rotation (clockwise) will not function if driven counterclockwise, as any small current delivered by the armature tends to demagnetize the fields and thus prevent the armature from delivering current. If conditions call for reversed rotation, connections should be made with either the armature leads transposed or the field leads transposed, but not both.

The polarity of a DC generator, with the accompanying direction of current in the several windings, is determined by the N and S polarity of the residual magnetism. Accidents or special manipulations may reverse this magnetic polarity; an unforeseen change may cause a disturbance or damage when it is connected to other generators or devices.

The direction of rotation of DC motors depends not on absolute polarity, but on a relative polarity between the field and armature. As no dependence is placed on residual magnetism for determining the initial direction of rotation, it is not necessary, when connecting DC motors, to regard the polarity of the DC supply connections. With a standard direction of rotation, if current is found to be flowing from terminal 1 to 2 in one of the windings, it will be found to flow in all of the other windings from 1 to 2. But, because of the disregard of polarity in connecting DC motors, the current is likely to flow from terminal 2 to terminal 1 in each of the several windings.

Reversal of rotation of a DC motor is obtained by a transposition of the two armature leads or by a transposition of the field leads (Figure B-14). With such reversed rotation (clockwise), when the polarity of the supply makes the direction of the current in the

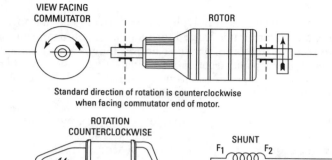

Standard direction of rotation is counterclockwise
when facing commutator end of motor.

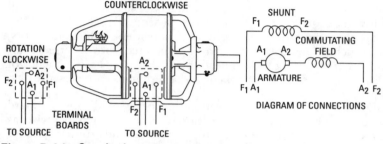

Figure B-14 Standard rotation, terminal markings, and terminal connections of a DC shunt motor. To reverse the direction, the field leads are interchanged as shown.

armature from terminal 2 to terminal 1, it will flow in the field windings from terminal 1 to terminal 3, and vice versa.

With synchronous converters, the practice of AC starting eliminates residual magnetism as the factor determining DC polarity. Proper polarity for connection to other apparatuses is secured either by separate excitation of the field or by special manipulation of a switch that permits the converter to reverse polarity, thus correcting a start with wrong polarity.

If a synchronous converter is to be operated with reversed direction of rotation (counterclockwise, viewing the commutator end), besides transposing the AC terminal connections, it is also necessary to make either a transposition of armature leads or a transposition of field leads.

Wiring Symbols

The symbols used on wiring diagrams to represent various electrical devices facilitate not only the reading and understanding but also the maintenance and installation of electrical equipment of all kinds. Various fundamental symbols commonly found on wiring diagrams and employed by representative electrical manufacturers are given in Appendix A. The fundamental symbols represent component parts such as coils, connections, contacts, instruments, transformers, and so on, and can be combined in various ways to represent complete electrical devices.

Circuit Wire Sizes

A motor must have the proper amount of current and voltage to operate in any prolonged condition. One way to make sure that the proper amount of power reaches the motor is to select the correct wire size for connecting the motor to the power source (Table B-2).

Table B-2 Circuit Wire Sizes for Individual Single-Phase Motor

Horse-power of Motor	Volts	Approximate Starting Current Amperes	Approximate Full-Load Current Amperes	Length of Run in Feet (from Main Switch to Motor)							
				Feet	25	50	75	100	150	200	300	400
1/4	120	20	5	Wire Size	14	14	14	12	10	10	8	6
1/3	120	20	5.5	Wire Size	14	14	14	12	10	8	6	6
1/2	120	22	7	Wire Size	14	14	12	12	10	8	6	6
3/4	120	28	9.5	Wire Size	14	12	12	10	8	6	4	4
1/4	240	10	2.5	Wire Size	14	14	14	14	14	14	12	12
1/3	240	10	3	Wire Size	14	14	14	14	14	14	12	10
1/2	240	11	3.5	Wire Size	14	14	14	14	14	12	12	10
3/4	240	14	4.7	Wire Size	14	14	14	14	14	12	10	10
1	240	16	5.5	Wire Size	14	14	14	14	14	12	10	10
1 1/2	240	22	7.6	Wire Size	14	14	14	14	12	10	8	8
2	240	30	10	Wire Size	14	14	14	12	10	10	8	6
3	240	42	14	Wire Size	14	12	12	12	10	8	6	6
5	240	69	23	Wire Size	10	10	10	8	8	6	4	4
7 1/2	240	100	34	Wire Size	8	8	8	8	6	4	2	2
10	240	130	43	Wire Size	6	6	6	6	4	4	2	1

Appendix C

Motor Connections

Figures C-1 through C-19 show schematics, block diagrams, and other drawings for motor connections.

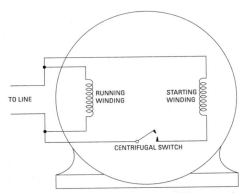

Figure C-1 Schematic wiring diagram of a split-phase motor.

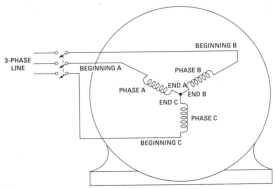

Figure C-2 Schematic wiring diagram of a star-connected, polyphase motor.

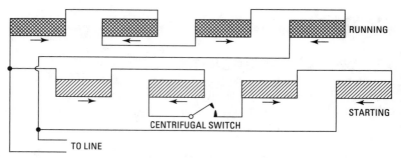

Figure C-3 Block diagram (extended type) of a four-pole, split-phase motor.

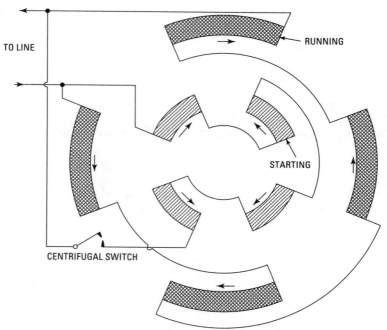

Figure C-4 Circular block diagram of a four-pole, split-phase motor.

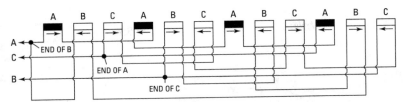

Figure C-5 Block diagram (extended type) of a three-phase, four-pole, series-delta motor.

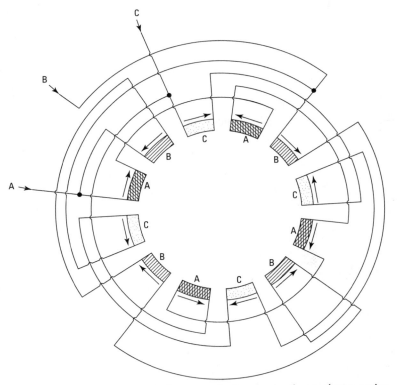

Figure C-6 Circular block diagram of a four-pole, three-phase, series-delta motor.

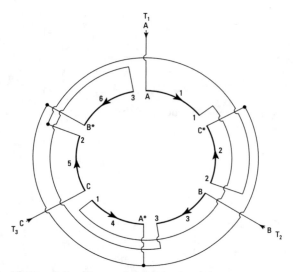

Figure C-7 Two-pole, three-phase, series-star connection.

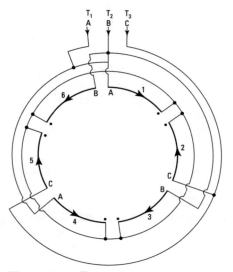

Figure C-8 Two-pole, three-phase, two-parallel-star connection.

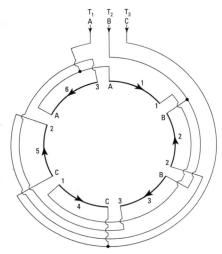

Figure C-9   Two-pole, three-phase, series-delta connection.

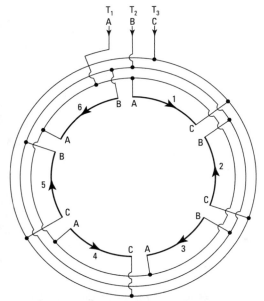

Figure C-10   Two-pole, three-phase, two-parallel connection.

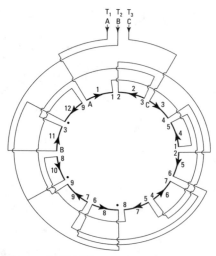

Figure C-11   Four-pole, three-phase, series-star connection.

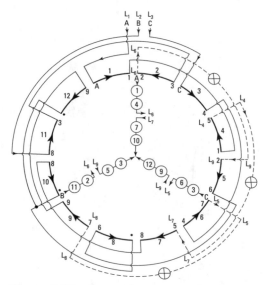

Figure C-12   Four-pole, three-phase motor, with three or nine leads for series-star or two-parallel-star connection.

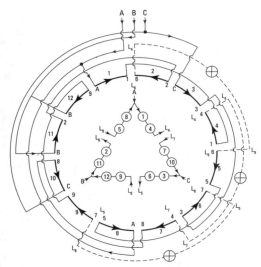

Figure C-13 Four-pole, three-phase motor, with three or nine leads for series-delta or two-parallel-delta connection.

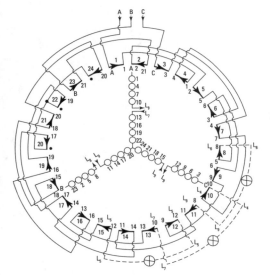

Figure C-14 Eight-pole, three-phase motor, with three or nine leads for series-star or parallel-star connection.

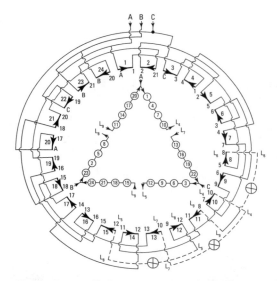

Figure C-15 Eight-pole, three-phase motor, with three or nine leads for series-delta or parallel-delta connection.

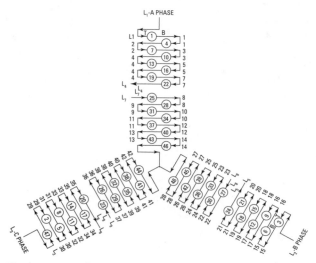

Figure C-16 Sixteen-pole, three-phase, nine-lead, series-star or two-parallel-star connection.

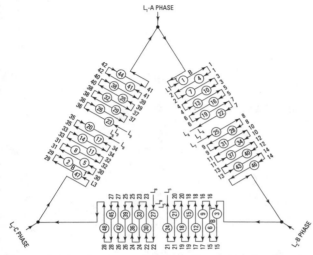

Figure C-17 Sixteen-pole, three-phase, nine-lead, series-delta, or two-parallel-delta connection.

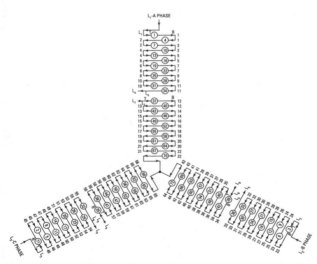

Figure C-18 Twenty-four-pole, three-phase, nine-lead, series-star, or two-parallel-star connection.

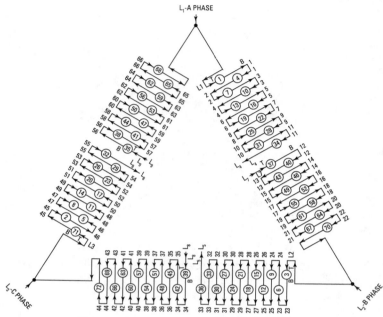

Figure C-19 Twenty-four-pole, three-phase, nine-lead, series-delta, or two-parallel-delta connection.

Glossary of Motor Terminology

Being able to order parts and select motors is entirely dependent upon being able to talk with or communicate with the people involved in the field every day. This glossary will enable you to become familiar with common terms.

Note
The Bodine Electric Company and the General Electric Company have been very helpful in furnishing information for this glossary.

Air Gap The space between the rotating and stationary member in an electric motor.

Air Over (AO) Motor intended for fan and blower service, and cooled by the air stream from the fan or blower.

Alternating Current (AC) The commonly available electric power supplied by an AC generator and distributed in one-, two-, or three-phase form. This is the standard type of power supplied to homes, businesses, industry, and farms.

Ambient The air surrounding the motor of air-cooled rotating machinery. The temperature of the space around the motor, the ambient temperature, should not exceed 40°C or 104°F.

Ampere The constant current that, if maintained in two straight, parallel conductors of infinite length and negligible cross-section and spaced 1 meter apart in a vacuum, produces 2×10^{-7} Newtons per meter of length. The unit of measurement for current. Ampere is abbreviated as A or, in some cases, amp.

Ampere Turn The magnetomotive force produced by a current of one ampere in a coil of one turn.

Angular Velocity Angular displacement per unit time, measured in degrees/time or radians/time.

Armature The portion of the magnetic structure of a DC or universal motor that rotates.

Armature Reaction Magnetic flux produced by the current that flows in the armature winding of a DC motor. This effect, which reduces the torque capacity, is called armature reaction and can affect the commutation and the magnitude of the motor's generated voltage. This magnetic flux is created in addition to that produced by the field current.

Basic Speed The speed a motor develops at rated voltage when rated load is applied.

Bearings (BRGS) A mechanical engine part that resembles a short length of bronze tubing with grooves to direct oil flow (Figure 20-6). Sleeve-type bearings (SLV) are preferred where low noise level is important. Ball bearings are used where higher load capacity is required or periodic lubrication is impractical (Figure GL-1).

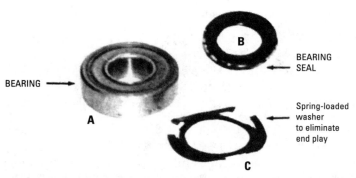

Figure GL-I End play is eliminated by using a spring-loaded washer (C) to hold the ball bearing (A) in place. The bearing seal (B) is used to prevent dust and dirt from getting into the bearing.

Braking Torque The torque required to reduce a motor's speed from running speed to standstill. The term is also used to describe the torque developed by a motor during dynamic braking conditions.

Breakdown Torque The maximum torque a motor will develop at rated voltage without a relatively abrupt drop or loss in speed.

Brush A piece of current-conducting material (usually carbon or graphite) that rides directly on the commutator of a commutated motor and conducts current from the power supply to the armature windings.

Canadian Standards Association (CSA) Organization that sets safety standards for motors and other electrical equipment used in Canada.

Cantilever Load A load that tends to impose a radial force (perpendicular to the shaft axis) on a motor or gearmotor output shaft.

Capacitor A device that, when connected in an alternating current circuit, causes the current to lead the voltage in time phase. The peak of the current wave is reached ahead of the voltage wave. This is the result of the successive storage and discharge of electric energy.

Center Ring The part of a motor housing that supports the stator or field core.

Centrifugal Cut-Out Switch A centrifugally-operated automatic mechanism used in conjunction with split-phase and other types of induction motors. Centrifugal cut-out switches will open or disconnect the starting winding when the rotor has reached a predetermined speed, and reconnect it when the motor speed falls below it. Without such a device, the starting winding would be susceptible to rapid overheating and subsequent burnout (Figure GL-2).

Figure GL-2 The centrifugal switch on this motor is mounted outside the frame with an arm extended into the housing to operate the switch.

Cogging A term used to describe nonuniform angular velocity. It refers to rotation occurring in jerks or increments rather than smooth motion. When an armature coil enters the magnetic field produced by the field coils, it tends to speed up; it tends to slow down when leaving it. This effect becomes apparent at low speeds. The fewer the coils, the more noticeable it can be.

Commutator A cylindrical device mounted on the armature shaft and consisting of a number of wedge-shaped copper segments that are arranged around the shaft and insulated from both the shaft and each other. The motor brushes ride on the periphery of the commutator and electrically connect and switch the armature coils to the power source.

Conductor Any material that tends to make the flow of electrical current relatively easy (copper, aluminum, gold, silver, and others).

Counter Electromotive Force (CEMF) The induced voltage in a motor armature caused by conductors moving through, or "cutting," field magnetic flux. This induced voltage opposes the armature current and tends to reduce it.

Direct Current (DC) Type of power supply available from batteries or generators (not alternators) and used for special-purpose applications. Current flows in one direction only; it does not alternate.

Drip-Proof (DP) Motor A type of enclosed motor in which ventilation openings in end shields and shells are placed so that drops of liquid falling within an angle of 15° from vertical will not affect the performance of the motor. Usually used indoors, in fairly clean, dry locations.

Duty Cycle The relationship between operating and rest time. A motor that can continue to operate within the temperature limits of its insulation system after it has reached normal operating (equilibrium) temperature is considered to have a continuous duty (CONT) rating. One that never reaches equilibrium temperature but is permitted to cool down between operations is operating under intermittent duty (INT) conditions.

Dynamic Unbalance A noise-producing condition caused by the nonsymmetrical weight distribution of a rotating member. The lack of uniform wire spacing in a wound armature or casting voids in a rotor or fan assembly can cause relatively high degrees of unbalance.

Eddy Current Localized currents induced in an iron core by alternating magnetic flux. These currents translate into losses (heat), and their minimization is an important factor in lamination design.

Efficiency In motors, the ratio of mechanical output to electrical input. It represents the effectiveness with which the motor converts electrical energy into mechanical energy.

Electrical Coupling The condition created when two coils are situated in such a way that some of the flux set up by either coil links some of the turns of the other.

Electromotive Force (EMF) A synonym for voltage, usually restricted to generated voltage.

Encapsulated Winding A motor whose winding structure is completely coated with an insulating resin (such as epoxy). This construction type is more designed for exposure to severe atmospheric conditions than is the normal varnished winding.

Enclosure (ENCL) The term used to describe the motor housing. Some of the more common types are drip-proof, explosion-proof, fan-cooled, nonventilated, open, and totally enclosed.

End Shield The part of the motor housing that supports the bearing and acts as a protective guard to the electrical and rotating parts inside the motor. This part is frequently called the end bracket or end belt.

Excitation Current A term applied to the current in the shunt field of a motor resulting from voltage applied across the field.

Explosion-Proof (EXP-PRF) Motor A special enclosed motor designed to withstand an internal explosion of specified gases or vapors and to allow the internal flame or explosion to escape. Usually used in smaller-rating (below ⅓ horsepower), nonventilated (EPNV) motors and in larger-rating, fan-cooled (EPFC) motors.

Fan-Cooled (TEFC) Motor Enclosed motor that includes an integral fan to blow cooling air over the motor.

Farad A unit of measurement for electrical capacitance. A capacitor has a capacitance of 1 farad when a potential difference of 1 volt will charge it with 1 coulomb of electricity.

Feedback, The voltage information received by a feedback circuit, as it generally relates to motors and controls. Depending on a predetermined potentiometer setting, a motor control can correct the voltage to deliver appropriate speed and/or torque.

Field A term commonly used to describe the stationary (stator) member of a DC motor. The field provides the magnetic field with which the mechanically rotating (armature) member interacts.

Field Weakening The introduction of resistance in series with the shunt-wound field of a motor to reduce the voltage and current, which weakens the strength of the magnetic field and thereby increases the motor speed.

Flux The magnetic field that is established around an energized conductor or permanent magnet.

Form Factor A figure of merit that indicates how much rectified current departs from pure (nonpulsating) DC. A large departure from form factor (pure DC) increases the heating effect of the motor and reduces brush life.

Fractional-Horsepower Motor A motor with continuous rating of less than 1 horsepower, open construction, at 1700–1800 rpm.

Frame (FR) Usually refers to the NEMA system of standardization of motor-mounting dimensions.

Frequency The rate at which alternating current reverses its direction of flow. Measured in hertz (Hz).

Full-Load Current The current drawn from the line when the motor is operating at full-load torque and full-load speed at rated frequency and voltage.

Full-Load Torque The torque necessary to produce rated horsepower at full-load speed.

Galvanometer An extremely sensitive instrument used to measure small current and voltage in an electrical circuit.

Gearhead The portion of a gearmotor that contains the actual gearing that converts the basic motor speed to the rated output speed.

Horsepower (hp) Power rating of the motor. It takes 746 W of electrical energy to produce 1 horsepower.

Hysteresis Loss The resistance to becoming magnetized offered by materials. Reduced by using silicon steel laminations.

Impedance The vectorial sum of both resistance and reactance in a motor. Total opposition to current flow. Measured in ohms. Z is the impedance symbol.

Inductance The characteristic of a coil of wire to cause the current to lag the voltage in time phase. L is the symbol for inductance. Inductance is measured in henrys (H).

Inertial Load A load (flywheel, fan, or the like) that tends to cause the motor shaft to continue to rotate after the power has been removed.

Insulation (INSUL) In motors, the covering or coating over the copper wire is the insulation, it usually is classified by maximum allowable operating temperature:

Class A, 105°C or 221°F

Class B, 130°C or 266°F

Class F, 155°C or 311°F

Class H, 180°C or 365°F

Insulator A material that tends to resist the flow of electric current.

Integral Horsepower Motor In terms of horsepower, a motor, built in a frame, that has continuous rating of 1 horsepower or more, open construction, at 1700–1800 rpm. In terms of size, an integral horsepower motor is usually greater than 9 inches in diameter, although it can be as small as 6 inches.

Line Voltage Voltage supplied by the power company or voltage supplied as input to the device.

Locked-Rotor Current Steady-state current (current that does not vary) taken from the line when the rotor is at standstill.

Locked-Rotor Torque The minimum torque that a motor will develop at rest for all angular positions of the rotor.

Magnetomotive Force (MMF) The magnetic energy supplied with the establishment of flux between the poles of a magnet.

Mechanical Degree The popular physical understanding of degrees (360° = 1 rotation).

Motor Types Classification by operating characteristics and/or type of power required for motors (Table GL-1). Induction motors include single-phase and three-phase motors. Direct-current motors are further classified as shunt, series, and compound.

Table GL-1   Motor Characteristics

		Duty	Typical Reversibility	Speed Character	Typical Start Torque*
Polyphase	AC	Continuous	Rest/Rot.	Relatively Constant	175% and up
Split-Phase, Synchronous	AC	Continuous	Rest Only	Relatively Constant	125–200%
Split-Phase, Nonsyn chronous	AC	Continuous	Rest Only	Relatively Constant	175% and up
PSC, Nonsyn-chronous, High-slip	AC	Continuous	Rest/Rot.[a]	Varying	175% and up
PSC, Nonsyn-chronous, Normal–slip	AC	Continuous	Rest/Rot.[a]	Relatively Constant	75–150%
PSC Reluctance-Synchronous	AC	Continuous	Rest/Rot.[a]	Constant	125–200%
PSC Hysteresis-Synchronous	AC	Continuous	Rest/Rot.[a]	Constant	125–200%
Shaded-Pole	AC	Continuous	Uni-directional	Constant	75–150%
Series	AC/DC	Int./Cont.	Uni-directional[d]	Varying[b]	175% and up
Permanent-Magnet	DC	Continuous	Rest/Rot.[c]	Adjustable	175% and up
Shunt	DC	Continuous	Rest/Rot.	Adjustable	125–200%
Compound	DC	Continuous	Rest/Rot.	Adjustable	175% and up

Table GL-1   *(continued)*

		Duty	Typical Reversibility	Speed Character	Typical Start Torque*
Shell Arm	DC	Continuous	Rest/Rot.	Adjustable	175% and up
Printed Circuit	DC	Continuous	Rest/Rot.	Adjustable	175% and up
Brushless DC	DC	Continuous	Rest/Rot.	Adjustable	75–150%
DC Stepper	DC	Continuous	Rest/Rot.	Adjustable	e

Percentages are relative to full-load rated torque. Categorizations are general and apply to small motors.

aReversible while rotating under favorable conditions, generally when inertia of the driven load is not excessive.

bCan be adjusted, but varies with load.

cReversible down to 0°C after passing through rest.

dUsually unidirectional; can be manufactured as bidirectional.

eDependent upon load inertia and electronic driving circuitry.

National Electrical Code (NEC) A code for the purpose of practical safeguarding of persons and property from the hazards arising from the use of electricity. It is sponsored by the National Fire Protection Institute, and serves as a guide for governmental bodies whose duty is to regulate building codes.

NEMA The National Electrical Manufacturers Association. This organization establishes certain voluntary industry standards relating to motors. These standards refer to operating characteristics, terminology, basic dimensions, ratings, and testing.

Nonventilated (TENV) Motor Enclosed motor that is not equipped with a fan for external cooling; depends on convection air for cooling.

Open Circuit In a motor, a defect that causes an interruption in the path through which the electric current normally flows.

Open (OP) Motor Enclosed motor with ventilation openings in end shields and/or shell to permit passage of cooling air over and around the windings. Locations of openings not restricted. For use indoors, in fairly clean locations (Figure GL-3).

Phase A term that indicates the space relationship of windings and changing values of the recurring cycles of AC.

Figure GL-3 An open-enclosure motor (OP).

Phase Displacement Mechanical and electrical angle by which phases in a polyphase motor or main and capacitor (or starting) windings in an induction motor are displaced from one another.

Plug Reversal Reconnecting a motor's windings to reverse its direction of rotation while running.

Polarities Terms (positive, negative, north, and south) that indicate the direction of current and flux flow in electrical and magnetic circuits at any given instant.

Power Factor (PF) A measurement of the time-phase difference between the voltage and current in an AC circuit. It is represented by the cosine of the angle of the phase difference. 0° has a power factor of 100 percent, which means the watts and volt-amperes are equal and there is nothing more than resistance in the circuit. Ninety degrees of angle represents nothing in the way of resistance and only inductance in the circuit. PF is also found by the formula

$$\frac{\text{True power (TP)}}{\text{Apparent power (AP)}}$$

Prony Brake A simple mechanical device, normally made of wood with an adjustable leather strap, that is used to test for the torque output of a motor. The prony brake loads the motor and a spring scale attached to it gives a relatively accurate measurement or torque.

Pull-In Torque The maximum constant torque that a synchronous motor will accelerate into synchronism at rated voltage and frequency.

Pull-Up Torque The minimum torque delivered by an AC motor during the period of acceleration from 0 to the speed at which breakdown occurs. For motors that do not have a definite breakdown torque, the pull-up torque is the minimum torque developed during the process of getting up to rated speed.

Rectifier An electronic circuit that converts alternating current into direct current.

Reluctance The characteristic of a magnetic material that resists the flow of magnetic lines of force through it.

Resilient Mounting A suspension system or cushioned mounting designed to reduce the transmission of normal motor noise and vibration to the mounting surface.

Resistance The degree of obstacle presented by a material to the flow of electrical current. Resistance is measured in ohms. R is the symbol for resistance.

Rotor The rotating member of an induction motor in a single-phase device. Current that is normally induced in the rotor reacts with the magnetic field produced by the stator. This produces torque and rotation (Figure GL-4).

Salient Pole Stator or field poles in a motor that are concentrated into confined arcs and wrapped within the winding (as opposed to poles that are distributed in a series of slots).

Secondary Winding A winding that is not connected to the power source but carries current induced in it through its magnetic linkage with the primary winding.

Semiconductor A material, usually silicon or germanium, that permits limited current flow.

Service Factor (SF) A measure of the overload capacity designed into a motor. A 1.15 SF means that the motor can deliver 15 percent more than the rated horsepower without injurious

Figure GL-4   This single-phase motor rotor has a centrifugal switch mechanism mounted with springs holding it in place. Note the three pressure pads on the part of the mechanism that moves outward to operate the switch.

overheating. A 1.0 SF motor should not be overloaded beyond its rated horsepower. Service factors will vary for different horsepower motors and for different speeds. Standard NEMA service factors for various horsepower motors and motor speeds are shown in Table GL-2, for easy reference.

Table GL-2   Service Factors

Horsepower	Synchronous Speed (rpm)			
	3600	1800	1200	900
¹⁄₂₀, ¹⁄₁₂, ⅛	1.40	1.40	1.40	1.40
⅙, ¼, ⅓	1.35	1.35	1.35	1.35
½	1.25	1.25	1.25	1.15
¾	1.25	1.25	1.15	1.15
1	1.25	1.15	1.15	1.15
1½ and up	1.15	1.15	1.15	1.15

Short Circuit A defect in a winding that causes part of the normal electrical circuit to be bypassed.

Skew Arrangement of laminations on a rotor or armature to provide a slight diagonal pattern of their slots with respect to the shaft axis. This pattern helps to eliminate low-speed cogging effects in an armature and minimize induced vibration in a rotor.

Slip The difference between the speed of the rotating magnetic field (which is always synchronous) and the rotor in a non-synchronous induction motor. Slip is expressed as a percentage of a synchronous speed and generally increases with an increase in load.

Slip Ring A conductor band, mounted on an armature and insulated from it. A conductor strip slides on the band as the armature rotates. The function of the slip-ring system is essentially the same as a commutator and brushes. Slip rings are also used to transmit current from the armature in a generator application.

Starting Torque The torque or twisting force delivered by a motor when energized.

Stator The part of an induction motor's magnetic structure that does not rotate. It usually contains the primary winding.

Synchronous Speed The speed of the rotating magnetic field set up by an energized stator winding. In synchronous motors, the rotor locks into synch with the field, and is said to run at synchronous speed.

Tachometer A small generator normally used as a velocity-sensing device. Tachometers are typically attached to the output shaft of DC servomotors that require close speed regulation. The tachometer feeds its signal to a control that adjusts its output to the DC motor accordingly (called closed loop feedback control).

Temperature Rise The amount by which a motor, operating under rated conditions, is hotter than its surrounding ambient temperature.

Thermal Protector A protective device, built into the motor, that disconnects the motor from its power source if the temperature becomes excessive for any reason.

Thermocouple A junction of two dissimilar materials that generates a minute voltage in proportion to its temperature. Such devices may be used as signal sources in indicating instruments and control equipment.

Torque Turning force delivered by a motor or gearmotor shaft, usually expressed in ounce-inches or Newton-meters. There are three types of torque associated with electric motors: starting torque, full-load torque, and breakdown torque.

Totally Enclosed (TE) Motor Motor with no openings in the motor housing (but not airtight). Used in locations that are dirty, oily, and so on. Some types are: totally enclosed, dust proof, with positive-pressure ventilation or the pipe-ventilation with a blower or fan delivering air. They may be totally enclosed, fan-cooled (TEFC) in addition to dust-ignition and pipe-ventilated. A totally enclosed water-air-cooled motor would also be approved by the NEC for Class II, Division I locations. Some motors may be allowed in Class II, Division 2 areas.

Underwriters Laboratories, Inc. (UL) An independent testing organization that sets safety standards for motors and other electrical equipment.

Variable Resistor A resistor, connected in series with a motor, that can be adjusted to vary the amount of current available and thereby alter motor speed.

Voltage The force that causes a current to flow in an electrical circuit. Analogous to pressure in hydraulics, voltage is often referred to as electrical pressure. Voltage is measured in volts (V).

Voltage Drop Loss encountered across a circuit impedance. Voltage drop across a resistor takes the form of heat released into the air at the point of resistance.

Watt The amount of power required to maintain a current of 1 ampere at a pressure of 1 volt. One horsepower is equal to 746 watts. The symbol for watt is W.

Index

speed control, 397, 489–490
frequency-change method of speed
control, 397, 403
full-complement needle bearings,
575
full-load armature current, 367
full-load current (FLC), 407, 453,
454, 462, 464, 492, 670
full-load efficiency, 366
full-load torque, 670, 678
furnace blower, 226, 550
fuses
AG (automobile glass), 540
cartridge, 540
dual-element, 408, 431, 465
overload, 538, 547
as overload protection, 408,
431, 465, 492, 538, 547,
553
plug, 540
purpose of, 462
two-element, 539
types of, 539–540

G
galvanometer, 7–8, 670
gearbox motors, 586
gearhead, 670
gearmotors, shafts in, 571, 573
gear puller, 508, 521
gear trains, noise and, 123, 125
General Electric (GE), 337, 665
nameplate, 218, 219, 237
shaded-pole motors
manufactured by, 145, 148,
149
generator(s)
alternating-current (AC), 33
compound, 29–31, 32, 53
constant current, 28
constant-potential, 32
constant-speed, 31
defined, 8–9, 53
differentially compounded, 30
direct-current (DC), 9–31, 53
field of, 33
flat-compounded, 30, 53, 59

overcompounded, 29–30, 53
for polyphase motors, 34–38
series, 26–28, 31–32, 53
shunt, 28–29, 32, 53
tachometer, 491
three-phase, 36–38
two-phase, 34–36
two-pole, 14, 20
undercompounded, 30, 53
graphite, 98
graphitized, self-lubricating sleeve
bearings, 577–578, 586
grinders, 153–154
grinding, 93–94
grounds
safety and, 350
testing for, 205–206

H
hacksaws, 94, 95, 106, 501–502,
512, 520
half-step, 434
hammers, 500–501, 520
ball-peen, 500, 520
claw, 500, 520
mallets, 500, 520
hand drill, 67–70, 71, 83, 85, 101
hand-reset overload relays, 193,
195–196
hand stoning, 92–93, 106
harmonics, 147
hazardous locations, 352
motors in, 159–160, 170
NEC classification of, 234–235,
238, 352, 452, 678
UL standards for, 234, 238, 351
headless screws. See setscrews
heater winding, 468
henrys, 671
hertz (Hz), 34, 670
hex-socket key sets, 504
high-flux density pole tips, 24
high resistance ground, 205
high-slip motors, 215, 233–234,
238, 311
high-torque, reversible AC motors,
255–257, 261